ASTROMY HACKS™

Other resources from O'Reilly

ASTRONOMY
HACKS™

Robert Bruce Thompson
and Barbara Fritchman Thompson

O'REILLY®

Beijing · Cambridge · Farnham · Köln · Paris · Sebastopol · Taipei · Tokyo

Astronomy Hacks™

by Robert Bruce Thompson and Barbara Fritchman Thompson

Copyright © 2005 Robert Bruce Thompson and Barbara Fritchman Thompson.
All rights reserved.
Printed in the United States of America.

Published by O'Reilly Media, Inc., 1005 Gravenstein Highway North,
Sebastopol, CA 95472.

O'Reilly books may be purchased for educational, business, or sales promotional use. Online editions are also available for most titles (*safari.oreilly.com*). For more information, contact our corporate/institutional sales department: (800) 998-9938 or *corporate@oreilly.com*.

Editor:	Brian Jepson	**Production Editor:**	Marlowe Shaeffer
Series Editor:	Rael Dornfest	**Cover Designer:**	Mike Kohnke
Executive Editor:	Dale Dougherty	**Interior Designer:**	David Futato

Printing History:

June 2005:   First Edition.

 This book uses RepKover™, a durable and flexible lay-flat binding.

ISBN: 0-596-10060-4
[C]

Contents

Credits

About the Authors

Robert Bruce Thompson became interested in astronomy when he was very young. In 1957, when he was four, his parents took him out in the front yard one autumn evening to look for Sputnik, the first artificial satellite. Someone shouted, "There it is!," and he watched, fascinated, as the bright pinpoint of Sputnik blazed its way across the night sky. Robert was hooked on science and space from that moment forward.

By the time he was 10, Robert was using an old binocular that his grandmother had given him to observe open star clusters and other wonders of the night sky. That Christmas, his parents gave him a small refractor that allowed him for the first time to observe Luna at the unprecedented magnification of 100X. That scope held his interest for a couple of years, but, struck by an early case of "aperture fever," he decided to grind his own mirror and build his own 6" Newtonian reflector telescope. In those days, even urban backyards were dark, and Robert spent the next several years exploring the night sky with that 6" scope. Most clear nights, he observed Luna, the planets, double stars, and many of the brighter deep-sky objects, including most of the Messier list.

Alas, life intervened. College, graduate school, jobs, marriage, family—everything seemed to get in the way of continuing the hobby. Then, one evening in late 2000, as Barbara and Robert were walking the dogs, Barbara expressed an interest in astronomy and suggested they buy a telescope. Robert ordered a 10" Dobsonian reflector and has never looked back. He spends his daytime hours writing books about computers (and now astronomy). But on clear, dark nights, you'll find him outdoors observing the night sky.

Barbara Fritchman Thompson can remember lying on the driveway with her dad back in the late 50s and early 60s, looking up at the stars and the Milky Way. They lived inside the city limits but, unlike today, the night sky was dark even in the city. Recently, her mom found an old junior high school project booklet she'd written, entitled "The Stars." Along with basic information about stars, their science, and astronomical tools, she included charts and drawings of how stars change and the major constellations of each season. After that time, around 1969, she seldom thought about the stars or telescopes again until the late 1990s.

As Barbara and Robert were walking the dogs one night, Barbara started noticing bright objects in the sky. She repeatedly asked Bob to identify whatever objects had attracted her attention. At first Robert tried to fool her, telling her stars and constellations had names like Calvin and Hobbes. However, when he realized she was becoming seriously interested in the night sky, they joined a local astronomy club and decided to buy a telescope.

At first Bob planned to grind a mirror and build their telescope. After doing some research, he realized good scopes were much less expensive than they had been in the middle 1960s. To Barbara's delight, the first scope they bought was an Orion XT10 10" Dobsonian. Several months later Bob ordered another scope, a 90mm refractor on an equatorial mount. Barbara hated using it and couldn't find a darn thing with it. Had he purchased that for her first true introduction to astronomy, she would not be sitting here writing this book. He also bought her a 50mm binocular.

During their early observing sessions, Barbara was not very successful with the scope and was an absolute failure with the binocular when it came to locating deep-sky objects. In order to make their observing sessions more productive and give her a structured framework to work within, Robert and Barbara founded the Winston-Salem Astronomical League (*http://www.wsal. org*) as an affiliate of the national Astronomical League (*http://www. astroleague.org*). Barbara started with the AL Urban Observing and Lunar Clubs. With Bob's coaching and a lot of practice, she became more confident in her abilities. Before long, she'd completed the AL Messier Club list, which requires identifying all 110 of the Messier objects and the AL Binocular Messier Club list. Barbara completed the AL Deep Sky Binocular Club, and is now pursuing the Hershel 400 and Caldwell Club lists.

Contributors

The following people contributed hacks, writing, and inspiration to this book:

- Gene Baraff is now learning the night sky. What a pity he waited because there was so much more of it back in the time when he used to observe in daylight only. Back then, like most teenage amateurs of his generation, astronomy meant grinding his own mirrors and building his own mounts. His first mirror was the standard 6" f/8, mounted in a Horseshoe-type equatorial mount. Its performance was very similar to that of the inexpensive imported equatorial mounts of today; i.e., it was inadequate in every detail.

  His second mirror, also 6" f/8 and ground after his first was stolen, was intended for solar use. It was mounted much more solidly. That mirror and mount, recently refurbished, are still usable today. The mirror was initially used unsilvered, to allow most of the Sun's rays to go out through the back of the scope. The diagonal, a prism mounted with its hypotenuse facing toward the mirror, allowed most of the light returned from the primary to pass harmlessly out of the front of the scope. Several cameras were built and used with this scope, and a brief note describing them appeared in the October 1947 issue of *Sky & Telescope* magazine.

  In the years since, Gene became a scientist, participated in the 1954 H-Bomb tests in the Pacific, got a doctorate in physics, worked for 40 years as a theoretical physicist for Bell Laboratories, and retired to go back to astronomy. Wife (the same one all these years), kids, grandkids, and the Skyquest-Telescope Forum take up major pleasurable amounts of his time.

- Mary Chervenak received a Ph.D. in organic chemistry from Duke University in 1995. After two years of post-doctoral research at the University of Alberta, she joined the Biocides group at Union Carbide in 1997. When Union Carbide was purchased in 1999, Mary accepted a position with The Dow Chemical Company's Antimicrobial group. She currently works in Technical Service and Development for UCAR Emulsion Systems (also The Dow Chemical Company). Although she spends far more time peering through a microscope than a telescope and her hobbies are not conducive to late-night stargazing sessions, she enjoys running under the stars and admiring the vast and startling beauty of the sky.

- Many amateur astronomers begin their hobby as teenagers. Not Steve Childers. He never even looked through a telescope until he was 51 years old. Spending most of his life under urban lights, Steve never paid any attention to the night sky. When he first considered purchasing a scope in 2002, Steve only knew two constellations by eye (the Big Dipper and Cassiopeia). He had no idea whether they appeared in the summer or winter, and, in fact, he would have been surprised to learn that constellations in general were not visible all year round. So, like many other people, his first purchase would probably have been a cheap department store telescope, spelling the end to a short-lived interest in astronomy. Fortunately, he received good advice from many helpful people, and purchased a 10" Dobsonian that provided a superb opportunity to enjoy amateur astronomy for the first time.

Owning his first telescope was exciting, but he quickly ran into problems. First, Steve couldn't find anything except the moon. Thanks (again) to good advice, he solved that problem by using a Telrad finder, a low-power, wide-field eyepiece, and a Palm Pilot as a live map of the night sky. Although he still doesn't know many of the constellations, with this setup he feels like he can find almost anything without resorting to computer-guided scopes. Another problem was a severe case of aperture fever that struck one night when Steve looked through a 20" Obsession Dob at a club observing session. There was only one cure for this miserable affliction, so over the next year he built a 17.5" truss Dob in his basement. Despite making every mistake in the book, Steve was shocked to discover the thing actually worked. In addition, he also has a small 80mm refractor, which spends most of its time connected to an H-alpha solar filter. Steve loves watching the changing face of the sun through H-alpha light. He also has a 6" Mak-Newt, which is used for Lunar and planetary photography. He's just beginning to dabble in astrophotography with a digital SLR camera, but he doesn't have much to show for it yet.

Despite this experience, Steve still considers himself a beginner in astronomy. In real life, he's a professor of Pharmacology, and conducts laboratory research in drug abuse. He's been married for 29 years (his wife believes anyone that goes out to look at stars in the middle of the night belongs in a straightjacket), and they suffer the indignities of two teenage kids.

- Paul Jones has degrees in chemistry from Oklahoma State University (1993) and Duke University (1998) and is currently an assistant professor of chemistry at Wake Forest University. Paul grew up under the dark skies of western Oklahoma and has been involved in amateur astronomy since 1983, when he attended a public observation in Tucson, Arizona that featured Comet IRAS-Araki-Alcock, Jupiter, and Saturn. Paul has authored several articles in amateur astronomy publications and designed and taught a workshop to introduce astronomy and astronomy activities to secondary school science teachers. Though he has longed to have an astrophoto published, it has turned out, inexplicably to him, that publishers prefer his writing (and even his sketching!) to his photography.

Acknowledgments

We are indebted to our contributors, Drs. Baraff, Chervenak, Childers, and Jones. In addition to contributing Hacks of their own, they also reviewed everything we wrote and made numerous helpful additions, suggestions, and corrections. Thanks to their help, this is a much better book than it would otherwise have been.

We also want to thank Mark Brokering at O'Reilly, who got the ball rolling and kept us on track. And without Brian Jepson, our editor, this book would never have seen the light of day (or the dark of night). Thanks, guys!

Thanks are also due to Al Nagler, the founder of Tele Vue Optics, Inc., who came to our rescue on very short notice when we desperately needed an image of a top-quality refractor for the front cover. The cover photograph is a Tele Vue-NP101 101mm f/5.4 apochromatic refractor on an ash-wood Tele Vue Gibraltar mount, a world-class instrument that any astronomer would be proud to own. For more information, visit the Tele Vue web site at *http://www.televue.com*.

Preface

Astronomy Hacks happened almost by accident. Our editor emailed us one day to say that O'Reilly was thinking about doing an astronomy book and to ask if we knew any amateur astronomers who might be interested in writing it. We sent a one-sentence reply, "Other than us, you mean?"

Of course, once O'Reilly realized that *we* were amateur astronomers, there never was much doubt about who would write the book. Robert was drooling at the thought of it. Barbara is too refined to drool, but she, too, was excited about the opportunity to write about our shared hobby. We had other high-priority books in progress, but this opportunity was too good to miss. So we dropped everything to write *Astronomy Hacks*. Writing about computers, our usual day job, is fun. We like computers, and we like writing about them. But we love astronomy.

There is something special about being out under the night sky. We look up and see the stars and constellations, just as our many-times-great grand-parents did hundreds and thousands of years ago. The stars provide a link across the generations, from remotest antiquity down to the present day. They establish an unchanging framework that places us in context within the universe.

We look at the Great Orion Nebula, for example, and realize that the light we see tonight began its journey about 1,550 years ago, when the Roman Empire was in its final days. Or we view the Andromeda Galaxy and realize that the photons striking our eyes left Andromeda about 2.9 million years ago, when early proto-humans were just coming down from the trees. And, as we look at the dim smudge of Andromeda and remember that that smudge is actually the light from nearly a thousand billion Suns, we wonder if someone there is looking back at us, wondering the same thing.

So, we jumped at the opportunity to write *Astronomy Hacks*. We had two goals in writing this book. First, of course, we wanted to convey our passion about astronomy to those many people who have some interest in the stars but have not yet begun their personal journeys of exploration and to encourage them along that road. But we also wanted to "pay forward." Over the years, many experienced amateur astronomers helped us learn the ropes. There's no way we can ever pay them back, but we can pay forward by helping others to learn and enjoy the hobby.

Why Astronomy Hacks?

The term hacking has a bad reputation in the press. They use it to refer to people who break into systems or wreak havoc with computers as their weapons. Among people who write code, though, the term hack refers to a "quick-and-dirty" solution to a problem, or a clever way to get something done. And the term hacker is taken very much as a compliment, referring to someone as being creative, having the technical chops to get things done. The Hacks series is an attempt to reclaim the word, document the good ways people are hacking, and pass the hacker ethic of creative participation on to the uninitiated. Seeing how others approach systems and problems is often the quickest way to learn about a technology.

When we were growing up in the 1960s, kids didn't sit around playing video games, listening to CDs, or watching DVDs. We *did* things. We ground our own telescope mirrors with blanks and grit purchased from Edmund Scientific (which is now, alas, a very pale shadow of what it once was), and built our scopes from Sonotube and pipe fittings. We built ham radio rigs from mostly scrounged components, and assembled rockets from salvaged conduit and army-surplus gyros (and concocted our own fuels). We tapped our girlfriends' telephone lines, learned how to pick locks, and built darkrooms in the basement. We forged drivers licenses, tore down and rebuilt car engines, repaired our relatives' televisions and appliances, constructed silencers for our .22 squirrel rifles, and made general nuisances of ourselves. In other words, we spent most of our free time hacking, although the term had not yet been invented.

Nowadays, most of what we did would quickly land us in a federal penitentiary. Back then, adults smiled, shook their heads, and told themselves that "boys will be boys" (even though some of us were girls). Most of those hacking opportunities are gone now, more's the pity. Hams buy most of their gear now rather than building it, and "rocket kits" and "chemistry sets" have been gutted to the point of worthlessness by companies fearful of litigation. Cops no longer have a sense of humor about teenagers' antics, and tearing down your car's engine will probably earn you a visit from EPA agents in black helicopters. What's a would-be hacker to do?

Well, there's still astronomy, which is one of the few remaining technical hobbies where hacking is not the exception, but the norm. Hacking is a time-honored practice among amateur astronomers, although most would not call it by that name. Many amateur astronomers still build their own telescopes—two of our contributors did—but even if that's a bit beyond your abilities, there are many other opportunities to hack.

But hacking doesn't just mean doing things; it means having a deep understanding of those things. At one level, it's possible to enjoy amateur astronomy without understanding any of the technical details. At its simplest, amateur astronomy requires nothing more than the night sky and your Mark I eyeball. Ultimately, though, you'll probably want to see more than is visible with your naked eye and to know more about what you're seeing.

That's where *Astronomy Hacks* comes in. Over the years, we've helped a lot of newbies over the hump, so we know the issues that beginning astronomers (and even more experienced ones) trip over. *Astronomy Hacks* is a collection of hard-won knowledge—everything from advice on choosing, using, and maintaining equipment to observing tips and tricks to short essays that explain the essential concepts you need to understand to more fully enjoy the hobby. *Astronomy Hacks* will help you get up to speed quickly, spend your money wisely, and bypass many of the frustrations beginners usually encounter. We tried to make this book the next best thing to having an experienced astronomer looking over your shoulder and offering advice as you learn the hobby.

How This Book Is Organized

You can read this book from cover to cover if you like, but each hack stands on its own, so feel free to browse and jump to the different sections that interest you most. If there's a prerequisite you need to know about, a cross-reference will guide you to the right hack.

The book is divided into four chapters, organized by subject:

Chapter 1, *Getting Started*
> Use the hacks in this chapter to move quickly from a standing start to full speed ahead. This chapter answers the most common newbie questions, including choosing the best telescopes and binoculars, and it helps you avoid the most common newbie mistakes. If you're just getting started, read this chapter through a couple of times before you do (or buy) anything. Even if you are a moderately experienced amateur

astronomer, it's worth taking the time to study this chapter in detail. You'll probably pick up at least a few tips and tricks that are new to you.

Chapter 2, *Observing Hacks*

This chapter covers observing activities, the heart of amateur astronomy. The hacks in this chapter explain everything from fundamental concepts, like stellar magnitude, to tips and tricks for seeing the most possible detail in very dim objects to organizing your records to sketching objects. Even very experienced astronomers are likely to pick up some tips and tricks from this chapter. We know, because some of the hacks we wrote about were new to us, suggested by our contributors and other experts.

Chapter 3, *Scope Hacks*

Use the hacks in this chapter to tweak, modify, upgrade, and maintain your telescope. This chapter focuses largely (although not exclusively) on the incredibly popular Dobsonian telescopes, which are not just amenable to hacking, but purely beg to be hacked. You'll learn how to improve your scope's motions (finally, a use for all those AOL CDs...), improve its optical performance, and otherwise tweak and tune it for maximum performance.

Chapter 4, *Accessory Hacks*

Accessorizing is half the fun in most hobbies, and amateur astronomy is no exception. This chapter explains what you need to know to choose, use, and maintain accessories properly. It's easy to go overboard when choosing astronomy accessories. If you follow our advice, you'll end up with only the accessories you need, at a price you can afford to pay.

Conventions

The following is a list of the typographical conventions used in this book:

Italics

Used to indicate URLs, filenames, filename extensions, and directory/folder names.

Color

The second color is used to indicate a cross-reference within the text.

You should pay special attention to notes set apart from the text with the following icons:

This is a tip, suggestion, or general note. It contains useful supplementary information about the topic at hand.

This is a warning or note of caution, often indicating that you or your equipment might be at risk.

The thermometer icons, found next to each hack, indicate the relative complexity of the hack:

    beginner   moderate   expert

How to Contact Us

We have tested and verified the information in this book to the best of our ability, but you may find that features have changed (or even that we have made mistakes!). As a reader of this book, you can help us to improve future editions by sending us your feedback. Please let us know about any errors, inaccuracies, bugs, misleading or confusing statements, and typos that you find anywhere in this book.

Please also let us know what we can do to make this book more useful to you. We take your comments seriously and will try to incorporate reasonable suggestions into future editions. You can write to us at:

O'Reilly Media, Inc.
1005 Gravenstein Highway North
Sebastopol, CA 95472
(800) 998-9938 (in the U.S. or Canada)
(707) 829-0515 (international/local)
(707) 829-0104 (fax)

To comment on the book, send email to:

bookquestions@oreilly.com

The web site for *Astronomy Hacks* lists examples, errata, and plans for future editions. You can find this page at:

http://www.oreilly.com/catalog/astronomyhks

For more information about this book and others, see the O'Reilly web site:

http://www.oreilly.com

To contact one of the authors directly, send mail to:

barbara@astro-tourist.net
robert@astro-tourist.net

We read all mail we receive from readers, but we cannot respond individually. If we did, we'd have no time to do anything else. But we do like to hear from readers.

We also maintain a messageboard, where you can read and post messages about astronomy topics. You can read messages as a guest, but if you want to post messages, you must register as a member of the messageboard. We keep registration information confidential, and you can choose to have your mail address hidden on any messages you post.

> *http://forums.astro-tourist.net/*

We each maintain personal journal pages, updated frequently, which often includes references to astronomy gear, our own observing sessions, upcoming astronomical events, and other things we think are interesting. You can view these journal pages at:

> Barbara: *http://www.fritchman.com/diaries/thisweek.html*
> Robert: *http://www.ttgnet.com/thisweek.html*

Got a Hack?

To explore Hacks books online or to contribute a hack for future titles, visit:

> *http://hacks.oreilly.com*

Safari Enabled

 When you see a Safari® Enabled icon on the cover of your favorite technology book, that means the book is avaiable online through the O'Reilly Network Safari Bookshelf.

Safari offers a solution that's better than e-books. It's a virtual library that lets you easily search thousands of top tech books, cut and paste code samples, download chapters, and find quick answers when you need the most accurate, current information. Try it for free at *http://safari.oreilly.com*.

Getting Started

Hacks 1–10

Getting started in amateur astronomy seems simple enough. Buy a telescope, take it out at night, point it at the sky, and you're good to go. Or are you?

Many thousands of people follow just this route every Christmas, and nearly all of them are disappointed. They overpay for an inferior scope at the mall or a big-box store. Once they get it assembled, they discover they can't figure out how to use it properly. They soon find that being outdoors with a telescope in wintertime gets cold fast, and decide they'd really rather watch television instead.

Even those who persist lose interest quickly. After they look at the Moon a time or two, and maybe Jupiter and Saturn, they decide there's really not much else to see. Perhaps they've bought a computerized go-to scope that claims to find objects for them automatically. If that's true, why are the objects invisible, even though the computer swears they're in the eyepiece? Where are all those brightly colored objects pictured on the telescope box? The new scope ends up gathering dust in the closet or for sale on eBay. It can all be very discouraging.

But it doesn't have to be that way. Astronomy can be a wonderful, life-long hobby, one that the entire family can enjoy together. Thousands of devoted amateur astronomers are outdoors on every clear night, observing the wonders of the night sky. You can join them, but you need to get started right. In this chapter, we'll tell you what you need to know to avoid the most common beginner mistakes.

H A C K **Don't Give Up**
#1 It's harder than it looks, but doable.

The night sky initially looks inviting—a big, black picnic blanket spangled with shine. Stellar objects are brilliant, easy to see, and seem to organize themselves into recognizable patterns. Vast, yes, but easily interpretable, welcoming.

Appreciating the beauty of a starry sky is easy. It's the next step that's hard. The night sky is the very worst kind of bully—the kind who punches you in the stomach, steals your lunch money, and then laughs at you when you cry.

When you first start looking at the stars through a telescope, the blanket shrinks to a napkin. The obvious becomes elusive and the elusive becomes invisible. Of course, the finding of things is part of learning to observe, but that knowledge is small comfort when you are unable to find the Andromeda Galaxy night after night after night.

What to say, really? Using a telescope can be frustrating. First, don't give up. Or rather, give up, but only for a while. Learning to squeeze an expanse of night sky into the eyepiece of a telescope and then to comb it degree by degree takes not only patience, but practice. Temporarily flinging up your hands and packing up the scope for the night is not only acceptable, it's necessary. When you're tired and angry, the fluid motion required to scan the sky become jerky and unpredictable; you're unlikely to find anything and you risk damaging your equipment.

Don't let the bully keep stealing your lunch money, though. Get your confidence back. The best tactic is a battle plan. Select a piece of sky, pull out a simple star map, and find your way around. Stars in the sky will look different from stars on paper, but once you've identified a few landmarks, go back to the scope. Wend your way through the familiar, retracing your steps until each star is a recognizable signpost.

The process is slow—arduous, even. But, eventually, as the stars stop looking like little blobs of light and start looking like a set of directions, the size of the viewing field matters less and less.

Now, go get that lunch money back.

—*Dr. Mary C. Chervenak*

Join an Astronomy Club

Meet others who share your interests, learn a few things from them, and maybe even play with their toys.

The first piece of advice newbies generally hear is to join an astronomy club. We couldn't agree more. Join an astronomy club. Join an astronomy club. Join an astronomy club. What we tell you three times is true.

Surprisingly few amateur astronomers belong to an astronomy club. For example, we live in Forsyth County, North Carolina, which has a total population of about 200,000. The Forsyth Astronomical Society has only about 50 members. The population of the United States is about 300,000,000. Credible estimates say that between 1,000,000 and 2,000,000 of them are interested in astronomy. On that basis, if all interested Forsyth County residents belonged to the club, it would have between 667 and 1,333 members, many times its actual membership. The same is generally true nationwide, which means a lot of people are missing out on one of the best resources available to learn and enjoy the hobby.

Joining your local astronomy club has many advantages, which usually include:

Help and advice from experienced members
> If you're just getting started, getting advice from more experienced club members can save you a lot of time, money, and aggravation. Unlike magazines—and even some web sites—that must please their advertisers, club members tell it to you straight. If something is junk, they'll say so, despite all the pretty full-color ads for it running that month in the astronomy magazines. And you can take that advice to the bank. Most astronomy clubs have a very strong focus on helping new members with everything from buying equipment to learning how to locate celestial objects. It's difficult to overstate the value of that help for a newbie.

A chance to try out equipment
> One of the biggest benefits of belonging to a club is that you can look at other people's stuff. If you're thinking about buying a $300 Nagler eyepiece, for example, but would like to see one first, a club is the place to be. Chances are, another club member already has just the eyepiece you're thinking about buying and would be happy to let you look through it. If you ask politely, he may even let you try it in your own scope. Local astronomy specialty stores are an endangered species in most cities, which means most astronomy gear is bought mail-order nowadays. If you're uncomfortable buying expensive items sight-unseen, get yourself to an astronomy club.

Access to dark-sky sites

As the blight of light pollution has continued to spread, it has become much harder to locate suitable dark-sky observing sites. Making matters worse, property owners' fear of liability and lawsuits has closed many sites that would otherwise be available to individual observers. Most astronomy clubs devote significant time and effort to locating and maintaining dark-sky observing sites, but for liability reasons those sites are often open only to club members. The small cost of an annual club membership gives you access to those club sites, which for many observers is by itself sufficient reason to join their local clubs.

Club observation sessions

Observing by yourself is a lonely pursuit. Most clubs schedule regular club observing sessions, usually at least on weekends near the new moon, but often at other times during the month as well. Many clubs also periodically schedule field trips to very dark sites, arranging for discounts on lodging and so forth. These club observations allow you to get together with other like-minded people. They're a great place to ask questions, try out other people's gear, and so on. Also, don't discount the safety aspect **[Hack #3]**. Most predators, two-legged and four-legged, avoid groups of people.

Presentations

Most clubs schedule regular presentations that range from astronomy-related videos to planetarium programs to lectures on timely topics. Many also hold periodic workshops on such topics as collimation, astrophotography, and building your own scope.

Access to the club library

Some clubs maintain a club library that contains field guides, catalogs, observing handbooks, back issues of astronomy magazines, and other reference materials. Quite often, the club library is the only local source for specialized reference books and similar items. Most clubs restrict use of club library resources to club members, so often the only way to get access to them is to join the club.

Loaner equipment

Some clubs maintain club scopes and other equipment that may be borrowed by members. There are often restrictions, such as requiring prospective borrowers to have been members for a certain time and to attend a training session. But if you haven't yet bought your own scope, a club loaner may tide you over until you do.

Discounts on magazines and books
>Many clubs offer discounts on astronomy magazines and books that are available to club members only. For example, *Sky & Telescope* magazine and *Astronomy* magazine both offer club discounts that are a significant percentage of many clubs' annual dues. Similar discounts are often available on specialty items such as the *Ottewell Astronomical Calendar*, yearbooks, and so on.

Finally, and not least, joining an astronomy club isn't just about getting. It's also about giving back. You can share your experience to help others learn and enjoy the hobby. If you're inexperienced, you may believe there's not much you can do to help, but that's not true. There's always someone with even less experience, and the best teachers are often people who are new enough at it themselves to empathize with the frustrations of a complete newbie. Ask around, find out what needs to be done, and pitch in.

You can locate a local astronomy club by calling the nearest observatory, planetarium, nature-science center, museum, or similar institution, or by contacting the physics/astronomy department of a local college. You can also use the Resources page of the *Sky & Telescope* web site (*http://skyandtelescope.com/resources/organizations/*) to search for local clubs.

HACK #3 Safety First

Take precautions to keep small problems small.

The days when most amateur astronomers observed from their own backyards are long gone. Light pollution forces most of us to seek observing sites far from cities, often in the middle of nowhere. But with civilization comes safety. If you have a problem at home, you dial 911 and the police or paramedics or animal control officers arrive in a few minutes. If you have a problem at a remote observing site, it's up to you to deal with it until help arrives, which may be some time.

If you're prepared, small problems tend to stay small. If you're not, a small problem can rapidly escalate into a dangerous emergency. Here's what we recommend to prepare yourself:

Observe in a group
>There really is safety in numbers. A lone observer may be victimized by two- or four-legged predators or may have a medical emergency. Bad things are less likely to happen when you observe with a group **[Hack #2]**, and if a problem does arise there are people there to help.

Never leave the last person alone

Use the buddy system. The last two vehicles remaining at the end of an observing session should leave together, particularly if the site is remote. If you have company, a flat tire or other breakdown is merely annoying. If you don't, it can be very inconvenient at best and dangerous at worst.

> Call us sexist, but we never leave a woman or even a group of women alone, even at a regular club observing site near civilization. It's just too risky. Most women appreciate having men stick around to protect them. For those who don't, we just pretend we aren't ready to leave until they start to pack up.

Carry a cell phone

A cell phone can be a lifeline in an emergency. Make sure the phone is charged and verify that you have a usable signal from your observing site. Store local emergency numbers for your observing site, not just 911, but the direct numbers for the local police and fire departments, hospital or emergency clinic, paramedic/rescue squads, and so on. Know the exact location of your observing site, and how to direct emergency services to locate it.

Notify someone of your location and expected return

Particularly if you are observing alone, make sure someone knows exactly where you are and when you expect to return. If the observing site is remote, make sure to provide a detailed map and/or GPS coordinates. If something happens to you while you are observing, it's good to know that someone will come looking for you sooner rather than later.

Carry a first-aid kit

Assemble or buy a first-aid kit such as the Johnson & Johnson Ready Organized First Aid Kit. At a minimum, have sterile dressings, adhesive bandages in various sizes, cotton balls or swabs, alcohol-based hand sanitizer, antibiotic ointment, burn ointment, eye-wash solution, aspirin, Imodium, and similar basic supplies. Also carry any medications you may need for acute emergencies, such as insulin, an inhaler, epinephrine injector, heart medication, and so on. Be sure to observe expiration dates for any medicines and replace them when they do expire. It's easy to lose track of how long it's been since you packed your kit.

Take first-aid and CPR training

Some of the members of your regular observing group—ideally all of them—should have at least basic first-aid and CPR skills. Contact the Red Cross for information about local classes.

Keep an eye out for severe weather

City dwellers don't fully appreciate the majesty of nature. It's one thing to sit out a severe thunderstorm huddled in your basement or hall closet. It's quite another to experience it up close and personal in the middle of an open field miles from nowhere. The spring and summer months are particularly hazardous because thunderstorms and tornados can pop up with little or no warning. Check the weather forecasts before you depart for an observing session, and know where the nearest shelter is.

Dress for the weather

Make sure your clothing is appropriate for the conditions. Dress warmly for cold weather sessions **[Hack #4]**. Disease-bearing mosquitos and ticks are a problem during warm weather in many areas. Use DEET-based insect repellant and pay particular attention to your legs. Wear high, thick socks, tight pant cuffs, and drench the clothing near your ankles with insect repellant. Lyme Disease or Rocky Mountain Spotted Fever is no fun at all. In poisonous snake country, wear high, snake-proof boots or snake pants.

If you use DEET, particularly in a high concentration, be careful with it around plastic, including plastic eyeglass lenses. DEET dissolves plastic.

Prepare your vehicle

Before you depart for an observing session at a remote site, make sure your vehicle won't let you down. Check the spare tire, battery, and oil level, and fill the gas tank. Carry a basic set of hand tools, jack, emergency tire inflator, jumper cables, fire extinguisher, and so on. A power inverter can also be very handy.

Carry an emergency kit

Pack an emergency kit in a duffle bag and leave it in your vehicle. Include blankets (traditional and space), a catalytic propane heater or other source of heat, a flashlight with spare batteries, a Swiss Army Knife, storable high-energy foods, and several liters of water stored in clean soft-drink bottles.

Arm yourself

Yes, we know this is controversial, but we consider it good advice, particularly if you observe at remote locations in areas where rabies is endemic or there are poisonous snakes or large predators. Robert generally brings a .44 revolver or a 12-gauge riot shotgun to observing sessions. If you don't own a firearm or if you are uncomfortable bringing it along, a whistle, air-horn, or other loud noisemaker may discourage bears and other four-legged predators.

Observing safety is mostly a matter of common sense. If you always keep safety in the back of your mind, you'll be fine.

HACK #4 Stay Warm

Hypothermia kills, and frostbite isn't much fun either. Learn how to protect your vital organs and keep all your extremities attached.

As odd as it sounds to non-astronomers, it's possible to become chilled while observing even during high summer. Observing is a sedentary activity, so you generate little excess body heat. The temperature of the night sky is very nearly absolute zero, and it serves as a gigantic heat sink that sucks the warmth out of you. Observing sites are often located at high elevations and unsheltered from the wind, which makes the problem worse.

When our astronomy club holds public observations at a local state park in July and August, it's easy to tell the experienced astronomers from the public visitors. The visitors are wearing the same t-shirts and shorts they wore during the heat of the day. The astronomers are wearing long-sleeve shirts and jeans, and often don jackets and caps as the night goes on. More than once, we've loaned blankets to visitors who were shivering, unable to figure out why they were cold. The night sky gets you every time.

For cold-weather observing sessions, the problem is much worse. The standard advice is to dress for temperatures 20°F to 30°F (11°C to 17°C) colder than the actual temperature, counting wind chill. For example, if the forecast low temperature is 30°F including the effects of wind chill, we dress for 0°F to 10°F, and are often none too warm at that.

Fortunately, there are steps you can take to remain warm even under severely cold conditions.

Avoid Alcohol

Even moderate alcohol consumption increases the risk and severity of hypothermia. Alcohol is dangerous in three ways. First, it increases the likelihood of hypothermia by increasing heat loss from your body. Second, it masks the effects of hypothermia by reducing shivering and other hypothermia symptoms. Third, and most dangerously, it gives you a false sense of feeling warm. Drinking alcohol before or during an observing session is always a bad idea, but it's a particularly bad idea for cold-weather observing sessions.

Dress in Layers

Two basic principles of keeping warm are that many thin layers of clothing are better than a few thick layers, and that loose layers are better than tight layers. Using several thin, loose layers of clothing traps your body heat most efficiently.

> When you purchase clothing for your observing wardrobe be sure to buy the outer layers such as sweatshirts and sweat-pants in a much larger size than you normally wear. This allows for layering without the fit being too snug. Remember, it is more important to be warm than to worry about how you look!

For a routine cold weather observing session, Robert wears underwear, long underwear, and jeans on the bottom, and long underwear, a t-shirt, a thin turtleneck, a sweater or sweatshirt, a flannel shirt, and a parka on top. For colder sessions, he wears underwear, long underwear, flannel-padded jeans, and sweatpants on the bottom, and adds two or three layers on top, including a hooded sweatshirt. Barbara dresses similarly, but dislikes jeans. She prefers to start with a heavy pair of sweatpants, paired if necessary with long underwear for added warmth. If it is very cold or windy, she adds a second pair of of sweatpants.

> Most people don't realize that long underwear comes in different ratings. Standard long underwear suffices for routine cold weather, but for observing in extreme cold you can get special "arctic" long underwear from outfitters such as Cabela's (*http://www.cabelas.com*). Outfitters also carry other specialty gear, such as full-body, zip-up "freezer suits" that are useful for observing in truly cold conditions.

Keep Your Head and Neck Warm

Your head is an amazingly efficient heat radiator, and your neck isn't far behind. If you dress warmly otherwise but do not protect your head, you will be cold. Even for summer observing sessions, it's worth wearing a baseball cap or similar light hat. For cool or moderately cold sessions, wear a knitted watch cap. From late fall to early spring, supplement the watch cap with a thin balaclava to cover your face and neck. During very cold weather, wear a fur-lined hat with ear flaps over the balaclava, and use a scarf to keep your neck warm.

Keep Your Feet Warm

It always amazes us to see some astronomers during cold weather observing sessions waddling around like penguins, bundled up in parkas, but wearing tennis shoes. Your feet are second only to your head in heat loss, and it's impossible to be warm if your feet are cold. To keep your feet warm, start with two layers of socks. Use a thin, cotton or polypropylene inner layer and a thick, woolen outer layer. For extreme cold, consider heated socks, which are powered by a battery pack and available from outfitters and hunting supply stores. Some observers place a scrap of carpet or a foam pad near their telescopes, standing on it to isolate their feet from the cold ground.

Choose your footwear according to the temperature. For cool or normally cold sessions, wear ankle-high hiking or walking boots. For very cold sessions, wear insulated calf-length hunting boots, such as those sold by L.L. Bean (*http://www.llbean.com*) and other outfitters.

> Never tease someone about being cold, any more than you would tease someone who was having chest pains. There is a certain macho outlook among some amateur astronomers, who would have you believe that Real Men don't feel the cold. That's a dangerous attitude. People vary in how they are affected by cold. One person may be in shirtsleeves while another has already donned jacket and earmuffs. One person may already be in the early stages of hypothermia while everyone else is quite comfortable.
>
> If you feel cold, do something to get yourself warm. Don't worry about what your observing buddies may think. Conversely, if you suspect one of your observing buddies is cold, don't tease him. Help him. Give him some of your hot coffee, break out one of your chemical heat packs for him, or fire up your propane heater. Hypothermia affects not only body temperature, but judgment. Don't take chances with it.

Keep Your Hands Warm

Although you lose less heat through your hands than through your head or feet, it's still important to keep your hands warm, not least to avoid dropping eyepieces and other equipment. A thin pair of cotton gloves or polypropylene glove liners may be all you need for moderately cold temperatures. In freezing temperatures, thin gloves keep your skin from freezing to eyepieces and other metal objects. For cold weather observing, supplement or replace the thin gloves or glove liners with heavy gloves or mittens.

Mittens are warmer than gloves but are cumbersome when you are handling equipment or jotting down something in your observing log. We found the best of both worlds in the hybrid fingerless glove/mittens shown in Figure 1-1. When you're writing in your observing log or otherwise need your fingers unencumbered, you simply flip up the mitten pouch portion to expose your fingers. When you're finished, flip down the mitten pouch to cover your fingers. You can even put a small chemical heating pack in the pouch to keep your fingers toasty warm.

Figure 1-1. Open-finger glove/mittens keep your hands warm while allowing full dexterity

Drink Plenty of Warm Fluids

For cool or cold weather observing sessions, always bring along a large Thermos of hot cider, coffee, tea, or cocoa. (Go light on the caffeine, which may exacerbate hypothermia.) Not only does drinking warm liquids help keep you warm, it helps prevent dehydration. Cold air, even if the relative humidity is high, contains very little moisture. Every time you inhale, your lungs warm and hydrate the cold, dry air. Every time you exhale, you lose that moisture to the outside air, where it is visible as it condenses as fog. Dehydration exacerbates the effects of hypothermia, and should be avoided at all costs. For very cold weather observing sessions, it's a good idea to take along a propane camp stove and use it to make hot drinks at least every couple of hours.

Screen Yourself from the Wind

Even a slight breeze can make cold air feel frigid and frigid air positively Antarctic. Screening yourself from wind makes a major difference in observing comfort. The most important step is to wear wind-resistant nylon outer clothing. All parkas have a nylon outer shell, of course, but don't overlook wind-resistant outer pants, hats, and glove shells.

If there is a breeze, choose where to set up your scope with that in mind. Moving just a few feet may allow you to screen your equipment and yourself with trees, outbuildings, or other objects. For cold-weather sessions, our club members sometimes park their vehicles in a laager to provide a central screened observing area. Even if you are alone, you can park your vehicle upwind of your equipment and observe in its lee.

If your vehicle is dark adapted [Hack #45], keep your charts, logs, and other working material inside the vehicle. You can study charts, record observations, and so on from the shelter of your vehicle, exposing yourself to wind only while you are actually locating and observing objects.

Use Supplemental Heat Sources

Regardless of how warmly you dress, you may need supplemental heat to stay warm. Many experienced cold-weather observers use one or more of the following solutions:

Chemical heat packs

Outfitters and sporting-goods stores sell disposable heat packs in various sizes. These small, flat packs produce heat by a chemical reaction that does not consume oxygen, and so can be used anywhere. They are wrapped in cellophane. To activate them, you simply remove the cellophane and shake them.

The smallest pack, about 2" × 3.5", is the right size for gloves, boots, and pockets, and sells for about $0.50 each. Once unsealed, it quickly warms to the 130°F to 150°F range, which is to say *hot*. It produces considerable heat for two or three hours, and continues to produce some warmth for several hours more. Larger packs, about 4" × 5", sell for $1 or so each, produce about the same temperatures but for longer periods, and are suitable to use inside your coat.

Drugstores sell reusable chemical heat packs that are larger and more expensive than disposable heat packs. These packs contain a solution of dissolved salts inside a heavy plastic pouch. You charge the pack by heating it for a few minutes in boiling water or a microwave oven. When charged, the solution is a clear liquid. When you activate the pack, the salts begin crystallizing out of solution, producing heat as they do so. When the pack is exhausted, the solution has completely solidified, and you can recharge the pack by heating it again. These reusable heat packs typically provide more heat but for a shorter time than disposable chemical heat packs.

> Chemical heat packs are also useful for warming inanimate objects. For example, once a heat pack has started to run down, don't discard it when you fire up a new one. Instead, put the partially depleted heat pack in your eyepiece case. It will provide just enough heat to keep your eyepieces warm enough that they won't dew up (or frost up) while you are using them.
>
> Similarly, cold weather is hard on laptop batteries. At very cold temperatures, a fully charged laptop battery that would provide several hours of use at normal temperatures may stop working in half an hour or less. Putting a partially used heat pack between the bottom of the laptop and the chart table keeps the battery warm and allows it to deliver its rated life. (We suggest wrapping the heat pack in a towel, both to isolate it from the cold chart table and to prevent the laptop from getting too warm.) Using a cardboard box to cover the laptop when it is not being used also extends battery life.

jon-e Handwarmers

Hunters have used jon-e Handwarmers for decades, and they're a good solution for astronomers as well. These devices look like an oversized cigarette lighter, and use a catalyst to burn liquid fuel without a flame. The Standard unit provides 8 to 12 hours of continuous heat on one filling, and the Giant unit runs for up to 36 hours per filling. You can simply keep a jon-e Handwarmer in your pocket to warm your hands occasionally, or you can buy a belt that holds two Standard jon-e Handwarmers over your kidneys to provide overall warming.

Propane heaters

Personal heaters are nice, but sometimes what you really want is to sit in front of a fireplace. Sadly, fireplaces are rare at observing sites, but there is a next-best solution. Portable catalytic propane heaters put out enough heat to make heat packs and other personal warmers seem tame by comparison. Catalytic heaters burn fuel without an open flame, and are safe to use inside a tent, vehicle, or other closed location as long as you provide some ventilation.

Don't buy a kerosene heater or a non-catalytic propane heater. These units put out a *lot* of heat, as much as 25,000 BTU/hour or more, but they produce much too much light to be usable at a dark observing site. They also cannot be used in a closed location because they produce deadly carbon monoxide gas.

Coleman (*http://www.coleman.com*) offers several models of catalytic propane heater that produce from 1,100 to 3,000 BTU/hour. (We recommend the 3,000 BTU/hour units for astronomy.) These heaters use disposable 16.4 ounce propane cylinders, which sell for a couple bucks at any hardware store and last from 8 to 18 hours, depending on the output of the unit. Catalytic propane heaters burn with a soft orange glow that is barely visible, even when you are fully dark adapted.

We don't want to overstate the amount of heat these units put out. You won't even notice the heat as you move around your observing site. In fact, the first time we fired ours up, we thought it hadn't started. It takes several minutes for the unit to start completely, and even once it's started, the heat is not obvious in the open air unless you are quite close to the unit. But 3,000 BTU/hour is a significant amount of heat if you concentrate and contain it. Figure 1-2 shows Robert warming himself with our catalytic propane heater. The blanket acts to trap the heat, and after a couple of minutes it becomes comfortably warm inside the impromptu tent.

If you or one of your observing buddies has a van, that provides the ideal solution. Simply place your catalytic heater in the van at the beginning of your observing session and start it running, leaving a window cracked a few inches for ventilation. Depending on how well the van is insulated, a 3,000 BTU/hour catalytic heater can raise the interior temperature by 20°F or more, providing a warm refuge for your observing group.

Don't forget that your coat and other cold-weather clothing are as good at keeping heat out as they are at keeping it in. When you're in the refuge, open or remove your coat to allow the heat to warm you.

If you use a propane or other heater at an observing site, remember that it produces a heat plume that roils the air over it and interferes with observing. Before you set up your heater, talk to other observers present and verify that where you plan to put the heater won't interfere with someone else's observing.

Figure 1-2. Robert warming himself with our catalytic propane heater

HACK #5 — Don't Violate Observing Site Etiquette

Rules exist for good reasons. Keep these simple ones in mind when you are observing, and you won't incur the wrath of your fellow participants.

An observing event can be anything from a large, formal star party with hundreds of participants to a small informal observing session among club members and friends who enjoy spending an evening together under the stars. Whatever the size of the event, behaving properly at the observing site allows you and others to enjoy the session.

A set of (usually) unwritten rules for observing site etiquette has developed. Most of the rules are based on common sense, but some may not be obvious to inexperienced observers. The rules vary with circumstances. If you are observing with a small group of close friends, you can be a bit more flexible. But if you are observing at a large star party, particularly if there are strangers present, it's best to follow the rules closely. Break a few rules, and you'll hear others muttering about the newbie. Break too many rules, and you may be asked to leave and not be invited back.

Don't Show a White Light After Dark

If you remember just one rule, remember this one: *do not show a white light after dark unless you have everyone's permission.* It takes half an hour or so for one's eyes to become fully dark adapted, and a brief flash of white light can instantly ruin that dark adaptation for everyone **[Hack #11]**.

If you must show a white light—and by "must" we mean for something really important, not just a matter of your own convenience—announce your intentions and get everyone's permission first. Once you have permission (which may be a long time in coming if someone is doing long-exposure imaging), give a timed warning—"White light in 30 seconds...15 seconds...5 seconds...white light is on." Limit the brightness of the white light—for example, by filtering the flashlight through your fingers—and keep it directed downward rather than waving it around. Finish what you need to do as quickly as you can, turn off the white light, and announce, "White light is off. Thank you."

Astronomers recognize three types of twilight, each of which occurs both in the evening and in the morning. Evening *civil twilight* is the period from sunset until the sun is 6° below the horizon. Evening *nautical twilight* begins when the setting sun is 6° below the horizon, and ends when the sun is 12° below the horizon. Evening *astronomical twilight* begins when the sun is 12° below the horizon, and ends when the sun is 18° below the horizon. Morning astronomical twilight begins when the rising sun is 18° below the horizon; morning nautical twilight begins when the rising sun is 12° below the horizon, and morning civil twilight begins when the rising sun is 6° below the horizon.

During civil and nautical twilight periods, the sky is still illuminated, and dark adaptation is not an issue. Dark adaptation starts with the beginning of astronomical twilight and is usually complete or nearly so by the end of astronomical twilight. Between the end of evening astronomical twilight and the beginning of morning astronomical twilight, the sky is as dark as it gets. Courtesy mandates using only red lights from the end of evening nautical twilight to the beginning of morning nautical twilight.

Arrive Before Dark

Schedule your arrival early enough to leave plenty of time to set up your equipment before full dark. If possible, plan to arrive at the observing site before sunset, but in no event should you arrive after the end of astronomical twilight. By then, everyone will be fully dark adapted, and some people may be imaging.

Plan Ahead When You Park

Access to observing sites varies, as do the rules for parking. Often, you can pull your vehicle right up to the observing area or pad. (We work directly out of our SUV at such sites.) Sometimes, you'll have to park at a distance and carry your equipment to the site. Whatever the arrangement, think about what you'll do if others are still observing when you want to leave.

Many astronomy clubs have specific rules, such as parking face-out so that you don't need to put your vehicle into reverse and thereby illuminate your backup lights. Depending on the peculiarities of your own vehicle, you may have to modify your actions. For example, our Isuzu Trooper SUV cannot be shifted into gear without depressing the brake pedal first. Although the brake lights are red, they are very bright. Accordingly, we try to point the rear of the SUV away from the observing area. Similarly, some vehicles with automatic transmissions cannot safely be left in Neutral, and must be shifted into Park. To shift from Park to Drive, you must pass Reverse, which illuminates your backup lights, if only briefly. Even that brief flash of white light is enough to ruin others' dark adaptation.

If you plan to leave when others are still observing, consider parking your vehicle temporarily while you unload, and then moving it far enough away from the observing site that you won't disturb anyone with your lights when you start your vehicle. Yes, that means you'll have to carry your gear to your vehicle when you tear down and pack up, but that's better than ruining everyone else's enjoyment.

Avoid Using Headlights After Dusk

Although you should always attempt to arrive at the observing site during daylight, if you must arrive after the end of nautical twilight, avoid using your headlights. Depending on circumstances, it may be acceptable to enter the observing area with your parking lights on. If in doubt, turn all of your lights off, walk into the observing area, and ask politely for a volunteer to lead you in with a red light.

When you depart, never use your headlights until you are well away from the observing site. Even the reflection of headlights is sufficient to ruin dark adaptation. Depending on circumstances, it may be acceptable to use your parking lights as you leave the observing site. At large, formal star parties held at dark sites, even parking lights are often forbidden. In that case, find a volunteer to lead you out using a red light. When you reach a safe point, give the volunteer at least a minute or two to get clear before you turn on your lights and ruin his night vision.

Dark Adapt Your Vehicle

If you observe frequently with others, consider modifying your vehicle [Hack #45].

Don't Crowd Others

When you set up your equipment, have consideration for those who are already set up. Leave plenty of space between you and those on either side of you. In addition to the scope itself, people need room for chart tables, chairs, and so on. In particular, giant Dobs need lots of working room, so if you're setting up near one always ask the owner if you're too close. (If you don't know what a giant Dob is, don't worry. You'll know one when you see it.)

Be Careful in the Dark

Particularly at very dark sites, be careful when you are moving around. You may be negotiating what amounts to an obstacle course in the dark, and there's a lot of expensive equipment at risk. Even if you are using a red flashlight, it's easy to walk right into someone's tripod or trip over a cable. Newbies sometimes keep silent for fear of appearing stupid, but experienced astronomers don't hesitate to ask, "Is there something here I might trip over?"

Conversely, do everything you can to help others avoid tripping over your own gear. If you are running power cables to your vehicle, for example, route them as much as possible out of traffic paths and tape or stake them down. Affix reflective tape to mounts and similar items, or install dim red LED flashers. When people approach your equipment, tell them what's in their way and use your own red flashlight to guide them around your gear.

Ask Before You Touch

Part of the fun of an observing session is having the chance to look through other people's scopes, try out each others' eyepieces and other accessories, and so on. But always ask before you touch anyone else's equipment or change anything about the setup. Treat other people's equipment as though it were your own. If you borrow an eyepiece, filter, or other accessory, return it promptly. If someone offers you a view through his scope, accept it, but tread lightly. Don't hog the eyepiece, and don't move the scope to a different object without getting permission first. If you don't know how to refocus the scope or keep it on the object, ask. Politeness counts.

Use Headphones for Music

You may think country-western (or rap, or rock, or classical) is the only type of music worth listening to. Others may disagree, sometimes vehemently. If you must listen to music while observing, use headphones to keep your music private.

Have Consideration for Others If You Smoke

The issue is not the smoke itself. Even non-smokers seldom object to someone smoking outdoors. The issue is the light produced when you light your cigarette, pipe, or cigar. When you light up, turn away from other observers and screen your match or lighter to avoid damaging their night vision.

Some clubs ban smoking entirely at observing sites. We think that goes much too far. Done properly, smoking need not interfere with others' enjoyment of the evening. However, if you are observing at a site that bans smoking, either honor that rule or find a different observing site.

Avoid Alcohol

Many clubs ban all alcohol consumption at their observing events, based on past unfortunate experiences. We certainly don't object to someone drinking a beer or two, but we avoid alcohol when we are observing. We suggest you do the same, whether or not it is banned by policy. Even moderate alcohol consumption severely hampers your ability to dark adapt and increases the risk and severity of hypothermia. Heavy alcohol consumption is a very bad idea when there is a lot of expensive equipment set up in the dark, not to mention a risk to your own safety [Hack #3].

Leave Young Children and Pets at Home

Unless young children and pets are specifically invited to an event, leave them at home. Many astronomy clubs run events specifically for young children, but only children old enough to be responsible—at least 10 or 12 years old—should attend general observing events.

It's always a bad idea to bring pets to an observing site. Even the best-behaved animal may misbehave when surrounded by strangers in the dark, and the last thing you want is to have Fido lift his leg on someone's tripod, or worse, spot a bunny, take off in hot pursuit, and knock over someone's scope.

Don't Be a Moocher

Come prepared with everything you need to enjoy the observing session, including—in addition to your equipment—warm clothing, food, warm drinks, and so on. Don't count on others to provide things you forgot to bring. Give as good as you get. If someone offers you a view through his scope, return the favor. If someone shares his coffee with you, offer him some of your munchies. If you borrow an eyepiece from him, offer him one of yours to try. Those who take and never give quickly become unpopular.

Have Consideration for Others' Time

Most amateur astronomers are social animals. They'll happily offer their scope to anyone who wants a look. But there are times when someone may be occupied, perhaps hunting down an elusive object or setting up an imaging session. If someone is clearly busy, don't bother him.

Don't Criticize Others' Equipment

You may have an $8,000 Starmaster Dob or a $5,000 apo refractor, but that doesn't give you the right to badmouth other people's equipment. We have seen some truly abominable behavior in this regard. At one public observation session, we watched a guy go from scope to scope, "star testing" them without asking permission and then telling the owners how bad their scopes were. (We would have bet this guy couldn't have run a star test to save his life, and the seeing that night was nowhere near good enough to judge much of anything.)

After you have looked through someone's scope, the proper response is "very nice" or something similar. If, and only if, the owner asks you for your opinion of his equipment should you say anything at all critical. Even then, be gentle. The guy may just want reassurance rather than an honest opinion. If it's truly clear to you that the owner wants you to be critical, then give him your honest opinion.

We make only one exception to this rule. SCTs and Newtonian reflectors are very often poorly collimated. Many inexperienced owners have no idea that their scopes ever need to
be collimated, let alone how to do it. We've seen some that
were so badly misaligned that they produced truly poor
images, much worse than the potential of the scope.

Miscollimation is easy and quick to fix, so it is a kindness to
make some non-judgmental remark, such as, "It looks like
your scope got knocked out of alignment when you set up
tonight." From there, it's an easy step to to to offer to collimate the guy's scope for him, and teach him how to collimate while you're doing it **[Hack #40]**.

Police Up Your Trash Before You Leave

Many club observing sites are on private property, and it's critical to maintain the goodwill of the property owner. But whether the site is on private or
public land, your goal should be to leave the site at least as clean as you
found it. Carry a trash bag or container and carry away everything you
brought in. If others have littered, pick up their trash as well.

Never Leave One Vehicle Alone

If only one other vehicle remains when you're ready to leave, wait until you
can leave together. We admit that we sometimes violate this rule, but never
at remote observing sites and never if the remaining person is a woman. Flat
tires, dead batteries, and other breakdowns are distressingly common. If one
vehicle is left alone, the results can be at best inconvenient and at worst
tragic **[Hack #3]**.

Do a White-Light Check If You Are the Last to Leave

If you're still there when the last group is ready to pack it in, your final step
before leaving should be to do a white-light sweep of the area to make sure
nothing is left behind. (We've found eyepieces, charts, and, in one case,
even an entire telescope that someone had forgotten to pack up.) In small
groups, someone will generally know to whom an item belongs. If you find
an overlooked item at a larger event, contact the event organizer or the club
president to make sure the item gets back to its owner.

HACK #6   Be Prepared

Carry what you need so you're ready to observe at any time.

Our regular observing gear could fill an SUV. In fact, it does. We have an Isuzu Trooper—Barbara calls it Astro Truck—dedicated to astronomy. It's always packed and ready to go on observing trips. Although it's not always feasible to carry that much gear, that doesn't mean we're not prepared to observe at any time and on no notice. The trick, as the Boy Scouts say, is to Be Prepared.

The amount and type of equipment you can have available depends on the situation; here's what we recommend.

Backpacking Kit

Weight and size limitations strictly limit how much equipment you can bring along. At a minimum, carry a 7X50 or 10X50 binocular, which is also useful for observing birds and other wildlife, and a planisphere. A planisphere, shown in Figure 1-3, displays the positions of the stars for any date and time you "dial in." (Planispheres are designed to work at a particular latitude, and they are generally produced in latitude increments of 10°. The one shown is designed for use at 40° N, but is usable from 30° N to 50° N. Choose the version that's closest to your own latitude.)

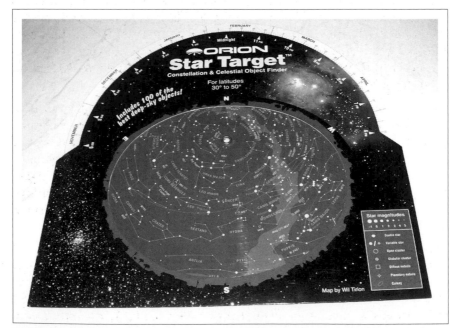

Figure 1-3. A planisphere

Better still, substitute a PDA with planetarium software installed for the planisphere [Hack #65]. The PDA weighs no more than the planisphere, is smaller, and packs a great deal more information. If you can afford the extra tonnage, substitute a giant binocular—12X60, 15X70, or so—for the standard binocular. Carry a couple of inflatable pillows that you can use while lying on the ground to brace yourself for stable views. A red LED flashlight and perhaps a narrowband filter, which you can hold between your eye and the binocular eyepiece, completes your kit [Hack #59].

An 80mm or 90mm short-tube refractor weighs the same or less than a giant binocular, but is much more flexible. With a low-power eyepiece, the scope has magnification and field of view similar to that of a binocular. With other eyepieces, the scope can provide much higher magnification. Inexpensive short-tube refractors are generally of poor optical quality and are limited to perhaps 75X or 100X. Better short-tube refractors—such as models from StellarVue, Tele Vue, Vixen, and others—can realistically support higher magnification.

Unfortunately, a short-tube refractor requires a mount of some sort. The best solution, if you or a companion can carry the weight, is a light tripod designed for astronomical use, such as the 7.5 pound Paragon HD-F2 sold by Orion. (Yes, it's heavy. Lighter tripods are available, but we know of none that are stable enough for a short-tube refractor except extremely expensive models made from exotic materials.) If even a light tripod is too heavy, consider carrying a table-top tripod. That requires you to lie on the ground to observe and may limit maximum elevation of the scope, but it provides sufficient stability for moderately high power viewing. Also consider using a monopod or mounting a tripod head atop your walking stick. It limits you to moderate magnifications, but that may suffice.

Although in general we are not fans of zoom eyepieces, they make sense for a lightweight kit built around a short-tube refractor. For low to moderate powers, the best choice is the $439 Pentax XL 8–24mm zoom eyepiece. If that's a bit rich for your blood, choose the $200 8–24mm zoom eyepiece made by Vixen and relabeled by Orion, Tele Vue, and others. For the higher powers supported by a quality short-tube refractor, either use a 3X Barlow, such as the Tele Vue 3X model, or use the $380 Tele Vue Nagler 3–6mm zoom eyepiece, depending on your budget.

For a scope-based kit, a planisphere really isn't adequate. You need real charts to find the objects that the scope can show you, and the lightest way to carry real charts is to carry a PDA with planetarium software installed [Hack #65]—of course, with red film covering the screen to save your night vision from the glare of the PDA [Hack #44]. With the PDA, you won't even need a red LED flashlight.

Car Kit

The opportunity to observe can occur any time, so it's a good idea to carry the essentials with you in your car at all times. We define a basic car kit as a 7X50 binocular and a planisphere. With just that much, you're prepared to grab any short observing opportunity that arises.

Our extended car kit is what we take along on trips. It's built around our grab-'n-go 90mm long-tube refractor. (After all, if we're traveling we don't need the grab-'n-go sitting in our library.) We bought a zippered, padded Cordura nylon case that holds the 90mm long-tube scope, an AZ-3 alt-azimuth tripod mount, the Telrad and optical finders, and the associated small parts. There's plenty of room left over for a small eyepiece case with a good selection of eyepieces and filters, a 7X50 binocular, red LED flash-lights, our *Sky Atlas 2000.0* charts, and so on. In fact, other than the limitations of only 90mm aperture, we're nearly as well equipped with our extended car kit as we are when we drive Astro Truck.

Airline-Portable Kit

You can ship even a large scope as packed luggage, but few people are com-fortable trusting their expensive optics to the tender care of the TSA and the airline baggage handlers. Accordingly, most consider "airline portable" to mean a scope they can take along as carry-on luggage, shipping only the tri-pod and mount as packed luggage. (Even the TSA goons can't do much to hurt a tripod and mount, or so we hope....) Of course, there's always FedEx.

For most people, the best choice in an airline-portable scope is a short-tube refractor. StellarVue, Takahashi, Tele Vue, Vixen, and others make suitable models that range in size from 60mm (2.4") to 102mm (4"). Most manufac-turers of airline-portable refractors also sell fitted cases that are guaranteed to meet airline size standards for under-seat stowage, and include space for eyepieces, Barlows, filters, and other accessories. Add a tripod with an equa-torial or alt-azimuth mount to your packed luggage, and you're good to go.

> If you need more aperture, there is an alternative. Teleport (*http://www.teleporttelescopes.com*) manufactures a 7" super-premium Dobsonian telescope that, incredibly, packs up small enough to fit under an airline seat. The optics are world-class, made by Carl Zambuto, and the scope itself is designed and constructed with the quality and attention to detail of a Rolls-Royce automobile or a Rolex watch. As you might expect, this is not an inexpensive scope. The Teleport 7" model costs $4,200. Tom Noe, the principal of Teleport, makes only about 15 telescopes per year, so the waiting list is quite long. Still, if you want a perfect airline-portable scope, this is the one to get.

Measure Your Entrance Pupil Size

HACK
#7

Use an Allen wrench to match your instruments to your eyeball.

Your pupils constrict in a bright environment to limit the amount of light that reaches your retinas. In the dark, your pupils dilate to admit as much light as possible as part of the dark adaptation process **[Hack #11]**. Astronomers refer to pupil diameter as *entrance pupil size* because it determines how much of the light from a binocular or telescope can enter your eyes. Measuring your entrance pupil size when your eyes are dark adapted gives you a key piece of information to help you select binoculars and eyepieces that are best suited to your own eyes.

Maximum dilated pupil size varies with age and other factors. A child under 10 years of age may reach maximum dilation of 8mm or slightly more when fully dark adapted. A young adult's entrance pupil may be as large as 7.5mm. As we age, our eyes may no longer dilate as fully as when we were young. By age 35 or 40, we may be limited to 6.5mm or less dilation, by 50 or 60 to only 6mm or less, and by 80 to only 5mm. (This is not invariably true; some 60-year-old eyes can still dilate to 7mm, and some younger eyes cannot dilate to a full 7mm.)

For maximum light gathering, you want the exit pupils of your binocular and telescope to be no larger than the entrance pupils of your dark-adapted eyes. This delivers the maximum possible amount of light and allows you to see the brightest possible image. If the exit pupil of the instrument is larger than your entrance pupil, you waste light.

To calculate the exit pupil of a binocular, divide the objective size by the magnification. For example, a 7X50 binocular delivers an exit pupil of 50/7=7.1mm, while a 10X50 binocular delivers an exit pupil of 50/10=5mm.

To calculate the exit pupil of a telescope, divide the focal length of the eyepiece in millimeters by the focal ratio of the scope. For example, a 25mm eyepiece used in an f/5 scope delivers an exit pupil of 25/5=5mm, while a 35mm eyepiece in the same scope delivers an exit pupil of 35/5=7mm.

If your fully dark-adapted entrance pupil is 5mm, for example, it's pointless to use a 7X50 binocular because it delivers a 7.1mm exit pupil. In effect, your 5mm entrance pupil stops down your 50mm objective lenses to 35mm. You could instead use a 7X35 or 10X50 binocular, either of which delivers a 5mm exit pupil that matches your entrance pupil, and the images would be as bright as those you see with the 7X50 binocular. Similarly, looked at solely from a light gathering perspective, it makes little sense to buy an eye-piece that has a focal length longer than the focal ratio of your scope multi-plied by your entrance pupil. If your entrance pupil is 5mm, for example, the longest eyepiece you should choose for use with an f/5 scope is $5 \times 5mm$ =25mm.

> You may choose to buy a longer eyepiece despite the fact that it "wastes light" because that longer eyepiece provides a wider field of view. For example, Robert, whose entrance pupil is about 6.5mm, routinely uses a 40mm Pentax XL eye-piece in his 10" f/5 scope. That eyepiece provides a huge 8mm exit pupil, and in effect turns the 10" f/5 scope into an 8" f/6.3 scope. That doesn't really matter, though, because Robert is seeing that larger field of view as brightly as it is possible for him to see it.

Your eye doctor can measure your fully dark-adapted entrance pupil for you, but you can also determine it for yourself. To do so, you'll need a set of met-ric Allen wrenches. Allow yourself to become fully dark adapted, which may take half an hour or more. Look directly at a bright star, and hold one of the smaller metric Allen wrenches along your cheek so that the long portion crosses your eye parallel to and near the eyeball, as shown in Figure 1-4. Move the wrench up and down until it is centered on your pupil. You'll see the star split into two stars, one on each side of the wrench. Substitute larger Allen wrenches until you reach a point where the star no longer splits, but is visible only as a single star on one side or the other of the Allen wrench. The size of that Allen wrench is the size of your fully dark-adapted entrance pupil.

If you observe frequently from a light-polluted site, repeat the experiment there. You may be surprised at the difference light pollution, particularly from nearby local sources, makes to your dark adaptation. For example, if your entrance pupil is a full 7mm at a truly dark site, it may be only 5mm at a brighter site. Your eyes operate on the same principles as any optical instrument. Light gathering ability varies with the square of the aperture. That means a 7mm entrance pupil admits nearly twice as much light (7^2 ver-sus 5^2) as a 5mm entrance pupil, which in turn means that you can see nearly one full magnitude deeper from the darker site [Hack #13].

HACK #8 Choose the Best Binocular
See more with both eyes.

Inexperienced astronomers often think of binoculars as poor substitutes for a telescope, something to use only if you don't have a scope. That's a mis-take. In fact, a binocular is essential observing equipment even if you own a

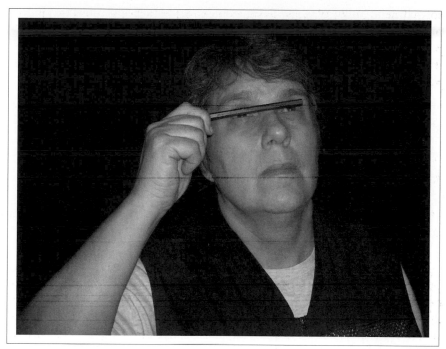

Figure 1-4. Barbara measures her entrance pupil using an Allen wrench

dozen top-notch telescopes. Experienced observers use binoculars to orient themselves within the constellation when they are locating objects they intend to view with their scopes—to place the object within its context of bright surrounding stars. A binocular is also useful for planning star hops **[Hack #21]** because the dimmest stars visible in a standard binocular are of similar magnitude to the dimmest stars visible in the optical finder of a typical telescope.

> Binocular is properly a single noun. Calling a single instrument a "pair of binoculars" is incorrect. "May I use your binocular" is grammatically correct; "May I use your binoculars" isn't, unless you mean two or more instruments. (Actually, asking to use someone's binocular is an etiquette faux pas among astronomers—nearly equivalent to asking to borrow someone's toothbrush—because most people have their personal binoculars adjusted to their own vision. If you do borrow a binocular, don't change the diopter adjustment.)

But binoculars are more than just an adjunct to telescopic observing. While the narrow fields of view of telescopes limit you to seeing just the trees, the wide fields of binoculars let you see the whole forest. For viewing large open star clusters, Milky Way star fields, comets, and other large objects, binoculars are often the best choice. On more than one occasion, we've set up our telescopes only to find at the end of the evening that we'd never used them. Instead, we'd spent the entire observing session using our binoculars to study the heavens.

Binoculars are also useful for more than just astronomical observing, of course. A good binocular has many daytime uses, including sports, birding, and other hobbies.

Key Binocular Characteristics

This section describes the important characteristics of binoculars. Understanding these issues allows you to choose a suitable binocular that fits your budget.

Magnification and Aperture

Binoculars are designated by their *magnification* and *aperture*. For example, a 7X50 binocular magnifies objects seven times and has 50mm objective lenses. The amount of magnification a binocular provides is important, but the size of the objective lenses is even more important because it determines how much light the binocular gathers.

Magnification

Magnification, also called *power*, is specified by a number followed by "X" and describes the amount of linear image enlargement provided by the binocular. For example, a 7X binocular makes an object appear seven times larger (or, alternatively, seven times closer) than if that object were viewed with the naked eye.

Binoculars are available with magnifications ranging from 4X to 20X or more in "giant" models. For hand-held binocular astronomy, the best choice is a binocular in the 7X to 12X range. All other things equal, binoculars with lower magnification show more of the sky, while those with higher magnification allow you to view more detail in celestial objects.

 Some models, called *zoom binoculars*, offer variable magnification, usually in the range of 5X to 8X on the low end to 15X to 30X on the high end. Although there are some (very expensive) exceptions, most zoom binoculars are junk—mechanically fragile and with very poor optics. We do not recommend zoom binoculars for astronomy, or indeed for any purpose.

Hand-holdability varies from person to person but is affected by the weight and magnification of the binocular. Most people can hand-hold a standard 7X binocular steadily; some can hand-hold a 10X binocular with acceptable steadiness; a few can hold-hold a 12X binocular steadily, although we consider 12X to be at the outside limit of hand-holdability.

> You can increase the steadiness of your binocular views by lying in a lawn chair or a partially inflated children's swimming pool. With such aid, we are able to handhold even 15X and 20X binoculars with acceptable steadiness. You can also tripod-mount a binocular to eliminate hand-held jitters. Even standard-size binoculars can be tripod-mounted; giant binoculars require it. Most binoculars intended for astronomy provide a standard 1/4X20 tripod socket, which is a feature worth having even if you intend to use your binocular hand-held most of the time.

Aperture

Aperture is the diameter of the objective lenses of the binocular, the large lenses at the front of the instrument (the end you don't look into). Larger apertures collect more light than smaller apertures, in an amount proportional to the square of their diameters. For example, a 50mm binocular has objective lenses twice the diameter of a 25mm binocular, but gathers 4 times as much light, because $2^2=4$. By the same calculation, a 100mm giant binocular gathers four times as much light as a 50mm binocular, and 16 times as much light as the 25mm model. Light gathering ability is essential because it allows you to see dimmer objects.

> Relative to the human eye, a 50mm binocular gathers from 50 to 100 times as much light. Because a 100X difference in brightness corresponds to five stellar magnitudes, using a 50mm binocular may allow you to see objects four to five magnitudes dimmer than those you can see naked eye. For example, if on a particular night from a particular site you can see stars down to magnitude 5.5 naked eye, the 50mm binocular might allow you to see stars down to magnitude 9.5 or dimmer **[Hack #13]**.

> To put this in perspective, for the same relative increase in light-gathering ability that you get from using a 50mm binocular instead of your naked eye, you would have to substitute a 14" to 20" telescope for the 50mm binocular.

Standard binoculars suitable for astronomy have objective lenses ranging from 35mm to 50mm. Oversized "giant" binoculars are popular with some astronomers. These giant binoculars have objective lenses ranging from 56mm up to 80mm or more, and are best used on a tripod or other solid mount. Such instruments are ideal for wide-field views of Milky Way star clouds and other large objects, but they are unsuitable for general use.

There are also so-called "super-giant" binoculars, which more resemble mounted pairs of refractor telescopes than they do standard binoculars. Super-giant binoculars have apertures ranging from 90mm to 6" (152mm) or more, and they are very specialized (and expensive) instruments. They use interchangeable eyepieces, optical finders, expensive mounts, drive motors, and other accessories more commonly associated with telescopes than with binoculars. The largest binocular we know of is the one built by a plucky Australian amateur astronomer using two 20" (508mm) telescope mirrors as objectives.

Exit pupil

If you hold a binocular up to a light source and look at the eyepieces from several inches or more away, you'll see bright circles in the eye lenses. Those bright circles are the exit pupils of the binocular, which define the point in the optical train where the light is delivered from the binocular into your eyes. You can calculate the exit pupil diameter simply by dividing aperture by magnification. For example, a 7X50 binocular provides an exit pupil of 50/7, or just over 7mm in diameter. A 10X50 binocular provides a 5mm exit pupil, as does a 7X35 or 8X40 model.

Exit pupil is important because the maximum entrance pupil of people's eyes varies in size with age and other factors **[Hack #7]**. Ideally, you want to match the exit pupils of the binocular to the entrance pupils of your dark-adapted eyes. By doing so, you deliver the maximum possible amount of light and see the brightest possible image. If the exit pupils of the binocular are larger than your entrance pupil, you waste light. For example, if your fully dark-adapted entrance pupil is 5mm, it's pointless to use a 7X50 binocular because it delivers a 7.1mm exit pupil. In effect, your eye is stopping down your 50mm objective lenses to 35mm. You could instead use a 7X35 or 10X50 binocular, either of which delivers a 5mm exit pupil that matches your entrance pupil, and the images would be as bright as those you see with the 7X50 binocular. Similarly, if you observe primarily from light-polluted locations, where your eyes can never fully dark adapt, a binocular with a 4mm to 5mm exit pupil may be the best choice.

Eye relief

The *eye relief* of a binocular is the distance between the outer surface of the eyepiece lens and where your pupil needs to be placed to view the image. Standard binoculars have eye relief ranging from only a few millimeters to 20mm or more. Long eye relief—at least 17mm to 20mm—is necessary if you wear glasses while using the binocular. Short eye relief is quite acceptable if you use the binocular without eyeglasses or with contact lenses. To decide how much eye relief you need, use the following guidelines:

- If your eyeglasses correct astigmatism or other non-symmetric vision problems, you'll need long eye relief so that you can wear your glasses while viewing. Look for a binocular that provides at least 17mm to 20mm of eye relief.

- If your eyeglasses correct only for near- or farsightedness, you can simply focus the binocular to accommodate your vision. Any otherwise suitable binocular should be fine, even if it provides minimal eye relief.

- If you are nearsighted in one eye and farsighted in the other, as both Robert and Barbara are, you can use the *diopter adjustment* on one of binocular eyepieces to adjust the binocular to your personal vision. (Make sure that the range of diopter adjustment available on the binocular you choose is sufficient.) To adjust the binocular to your personal vision, close the eye on the side with the diopter adjustment and focus the binocular sharply on the stars using the other eye. Then, close that eye, and using only the eye on the side with the diopter adjustment, turn the diopter adjustment until the focus in that eye is also sharp.

Many observers who routinely wear contact lenses find them unsuitable for astronomical observing and so wear their glasses instead. If you intend to buy a binocular to use while wearing contact lenses, try using another binocular first while wearing your contacts.

Field of view

Field of view (FoV) quantifies the angular range of the image visible in the binocular eyepieces. FoV is determined by the optical design of the binocular, including its focal length and the type of eyepieces used. FoV may be specified in two ways. Binoculars marketed to astronomers usually specify the FoV angularly, in degrees. Binoculars marketed for birding and other terrestrial uses may specify FoV linearly, in terms of the number of feet visible at 1,000 yards or the number of meters visible at 1,000 meters. (In fact, binoculars sold for astronomy are usually quite suitable for terrestrial use and vice versa, so don't let that worry you.)

> To convert linear FoV to angular FoV, divide the linear FoV by the conversion factor 52.3598 if the FoV is given in feet at 1,000 yards, or the factor 17.4533 if the FoV is given in meters at 1,000 meters. For example, to convert a FoV of 314 feet at 1,000 yards to degrees, divide 314 by 52.3598 to yield a 6.0° FoV. To convert a FoV of 131 meters at 1,000 meters to degrees, divide 131 by 17.4533 to yield a 7.5° FoV.

For astronomy, a wide FoV is desirable—you can see more of the sky with a wider field—but a very wide FoV may be problematic. Increasing the FoV beyond a certain point requires optical compromises that cause blurred images at the edges of the field, distortion, short eye relief, and other problems. For 7X or 8X binoculars, something in the 6.5° to 8.5° range is reasonable. For 10X binoculars, look for something in the 5.0° to 7.0° range, and for 12X models something in the 4.5° to 6.0° range.

> All other things being equal, a binocular with a FoV on the low end of our recommend range will probably provide better edge performance and eye relief than a similar model with a wider field. High-end binoculars can push the limits by using complex (and expensive) eyepiece designs that provide wider fields while maintaining optical quality and long eye relief.

Interpupilary distance

Interpupilary distance is the distance between the centers of your two pupils. Standard binoculars are adjustable to accommodate different interpupilary distances, typically within a range of 60mm to 75mm. That suffices for most people, but some women and many children have interpupilary distances too short for a standard binocular to accommodate. Unfortunately, the only answer is often compact binoculars, which are unsuitable for astronomy because of their small objective lenses.

Prism type

Binoculars use prisms to present a correct-image view, right-side up and not reversed left-to-right. Two types of prism are commonly used, *Porro prisms* and *roof prisms*. Ironically, although roof prisms are generally considered "better" than Porro prisms, we consider roof-prism binoculars less than ideal for astronomical use.

It's easy to identify the two types of binocular visually. Roof-prism binoculars align the objective lenses and eyepieces on a common center optical axis, giving roof-prism binoculars their characteristic single-piece "H" shape. Porro-prism binoculars are available in the two-piece "Z" or "German" shape or the single-piece "B" shape. All Porro-prism binoculars, regardless of shape, align the objective lenses and eyepieces in a stepped or offset arrangement, with the objective lenses farther apart than the eyepieces. **Figure 1-5** shows two inexpensive Orion Scenix 7X50 binoculars, which are typical Porro-prism models.

Figure 1-5. Orion Scenix 7X50 binoculars, typical Porro-prism models

Roof-prism binoculars are generally much more expensive than Porro-prism binoculars of equal optical and mechanical quality. That's true both because the roof prisms themselves are more expensive to produce than Porro prisms and because roof-prism binoculars are much less forgiving of collimation (alignment) errors. Roof prisms are typically collimated to an angular accuracy of two arcseconds. For equivalent image quality, Porro prisms must be collimated to an accuracy of only 10 arcminutes, or 1/300th as precise. These strict tolerances mean that roof prisms are ordinarily affixed permanently to a solid metal plate and laser or interference collimated before assembly, which is a time-consuming and expensive procedure. The upside is that because roof prisms are permanently affixed to a solid substrate, they seldom require recollimation.

With very few exceptions, binoculars must be collimated at the factory or a service center. Binoculars are generally pretty rugged. We've dropped numerous binoculars through the years and have never knocked one out of alignment. If you do knock a binocular hard enough to require recollimation, though, expect an expensive trip back to the factory and a long wait before you see the binocular again.

The possibility of miscollimation is a good reason to buy your binocular locally, or at least from an online source that has a no-questions-asked return policy. Even premium binoculars are sometimes miscollimated when they leave the factory, and that problem is even more likely to occur with inexpensive models.

The other disadvantage of roof prisms for astronomy is that they transmit less light than Porro prisms. The optical design of roof-prism binoculars requires that one surface of a roof prism be semi-silvered to transmit part of the light and reflect part of it. This arrangement causes light loss of 12% to 15%, which doesn't matter for daytime use but is a severe drawback for astronomy, where every photon counts when viewing dim objects.

Accordingly, we recommend avoiding roof-prism binoculars for astronomy. Stick to the tried-but-true Porro-prism binoculars. You'll see brighter images and save money as well.

Two types of glass are used to make Porro prisms. Cheap Porro prisms are made from inferior BK-7 borosilicate flint glass. Better Porro prisms are made from superior BaK-4 barium crown glass. Although nearly any binocular that uses BaK-4 prisms advertises that fact, it's easy enough to check for yourself. Simply hold the binocular with the eyepieces several inches from your eyes, pointing at the sky or another evenly lit light source, and look at the exit pupils. If the prisms are of BaK-4 glass, the exit pupils are round and evenly illuminated. If the prisms are of BK-7 glass, you'll see a square inside the circle, with the area inside the square brightly illuminated and the area outside the square dimmer.

Because of their design, roof-prism binoculars, can use BK-7 glass without suffering the edge dimming and vignetting that occurs with BK-7 Porro prisms.

Coatings

A typical binocular has from 6 to 10 lenses and prisms in the optical path between the object you are observing and your eye. Each of those optical elements has at least two surfaces, and each surface reflects a small percentage of the light rather than transmitting 100% of the light that strikes it.

An uncoated surface may reflect as much as 4% of the light that strikes it, which doesn't sound excessive until you run the numbers. If a binocular has 10 optical elements, each with two 4%-reflective surfaces, its overall light transmission is 0.96 raised to the 20th power, or 0.442. In other words, that binocular transmits only 44.2% of the light that strikes its objective lens. The remaining 55.8% of the incident light is reflected and scattered, which reduces both brightness and contrast dramatically.

Optics makers use *interference coatings* on lenses and prisms to reduce these reflections. The simplest and least expensive coating is a thin single layer of magnesium fluoride (abbreviated MgF or MgF_2), which can increase transmission to 0.985. A 10-element binocular with single MgF coatings on all surfaces transmits $(0.985)^{20}$, or about 73.9% of the incident light, a great improvement over an uncoated binocular.

Reflection can be further reduced by using multi-layer interference coatings, a process known as multicoating. Although a multicoated optical element may have literally hundreds of coatings, three to seven layers is more typical. A properly multicoated surface may have transmission ranging from 0.995 to 0.999, which translates to overall transmission of 90.5% to 98.0%. Obviously, good multicoatings make a huge difference, even compared to standard coatings.

But coating costs money, and multicoating more so. To cut costs, some optics makers coat (or multicoat) only some of the surfaces. A standard terminology has arisen to describe the levels of coatings used on binoculars and other optical equipment.

Coated

> Single-layer MgF coatings have been applied to some but not all of the optical surfaces, typically only to the external surfaces of the objective lens and eyepiece lens. We consider a coated binocular unsuitable for serious astronomical use. Only the cheapest binoculars—we are tempted to call them toys—are in this category.

Fully coated

> Single-layer MgF coatings have been applied to all of the optical surfaces. We consider a fully coated binocular the minimum acceptable for serious astronomical use. Most of the inexpensive binoculars sold by Wal*Mart and similar big-box retailers are in this category.

Multicoated

> Multi-layer coatings have been applied to some but not all of the optical surfaces, typically only to the external surfaces of the objective lens. Usually, but not always, the other surfaces have had single-layer MgF coatings applied, which is sometimes described as "fully coated and multicoated." Most inexpensive binoculars suitable for astronomical use, such as the Orion Scenix models (*http://www.telescope.com*), are in this category.

Fully multicoated

> Multi-layer coatings have been applied to all of the optical surfaces. Binoculars in this category are the best choice for astronomical observing, but are not inexpensive. Low-end, fully multicoated models, such as the Orion UltraViews, start at $150 and go up from there.

 All coatings are not of the same quality. Applying top-notch multicoating is a very expensive process that, if done properly, can exceed the cost of making the lenses themselves. On the other hand, if all you care about is being able to claim that your optics are multicoated, you can slap on multicoating relatively cheaply. Optics from Zeiss, Nikon, Fujinon, or Pentax (to use just a few examples) have superb multicoating. Optics from second-tier Japanese makers, such as Vixen, have good multicoating, but not as good as that of first-tier makers. Cheap multicoated optics from Chinese makers, well, they're multicoated, but that's about the most you can say about them. Avoid any binocular with "ruby red" or other strange coatings.

Recommended Binoculars

Binoculars are available in a wide range of prices. For convenience, we classify standard-size binoculars by price range as inexpensive (<$75), midrange ($75 to $250), premium ($250 to $500), and super-premium (>$500). As you might expect, it's not difficult to get a good binocular if you are willing to pay premium or super-premium prices. Even midrange binoculars are generally excellent.

In the sub-$250 range, each doubling of price generally buys you substantially better optical and mechanical quality, all other things being equal. For example, a $100 7X50 binocular is not twice as good as a $50 7X50 binocular, but it is likely to be built more solidly and to provide noticeably superior images. A $200 7X50 binocular provides a similar relative improvement over a $100 binocular. Once you get into the $300 range, although you can spend much more, the improvements become incremental, not to say invisible. For example, other than by examining the name plate, few people would be able to detect any difference in optical or mechanical quality between a $300 10X50 binocular and a $600 10X50 binocular.

Even some inexpensive binoculars provide surprisingly high bang-for-the-buck. When Barbara first became interested in astronomy several years ago, Robert bought her a $90 Orion Scenix 7X50 binocular, figuring that if she lost interest in astronomy it wouldn't be any great loss. Robert had been on a 20-year hiatus from active observing—college, jobs, and "real life" intervened—but was determined to get back into the hobby. At first, he planned to buy a premium Zeiss, Leitz, or Swarovski binocular, but when he saw how good Barbara's inexpensive Orion Scenix binocular was, he bought a Scenix for himself. Is the Scenix as good as premium models? No, but for one-fifth to one-tenth the price, it's astonishingly good.

- If your budget is very tight, nearly any inexpensive binocular is better than nothing. Avoid gimmicks, such as "instant focus," red coatings, and so on. Look for a binocular that uses BaK-4 prisms and is at least fully coated. A 7X35 model is acceptable, and in fact may be a better choice than a 7X50 or 10X50 model at the same price.

- For standard binoculars in the $75 to $250 range, most models from Bausch & Lomb, Celestron, Minolta, Nikon, Olympus, Orion, Pentax, Pro Optic, and Swift are reasonable choices. At the low end of this range, we think the 7X50 and 10X50 Orion Scenix models are the stand-out choices. (We'd avoid the 12X50 Scenix and particularly the

8X42 model.) At the upper end of this range, we think the Orion Vista binoculars in 7X50, 8X42, and 10X50 offer very high value, as do the similar but more expensive Celestron Ultima models.

> Some models in this price range include built-in digital cameras, which we consider a drawback. Too much of the price goes to the digital camera and too little to the optics. Avoid any "gimmick" binocular. It almost certainly has trash optics and will be none too durable.

- For standard binoculars in the $250 to $500 range, there are many suitable candidates. At the lower end of the range are the excellent Celestron Ultima models. At the middle and upper parts of this range, you will find many suitable models from Alderblick, Celestron, Fujinon, Nikon, Pentax, and Steiner.

> Canon image-stabilized binoculars also fall into this range. Although they are popular with some astronomers and of excellent optical quality, we consider them a poor choice for astronomy. Their apertures and exit pupils are small, and they are expensive. Also, many people find their electronic image stabilization makes for an odd viewing experience. All things considered, if you're concerned about image stability, we think a standard binocular on a decent tripod is a better choice.

- For standard binoculars in the stratospheric super-premium price range, nearly any Porro-prism model with provision for tripod mounting is an excellent choice. This is the realm of world-class optics from companies like Leitz, Swarovski, and Zeiss. Premium models from companies like Fujinon, Nikon, Pentax, and Steiner are also in this price class. Binoculars simply don't get any better than this.

> If you are considering a mid-range or better binocular, buying a used model can save you money. Decide which brands, models, and price ranges suit you, and then check Astromart (*http://www.astromart.com*) to see what's available (the selection varies day to day and even hour to hour). Depending on age and condition, used mid-range and premium binoculars generally sell for 50% to 80% of what you'll pay for the same model new. Don't buy an inexpensive binocular used, though. Chances are good you'll just be buying someone else's problem.

Like standard binoculars, giant and super-giant binoculars are available in various sizes and price ranges. Although we do not recommend a giant binocular as your first (or only) binocular, they have their place. In fact, some observers eschew telescopes entirely and spend all of their observing time with a good tripod-mounted giant binocular.

We classify giant binoculars as inexpensive (<$250), midrange ($250 to $1,000), premium ($1,000 to $5,000), and super-premium (>$5,000). Some inexpensive giant binoculars are surprisingly good for their price. You won't mistake them for premium units, but they are quite usable. They are generally sharp at the center of the field, but with noticeable softness in the outer 15% to 30% of the field. Inexpensive giant binoculars often show significant chromatic aberration (false color) on very bright objects such as Luna and bright stars, but they are not really intended for observing those types of objects. For doing what they do best—scanning Milky Way star fields and viewing open star clusters—they serve quite well. If you can afford better, you'll find that spending twice as much delivers noticeably better image quality and mechanicals. As with standard-size binoculars, there are premium and super-premium models available for those who can afford them (large Fujinon, Leitz, Takahashi, and Zeiss models sell for more than $10,000, sometimes much more).

- If you're on a tight budget, the Chinese-made Celestron SkyMaster series is a good choice in a giant binocular. Three models are available, 15X70, 20X80, and 25X100. All use BaK-4 Porro prisms and are multi-coated. Eye relief is acceptable, at 18mm in the 15X70 model, and 15mm in the two larger models. The 15X70 model sells for well under $100, and even the 25X100 model can sometimes be found on sale for under $250. All models have provision for tripod mounting, which is necessary for these large instruments. The 15X70 model is water resistant, and the two larger models are waterproof. Image quality is mediocre, particularly at the edges, but is surprisingly good for the price. We think most occasional users will be quite pleased with a Celestron Sky-Master binocular.

- For inexpensive semi-giant binoculars, we think the stand-out choice is the Orion Mini-Giant series. These fully multicoated Japanese-made optics are available in 8X56, 9X63, 12X63, and 15X63 models, ranging in price from $159 to $219 and with fields of view from 5.8° in the 8X model to 3.6° in the 15X model. Eye relief across the line is excellent, from 17.5mm to 26mm. The smaller models are hand-holdable, although all models have tripod sockets. Image quality is, if not quite up to the level of the premium brands, more than acceptable to most people.

We can't make other recommendations because our experience with giant binoculars is very limited. However, we will say that Fujinon offers several giant binoculars that are extremely popular with amateur astronomers and receive uniformly excellent reviews. If we wanted to buy a premium or super-premium giant binocular, we'd look at Fujinon models first.

HACK
#9

Choose the Best General-Purpose Telescope

Understand the advantages and disadvantages of popular scope types.

If you want to start a war, just ask a group of astronomers what the best type of scope is. Everyone agrees that "junk" scopes should be avoided, of course, but that's about the extent of the agreement. There are many types of scopes, all of which have advantages and drawbacks relative to the other types. Each type of scope has proponents and detractors, and the debate can become quite heated. In this hack, we'll attempt to provide unbiased advice about the strengths and weaknesses of each type of scope.

Just so that you're aware of it going in, we confess that we're "Dob bigots." We tried to get an "SCT fanatic" we know to help us with this hack, but he's not speaking to us. Neither is a "refractor maniac" we know. And they're not speaking to each other. We're only kidding, of course, but feelings do run high when people start debating the best scope type. The Coke–Pepsi, PC–Mac, and Linux–Windows wars are nothing compared to the scope-type wars.

Scope Characteristics

Here are the three most important characteristics of a telescope:

Aperture
> The aperture of a telescope is the diameter of its primary mirror or objective lens, which may be specified in inches or millimeters. Amateur telescopes have apertures ranging from 60mm (2.4") to 30" or more. Aperture determines the amount of light a scope can gather, the fineness of detail it can resolve, and the maximum and minimum useful magnifications for the scope.

> Light gathering is proportional to the square of the aperture. For example, a 10" scope gathers four times as much light as a 5" scope. The amount of light gathered determines how "deep" the scope can go. Larger aperture allows you to see dimmer objects (and more detail in all objects) than a smaller aperture.

Resolution is proportional to the aperture. For example, a 10" scope can resolve detail twice as fine as a 5" scope (assuming equal optical quality and steady seeing conditions).

Aperture

See above.

Aperture

See above.

Above all, aperture rules. Here are some other important characteristics:

Focal length

The *focal length* of a scope is the actual or virtual distance from the optical center of its primary mirror or primary objective lens at which it brings an object located at infinity to focus. The focal length of amateur telescopes ranges from 400mm (~16") for short-tube refractors and other small scopes through 4,000mm (~160") or more for the largest amateur instruments.

For a given focuser size, focal length determines the maximum possible true field of view (TFoV) of a scope, which is to say how wide a swath of sky is visible in that scope. For example, a scope of 400mm focal length with a 2" focuser has a maximum possible TFoV of about 7°, while a scope of 2,800mm focal length with a 2" focuser can never show more than about 1° of sky.

Conversely, a short focal length makes it difficult to achieve high magnification. Magnification is calculated by dividing the focal length of the scope by the focal length of the eyepiece. For example, using a 14mm eyepiece with a scope of focal length 2,800mm yields 200X because 2,800/14=200. To get that same 200X magnification in a scope of only 400mm focal length, you'd need a 2mm eyepiece (400/2=200). But such short focal length eyepieces have problems of their own—typically including tiny eye lenses and very short eye relief—that make them uncomfortable to use.

Accordingly, short focal length telescopes are best for low-power, wide-field applications, such as scanning Milky Way star fields—while long focal length telescopes are best for high-power, narrow-field applications—such as Lunar and planetary observing. Typical general-purpose amateur telescopes have focal lengths ranging from about 1,000mm to 2,500mm, which allow using high magnifications easily while maintaining reasonably wide fields of view.

Focal ratio

The *focal ratio* of a scope is the ratio of its focal length to its aperture. For example, a scope of 250mm (~10") aperture and 1,250mm focal length has a focal ratio of 1,250/250=5, expressed as f/5. A typical 8" (203.2mm) SCT (Schmidt-Casssegrain Telescope) has focal length of 2,032mm, and accordingly a focal ratio of f/10. Mainstream amateur telescopes have focal ratios ranging from f/4 to about f/16, with the vast majority in the lower half of that range. A scope with a focal ratio of f/6 or less is considered to be a "fast" scope. An f/6 to f/9 focal ratio is considered medium, and a larger focal ratio is considered slow.

Fast, medium, and slow as applied to telescope focal ratios is a holdover from photography, where a lens with lower focal ratio allows a shorter exposure time than a lens with a higher focal ratio. Focal ratio has no relationship to the brightness of the image a scope provides when used visually. For example, an 8" f/5 scope provides the same image brightness as an 8" f/10 scope, if the two scopes are used at the same magnification.

Focal ratio is important because, for standard reflector and refractor telescopes, the focal ratio determines the length of the optical tube assembly (OTA). For example, a 10" f/5 scope has a tube about 50" long, whereas a 10" f/10 scope has a tube 100" long. Long tubes are heavy, hard to transport, and require heavy, expensive mounts. All other things being equal, then, a fast focal ratio is clearly desirable.

Unfortunately, all other things are not equal. Fast focal ratios require deeply curved mirrors and lenses, and these deep curves are much harder and more expensive to produce to the required level of precision than are the shallower curves used by instruments with longer focal ratios. Fast focal ratios are also hard on eyepieces. Nearly any modern wide-field eyepiece provides excellent image quality in an f/8 or slower scope, but only modern, complex (read "expensive") eyepiece designs can provide a wide-field image that's sharp edge to edge in a fast scope. In exchange for greater portability, those who buy fast focal ratio scopes resign themselves to buying expensive premium eyepieces or putting up with soft edges in inexpensive wide-field eyepieces.

Optical quality

As odd as it sounds, optical quality is not a major issue with modern commercial telescopes. There are differences, certainly. Premium telescopes have optics as accurate as it is humanly possible to make and are priced accordingly. But even mass-produced Taiwanese and Chinese scopes have surprisingly good optics. So good, in fact, that only a very experienced observer on a night when the atmosphere is extremely stable will be able to tell the difference in optical quality between a premium scope and a mass-produced model. For most observers in most locations on most nights, atmospheric turbulence will be the limiting factor, not the quality of the optics.

There are a couple of caveats about mass-produced scopes. There is, of course, a great deal more variation in optical quality among mass-produced scopes than among premium models. When you buy a premium scope, you get superb optics, period. That's part of what you're paying for. When you buy a mass-produced scope, the optics may be anything from mediocre to excellent. That variability is part of the reason the scope costs much less than a premium model. Quality control costs money.

Also, when we sing the praises of mass-produced scopes, we're not endorsing all of them. Some models are excellent, but there are many "junk" scopes available—including some with well-known names like Meade and Celestron—that have simply terrible optics. The trick is knowing the difference, but then that's why you're reading this.

Mount

Every telescope requires a mount of some sort. If you've ever used a binocular, you know how unsteady the image can be, even at only 7X or 10X magnification. Telescopes typically operate at magnifications in the range of 50X to 300X, which makes a stable mount imperative.

There are two broad classes of telescope mount. An altitude-azimuth (alt-az) mount is simple, inexpensive, light, and intuitive to learn and use; however, it is not designed to track the apparent motions of the stars unless you add supplementary equipment. An equatorial (EQ) mount is complex, more expensive, heavier, and difficult to learn to use properly; but it is designed to track the stars.

Unfortunately, as scopes have gotten better and less expensive, equatorial mounts have gone in the opposite direction. The typical cheap Chinese equatorial mounts supplied with low-end scopes are much too light to support the weight of the scopes they're bundled with for anything more than rudimentary visual observing. They're crude, shaky, and shoddily constructed. Most beginners who attempt to use such mounts believe the problems they have are their fault. They're not. The mounts themselves are the problem. Even an experienced astronomer can't do much with them. For an inexpensive scope, an alt-az mount is usually a better choice.

If you are thinking of buying an inexpensive equatorial mount because you want to do astrophotography, disabuse yourself of that notion. Cheap EQ mounts have neither the accuracy nor the precision needed for long-exposure astrophotography. The least expensive EQ mount suitable for serious astrophotography is the Vixen GP-DX, which costs $1,300. That's for the mount only—no scope included.

Some alt-azimuth and equatorial mounts provide a go-to feature, either standard or as an optional upgrade. Go-to mounts include drive motors and a hand controller with electronics that calculates the current position of a specified object and automatically moves the scope until it is pointed at that object.

To use a go-to mount, you initialize it by choosing two or three bright "guide stars." (A go-to scope can usually be initialized against any of 25 or more guide stars, so there are always guide stars available from any location and at any time of year.) As you point the scope to each guide star, you press a button to indicate that the selected guide star is centered in the eyepiece. After the scope is initialized, you simply choose an object on the hand controller, and the scope moves automatically to that object.

A go-to mount can be very useful, as long as you don't use it as a crutch to avoid learning the night sky. On nights when you're more interested in looking at objects than pursuing the challenge of locating them manually, a go-to scope allows you to spend your time looking instead of finding. A go-to scope is also very useful for **urban observing [Hack #10]** because it can be very difficult to locate objects manually under bright urban skies.

Unfortunately, inexpensive go-to scopes are typically quite unreliable, both mechanically and in terms of locating objects. They seldom center the desired object in the eyepiece, and quite often fail entirely to put the object anywhere in the field of view. They use cheap motors, plastic gears, and other low-end components that are likely to fail sooner rather than later. Because so much of the cost of an inexpensive go-to scope goes toward the motors and electronics, the optics are usually small and of low quality, which means that even if the go-to succeeds in locating the object, you won't be able to see much detail, if the object is visible at all. Unless you are willing to spend at least $1,500 to $2,000 on a go-to scope, we suggest you avoid go-to.

Some telescopes, notably Orion IntelliScope Dobsonians, include *digital setting circles* (DSCs), which serve a similar purpose—helping you locate objects automatically. Unlike go-to scopes, these "push-to" scopes don't have motors, so they can't move to the object on their own. Instead, when you enter the object to be located, the hand controller displays arrows to tell you which direction to move the scope. You push the scope in the direction indicated, and when it's pointed at the object, the hand controller display "zeroes out" to tell you that you've arrived.

Portability

Don't underestimate the importance of portability, which is determined by the size, weight, and bulkiness of the various parts of the telescope and mount, as well as the ease or difficulty of setting up and tearing down the scope. A scope that is light, portable, and can be set up quickly and easily will be used much more than one that is heavy, awkward, and requires more time and effort to set up and tear down.

Some scopes are obviously very portable, for example small to mid-size refractors. Other scopes are just as obviously difficult to transport, such as large tube Dobsonians. A 16" tube Dobsonian, for example, may require a van to transport and two or three people to set up and tear down, both because of its bulk and its weight. Although mid-size (8" to 10") SCTs can be handled by one person, large SCTs suffer from portability problems. An 11" to 14" SCT really needs two people to set up safely, and the 16" and larger models might as well be considered observatory instruments.

Despite their bulk, 10" and smaller tube Dobsonian scopes are very portable, assuming your vehicle is large enough to accommodate a tube that's a foot or so in diameter and four feet long. Tube Dobs have only two parts—the optical tube and the base—and can easily be set up and torn down by one person in a minute or less. We consider 12" to 12.5" the maximum practical size for a tube Dob, although tube Dobs are made in sizes up to 17.5". For larger Dobs, tubes are simply too heavy, large, and unwieldy. Fortunately, various companies manufacture truss-tube Dobs, which replace the single large solid tube with a structure of thin tubes. Truss Dobs are made in sizes as large as 30", and even 20" models can be assembled and torn down by one person. Unfortunately, truss Dobs are very expensive compared to a tube Dob of the same size.

Cool-down time required

In order to provide its best images, the mirrors and/or lenses in a telescope must be allowed to equilibrate to ambient temperature. (For some reason, amateur astronomers call this process "cool down," even when the telescope starts out cooler than the outside air.) Until the optics reach air temperature, the scope does not provide its best resolution and, in some types of scopes, tube currents cause wavy images, distortion, and other visual anomalies. All other things being equal, a larger scope always takes longer to cool down than a similar smaller model, but different types of scopes have different cool-down characteristics.

In general, small refractors cool quickly. Unless the temperature differential is large, a refractor is ready for use within a few minutes after it is set up. A 6" to 10" Newtonian reflector (such as a Dob) may require 30–60 minutes to cool down, depending on the temperature differential, and less if fans are used to expedite cooling. An 8" to 10" SCT (Schmidt-Cassegrain Telescope) might require between one and two hours to cool completely, and a 5" MCT (Maksutov-Cassegrain Telescope) might require two to three hours to cool fully. Large scopes, including 12" or larger SCTs and large Newtonian reflectors, may never cool fully, even with fans, because their massive mirrors simply cannot lose heat fast enough to keep up with the decline in air temperature over the course of the evening.

Proper cool down is particularly important when you view Luna or the planets. For observing faint fuzzies, cool down is less critical. You can't see much fine detail in them anyway, so the resolution degrading effects of an uncooled mirror don't matter much. We often observe DSOs **[Hack #22]** while we wait for our mirror to cool down.

All of these characteristics are important, and all of them pertain to any scope of any type you might buy. In the following sections, we examine the details of the various types of scopes available.

Scope Types

There are actually dozens of different optical designs used in scopes, but all of them fall into one of three broad categories:

- *Refractor* telescopes use only lenses to form the image that is delivered to the eyepiece.
- *Reflector* telescopes use only mirrors.
- *Catadioptric* telescopes use both lenses and mirrors.

In the following sections, we examine each of those categories and explain the advantages and disadvantages of each.

Refractors. When most people hear the word "telescope," a refractor is what they think of (which is why we used a refractor on the front cover). When Galileo first turned his telescope to the heavens in 1610, it was a refractor, and refractors have remained popular with amateur astronomers ever since. Figure 1-6 shows a typical refractor. Galileo would have felt right at home with it. (Actually, he would have killed for one anywhere near this good.)

Figure 1-6. A typical refractor: our Orion 90mm f/11.1 "long-tube" on an alt-az mount

In its simplest form, a refractor comprises a tube with an *objective lens* on one end and a focuser with an eyepiece on the other. The objective lens gathers and focuses the light to a point about midway in the focuser's travel. The tube of a refractor is therefore roughly as long as the focal length of the objective lens, plus the length of the lens shade on the front end and the focuser mechanism and drawtube on the rear. Our 90mm f/11.1 refractor, for example, has a focal length of 1,000mm. The tube, including lens shade and with the focuser mechanism fully extended, is roughly 1,200mm long.

For practical reasons, most modern traditional-style ("long-tube") refractors have focal lengths of 1,000mm or less, regardless of their apertures. This is true because tubes much longer than one meter are unwieldy and require tall (and expensive) mounts. Accordingly, most manufacturers design their refractors to have a focal length of no more than 1,000mm. But if the aperture varies and the focal length remains constant, the focal ratio of the scope must also vary. And that is exactly how manufacturers handle the problem. For example:

- A 60mm (2.4") refractor has a focal ratio of f/10 to f/16, and a focal length of 600mm to 1,000mm.
- A 70mm (2.8") refractor has a focal ratio of f/10 to f/14, and a focal length of 700mm to 1,000mm.
- An 85mm (3.3") refractor has a focal ratio of about f/12, and a focal length of about 1,000mm.
- A 100mm (4") refractor has a focal ratio of about f/10, and a focal length of about 1,000mm.
- A 127mm (5") refractor has a focal ratio of about f/8, and a focal length of about 1,000mm.
- A 152mm (6") refractor has a focal ratio of about f/6.7, and a focal length of about 1,000mm.

Notice the pattern? As the aperture increases, the manufacturer reduces the focal ratio to keep the focal length no longer than about 1,000mm. (Actually, a few 6" f/8.3 refractors are made, with focal lengths of about 1,250mm and correspondingly longer tubes, but these scopes are extremely unwieldy because of their length and weight.) It might seem that the manufacturers could continue that game forever. Why not, for example, make an 8" f/5 refractor or even a 10" f/4 model? Either would have the same 1,000mm focal length, and a reasonably short, albeit heavy, tube.

Alas, the laws of optics don't allow it. The problem is *false color* (otherwise known as *chromatic aberration*), which manifests as a colored fringe around bright objects, or even as an overall color cast. False color is not just esthetically displeasing; it actually reduces the amount of visible detail significantly. And, in any refractor design, false color increases as you increase the aperture and as you decrease the focal ratio. Our imaginary 8" f/5 or 10" f/4 refractor would be hit by a double whammy: much too much aperture, and much too short a focal ratio. (Actually, the false color in a standard 6" f/6.7 refractor is hideous. Most people consider the false color even in a standard 5" f/8 refractor unacceptably high. The false color in our imaginary 8" or 10" refractor would be positively kaleidoscopic.)

The subject of false color is always on the minds of refractor owners. Even people who own $5,000 and $10,000 premium refractors debate whether this or that model has a bit less false color than an equally expensive competing model. The best refractors show only slight false color even on bright objects, but some false color is inherent in the optical design of any refractor.

Broadly speaking, there are two classes of refractors:

Achromatic refractor

An *achromatic refractor* (or *achromat*) typically uses a two-element objective lens, with elements of crown and flint glass. An achromat is reasonably well corrected for most optical aberrations, but it shows noticeable chromatic aberration on bright objects such as Luna, the planets, and bright stars. False color in an achromat can be nearly eliminated by using a longer focal ratio (and thereby increasing the length of the tube), but that is impractical for any aperture larger than 90mm to 100mm because the length of the tube becomes excessive.

Apochromatic refractor

An *apochromatic refractor* (or *apochromat*) typically uses a three-element objective lens, or a two-element objective made from expensive rare-earth glasses or calcium fluoride. An apochromat is well corrected for optical aberrations, including chromatic aberration.

> Whether the scope is an achromat or an apochromat, chromatic aberration increases with increasing aperture and decreasing focal ratio (although the level is much less in an apochromat of the same aperture and focal ratio.) For example, a 5" refractor shows more chromatic aberration than a 3.5" refractor of the same focal ratio. Conversely, an f/5 refractor has much more chromatic aberration than an f/10 refractor of the same aperture.

Achromats are inexpensive to moderately priced for their aperture, typically $35 to $100 per inch of aperture. Small achromats—those in the 60mm to 100mm range—are popular beginner scopes, although we believe they are seldom a good choice.

Apochromats are very expensive, typically $250 to $1,000+ per inch of aperture, with the cost per inch climbing rapidly as aperture increases. Apochromats in the 60mm to 127mm (2.4" to 5") range are popular among advanced amateur astronomers, particularly those who do imaging. Apochromats of 6" to 10" aperture are available, although in larger sizes the prices become—dare we say it?—astronomical. For example, a 6" apochromat might cost $7,500 (tube only), and a 10" model might cost $40,000.

Some refractors are described by their manufacturers as "semi-apos" or "neo-achros." These are marketing terms rather than technical categories. Both indicate a scope that isn't quite well corrected enough to be honestly described as apochromatic, but has better color correction than standard achromats. A refractor that does not claim to be an apochromat but is described as using "ED glass" or a "fluorite element" is usually a semi-apo, although true apos also use ED glass and/or fluorite elements.

The classic or long-tube refractor has been around for hundreds of years. Recently, the *short-tube refractor*, shown in Figure 1-7, has become quite popular. These scopes are typically 70mm to 90mm in aperture with focal ratios of f/5 to f/6, although some models are as large as 150mm. The short focal lengths provide wide fields of view and a short, easily mounted optical tube, both of which are desirable features. Short-tube refractors are popular as grab-'n-go scopes **[Hack #10]** and for scanning Milky Way star fields, open star clusters, and other large astronomical objects. Short-tube refractors are generally a poor choice for high-magnification viewing, such as Lunar and planetary observing.

Refractors have the following advantages:

Simplicity and durability
  Assuming that you take reasonable care, there's not much that can go wrong with a refractor. The optics are collimated at the factory and seldom if ever need to be recollimated **[Hack #38]**. There's no twiddling necessary with a refractor. You simply install it on its mount and start viewing.

Portability
  Small refractors, particularly short-tube models, are extremely portable. They are relatively short and light, and so they do not require a heavy or complex mount. Easy portability is one reason why refractors are very popular as grab-'n-go scopes.

Fast cool down
  Small and mid-size models require little or no cool-down time to provide their best images. You can simply set them up and start observing.

Pristine image quality
  Refractors have no secondary mirror or other central obstruction in the light path to produce diffraction spikes and reduce contrast. A well-made refractor provides bright pinpoint stars on a velvet black background. Most astronomers agree that refractors provide the most esthetically pleasing images of any telescope type.

Figure 1-7. The StellarVue 80mm f/6, a typical short-tube refractor

Usable for terrestrial viewing

Astronomical telescopes, including refractors, provide an image that is flipped left-to-right and/or inverted, and so is useless for terrestrial observing. However, you can convert an astronomical refractor to a terrestrial scope simply by installing a correct-image diagonal or eyepiece, allowing the scope to serve two purposes.

Ideal for astrophotography

Refractors, particularly apochromatic models, are well suited for astrophotography and are the scope of choice for many professional astrophotographers. Their absence of diffraction effects and high contrast makes refractors an ideal match for photography.

Most astronomers feel an urge to photograph the heavens, but astrophotography is an expensive hobby that requires total dedication. Don't expect to take good astrophotographs with inexpensive equipment. Those "amateur" astrophotographs you see in the astronomy magazines often represent weeks or months of effort using equipment that costs $5,000 to $50,000. Failing at astrophotography is one of the main reasons people leave the hobby. If you are determined to shoot high quality astrophotographs, plan to spend a lot of money on equipment and months or years learning how to do it.

Refractors have following disadvantages:

Small aperture
Practical refractors have apertures of 60mm to 150mm. These relatively small apertures limit both the light-gathering ability and the resolution of refractors. If your observing is limited to Lunar, planetary, and double stars, a refractor may be a good choice. But if you have any interest in observing DSOs, you need a larger scope.

False color
As we said earlier, all refractors exhibit false color to some extent. You can limit the problem by choosing a scope of relatively small aperture and long focal ratio or by spending the money necessary to get an apochromat.

Inconvenient eyepiece location
Because the eyepiece of a refractor is on one end of a long tube, the eyepiece position changes dramatically with the elevation of the scope. If you are observing an object near zenith, you may find yourself lying on the ground to get your eye low enough to see into the eyepiece. Conversely, if you are observing an object near the horizon, you may find yourself standing erect, or even on a stepladder or stool **[Hack #60]**, depending on the height of the mount.

High price
Refractors, particularly apo models, are the most expensive type of scope in terms of dollars spent per inch of aperture.

Dobsonian reflectors. In the 1970s, John Dobson started the Sidewalk Astronomers (*http://www.sidewalkastronomers.com*) in San Francisco. Dobson set out to bring telescopes to the people. His goal was to build large telescopes at low cost. He achieved that goal by scrounging, begging, and recycling materials to build his scopes. He ground his primary mirrors, for example, from salvaged ship portholes, and recycled old binoculars for finder scopes.

The real problem was the mount. Traditional mounts for scopes the size of those Dobson was making would have cost thousands of dollars, and it was impossible to produce home-made versions of these mounts with adequate precision. In a moment of inspiration, Dobson came up with a simple but brilliant idea. Instead of using a traditional tripod or pier mount, Dobson designed a simple alt-azimuth box mount that sat flat on the ground and rode on Teflon bearings. Such mounts could be produced cheaply from inexpensive, easily worked materials such as plywood and kitchen counter laminate, and were stable enough to support even the largest scopes.

Commercial telescope makers grabbed the ball and ran with it, and nowadays Dobsonian reflector telescopes are ubiquitous. If you attend a large star party [Hack #2], you'll probably see more Dobs than all other types of scopes combined. Figure 1-8 shows our 10" f/5 Orion XT10, a typical commercial Dobsonian telescope. Similar models are produced or sold by numerous companies, including Orion, Celestron, and others.

Figure 1-8. A typical tube Dobsonian telescope

Dobsonian-style scopes were responsible for the proliferation of large-aperture scopes among amateurs. When Robert started observing in the mid-60s, the standard amateur instruments were commercial 60mm refractors and 6" home-made reflectors on home-made equatorial mounts. People would drive for hours for a chance to look through an 8" scope, and if you had a 10" scope you probably had one of the largest amateur instruments in the state. Nowadays, 10" and even 12" Dobs are considered mid-size instruments, suitable even for beginners, and many serious amateurs own 15", 20", and even 30" Dobs. And it's all thanks to John Dobson.

As scope sizes began to increase, a problem became apparent. In Dobs up to 10" or 12", the scope tube is awkward but manageable. Most people can, with little or no assistance, handle a 4- or 5-foot tube that's 12" or 14" in diameter and weighs 30 to 50 pounds. In larger apertures, though, a solid tube becomes impractical. The largest tube Dobs have tubes 8 feet long that weigh more than 300 pounds. You need a crane to move them, or at least a couple of strong friends.

With the tube putting a practical upper limit on aperture, some alternative was needed if Dobs were to continue growing larger. That alternative is called a truss Dob. In a truss Dob, the solid tube is eliminated, replaced by a structure of light aluminum tubes that connect the focuser cage to the mirror box. Figure 1-9 shows a typical truss Dob; this one is a 17.5" f/5 model built by contributor Steve Childers (at left), with Paul Jones center and Robert on the right.

To give you an idea of the scale of large truss Dobs, all of us are about 6'4" tall, and this is "only" a 17.5" model. So-called "monster Dobs" are 30" to 40" in aperture. To see through the eyepiece when these huge scopes are pointed near zenith, you must stand atop an 18' or 20' stepladder. (Standing on a stepladder two stories off the ground in pitch blackness—now there's our idea of a good time.)

Dobsonian reflectors have the following advantages:

Price

Dobsonian reflectors are the least expensive type of telescope in terms of dollars spent per inch of aperture. Mid-size tube Dobs—those in the 6" to 12" range—typically sell for $50 to $75 per inch of aperture, a small fraction of the cost of other types of mid-size to large telescopes. With a tube Dob, most of what you pay goes for the optics rather than the mount, motors, and electronics for which you pay a high price with other types of scopes.

Figure 1-9. A typical truss Dobsonian; this one is a home-built 17.5" f/5

Truss Dobs, although they cost considerably more than tube Dobs of similar size, are also a bargain. Although $3,000 or more for a 15" truss Dob sounds expensive, when you compare that to the cost of a traditionally mounted 15" scope, the cost advantage of the Dobsonian mount becomes apparent.

Stability

A well-built Dob is inherently immensely stable. The optical tube sits atop a rigid plywood rocker box that rests on a groundboard, which sits flat on the ground. The center of gravity is very low, and the weight of the scope is distributed to three widely spaced feet on the groundboard. With other inexpensive mounts, just touching the focuser causes the image to bounce around for several seconds. With a Dob, vibrations damp out almost instantly.

Intuitiveness

Nearly everyone intuitively understands the up-down-left-right motions of a Dob. When you want to point a Dob at a different object, you just grab the tube and move it to the object. Even a complete newbie can learn to use a Dob in about one minute flat. (Of course, that doesn't mean a newbie can learn to locate objects that quickly, but intuitive operation is a real advantage nonetheless.)

Portability

Despite its large tube, a tube Dob is easily portable, at least if the tube fits your vehicle. A tube Dob has only two parts—the tube and the base—so setup is a simple matter of placing the base on the ground and putting the tube on the base. Tear-down is equally quick and easy. It takes us literally one minute to unpack and set up our 10" Dob at the beginning of a session, and another minute at the end of the session to tear it down and repack it.

Truss Dobs are even more portable than tube Dobs, although they do take longer to set up and tear down. We know owners of 15" to 20" truss Dobs who carry their scopes in sub-compact cars. Depending on size, a truss Dob takes 5 or 10 minutes to set up and a similar time to tear down. Truss Dobs 20" and smaller can usually be set up and torn down by one person. Larger models may require a helper.

Fast cool down

Newtonian reflectors, including Dobs, cool down faster than any other type of scope except a refractor. Our 10" Dob, for example, normally stabilizes within 30 minutes or so. When the differential between the scope temperature and the ambient air temperature is extreme, a small or mid-size Dob may require an hour to cool, versus two to four hours for a catadioptric scope of similar size. Because the primary mirror on a reflector is exposed, it's easy to add fans to larger models to shorten cool-down time.

High image quality

Although the secondary mirror and spider vanes on a Newtonian reflector inevitably contribute diffraction effects, the image quality of a good Newtonian is second only to a refractor. Well-designed Newtonians with long focal ratios and correspondingly small secondary mirrors provide refractor-like images, with few diffraction effects and very high contrast. Even faster Newtonians, such as f/5 and f/6 Dobsonian models, exhibit contrastier images and finer detail than catadioptric scopes of similar aperture.

Dobsonian reflectors have the following disadvantages:

Frequent collimation

Collimation is the process of aligning the mirrors and/or lenses in a telescope so that they share a common optical axis. All scopes must be properly collimated to provide their best image quality, but reflectors (including Dobsonians) require more frequent collimation than do other types of scopes, and the collimation process is somewhat more complex for reflectors.

Beginners sometimes shy away from reflectors because they fear they will be unable to collimate properly. In fact, a full collimation **[Hack #38]** of the secondary mirror and primary mirror takes only a few minutes, and it is usually required only when you first assemble the scope. You collimate the primary mirror **[Hack #39]** every time you set up the scope, but that takes only a minute to do. Finally, you tweak the scope into perfect alignment **[Hack #40]**.

As evidence of how trivially easy it is to collimate a reflector, we forgot to include collimation as a disadvantage in the first draft of this chapter, and we had to come back and add it during an editing pass.

Lack of tracking

By design, a Dob is an unmotorized alt-azimuth mount, which means the scope doesn't track the apparent motion of the stars automatically. You have to move the scope manually to keep objects from drifting out of view. If your Dob has smooth motions **[Hack #42] [Hack #43]**, it's no problem to track manually, even at high power. However, the absence of motorized tracking does make it difficult, for example, to sketch at the eyepiece.

There are ways to add tracking to a Dob. You can install a commercial motorized tracking system such as the Dob-Driver (*http://homepages.accnorwalk.com/tddi/tech2000*) or the ServoCAT (*http://www.stellarcat.biz*), or you can use an equatorial platform **[Hack #63]**.

Unsuitability for astrophotography

Even if equipped with motorized tracking, Dobsonians are generally poorly suited for astrophotography other than perhaps Lunar and planetary photography using eyepiece projection. The focusers on most Dobsonians have insufficient in-travel to accommodate even CCD cameras at prime focus, and film cameras are usually out of the question. Just as important, serious astrophotography requires hanging a lot of equipment on the scope—camera, autoguider or guidescope, and so on—and that often causes insurmountable balance problems with a Dob. If you want to do serious astrophotography, a Dob is about the worst possible choice of scope.

Equatorially mounted reflectors. Although all Dobs are Newtonian reflectors, not all Newtonian reflectors are Dobs. Rather than mounting a Newtonian reflector optical tube on a Dobsonian base, it's possible to install it on a traditional equatorial mount supported by a tripod or pier. In fact, until the advent of the Dobsonian, an equatorially mounted reflector was the most common amateur instrument.

An *EQ Newt* shares the optical advantages of a Dobsonian—large, high-quality aperture, fast cool down, and so on—and also has the advantage of motorized tracking (as an extra-cost option, if not always as a standard feature). Despite these advantages, EQ Newts aren't very popular nowadays because the EQ mount itself adds significantly to their price relative to that of a Dob of similar size. For example, Orion sells their 6" XT6 Dobsonian for $249. Their similar EQ version, the SkyView Pro 6LT Equatorial Reflector, sells for $548. Similarly, their 10" XT10 Dobsonian sells for $499, while their Atlas 10 EQ Reflector—which is essentially the same optical tube mounted on an EQ mount—sells for $1,299.

Faced with that trade-off—motorized equatorial tracking versus much more aperture for the same money—most people choose the latter. Rightly, we think. For example, given the choice of a 10" Dob for $500 or a 6" EQ reflector for $550, we think most people will be happier with the Dob. The larger aperture of the Dob allows you to see more than one full magnitude deeper—which is an immense difference—and the Dob is much easier to set up and use than the EQ Newt.

For the price differential between the 10" Dob and 10" EQ models, we'd sooner buy the Dob and spend part of that extra $800 on an equatorial platform **[Hack #63]**. In fact, we'd probably buy the $949 12" IntelliScope Dob instead and spend the remaining $350 buying the parts to build a top-notch equatorial platform.

For larger EQ-mounted scopes, the other consideration is weight, which directly affects portability. The Orion 10" Dob, for example, weighs 55 pounds complete, divided about equally between tube and base. The Atlas 10 EQ Reflector optical tube weighs 27 pounds, but the mount weighs 90 pounds. Most healthy adults can handle the 10" Dob without assistance, but setting up the Atlas mount may require a helper.

If you hadn't guessed, we're not big fans of equatorially mounted reflectors. We're apparently not alone, as few amateurs buy these scopes nowadays.

Catadioptrics. A *catadioptric telescope* ("cat" for short) employs both mirrors and lenses. There are many types of catadioptric telescope, including the *Schmidt-Cassegrain Telescope* (SCT), the *Maksutov-Cassegrain Telescope* (MCT or *Mak-Cass*), the *Maksutov-Newtonian* (MN or *Mak-Newt*), and the *Schmidt-Newtonian* (SN). There are other types of catadioptric telescope, but none are commonly used by amateurs.

The Schmidt/Maksutov and Newtonian/Cassegrain nomenclature confuses a lot of people, but it's really not that difficult to understand. Most catadioptric scopes use a Newtonian or Cassegrainian primary mirror at the back of the tube and a Schmidt or Maksutov full-aperture corrector plate at the front of the tube. The corrector plate, as its name implies, is not simply a flat piece of glass. It is an actual lens, with curves calculated to reduce the aberrations produced by the primary mirror.

- Cassegrain variants use a Cassegrain primary mirror, which has a central hole. Incoming star light reflects from the primary mirror to a convex secondary mirror mounted on the rear of the corrector plate, which reflects the light straight back (180°) toward the rear of the tube, out of the central hole in the primary mirror and into the focuser and eyepiece. Because the secondary mirror is convex, it causes the fast-converging light cone from the primary mirror to diverge, which has the effect of increasing the apparent focal length of the scope.

> Because Cassegrain variants use a convex secondary and "fold" the light path, they are physically much shorter for their focal lengths than other types of scopes. For example, a typical 8" f/10 SCT has a focal length of 2,032mm. The primary mirror is actually f/2, which means its native focal length is about 400mm. But the convex secondary mirror functions, in effect, as a 5X Barlow, causing the rapidly converging f/2 light cone to converge more slowly as an f/10 light cone. The severe aberrations, particularly coma, that are inherent in any fast mirror—let alone one operating at f/2—are corrected by the corrector plate.

- Newtonian variants use a solid Newtonian primary mirror. Incoming star light reflects from the primary mirror to a flat secondary mirror mounted diagonally on the rear of the corrector plate, which reflects the light at a 90° angle out the side of the tube and into the focuser and eyepiece.

- Schmidt variants use a relatively thin corrector plate with shallowly curved aspheric surfaces.

- Maksutov variants use a much thicker corrector plate with deeply curved spherical surfaces. This thick glass corrector plate retains heat, and Maksutov telescopes are notorious for taking long periods to cool down.

So, for example, a Maksutov-Newtonian Telescope uses a Newtonian mirror—which means you view from the front of the scope—and a Maksutov corrector plate, which means the scope takes a long time to cool down. Conversely, a Schmidt-Cassegrain uses a Cassegrain mirror—which means you view from the rear of the scope—and a Schmidt corrector plate, which means the scope cools down much faster than a Maksutov variant, although still slower than a refractor or a reflector.

Schmidt-Cassegrain

SCTs are by far the most popular type of catadioptric scope used by amateurs. In fact, they come close to Dobs in overall popularity. SCTs are the proverbial jack of all trades and master of none. They do nothing that some other type of scope doesn't do better. On the other hand, they do just about everything reasonably well. SCTs pack a lot of aperture into a physically compact package. They are available in apertures ranging from 4" to 20". Models up to 12" are reasonably portable, but larger SCTs are for all practical purposes limited to fixed observatory mountings. Figure 1-10 shows a typical SCT; this one is an 8" Celestron model mounted on a Vixen Super Polaris equatorial mount.

The two leading SCT producers, Celestron and Meade, both produce SCT tubes in various apertures and offer them on different mounts of widely varying quality and cost. For example, Celestron offers the 11" C11 optical tube on four different mounts:

- The $1,700 C11-S uses a crude, Chinese-made CG-5 equatorial mount. Although Celestron has enhanced the CG-5 tripod with heavy tubular stainless steel legs, the 11" OTA is grossly under-mounted on a CG-5. Most experienced astronomers consider the CG-5 to be an adequate visual mount for an 8" SCT, marginal for the 9.25" SCT, and utterly insufficient for an 11" SCT, even for visual observing. Incredibly, Celestron claims the C11-S is suitable for astrophotography. We'll be kind here, and say we think they're exaggerating.

Figure 1-10. A typical SCT

The $2,000 C11-SGT is a C11-S with go-to functionality added. The extra $300 buys you a hand controller. You punch in an object on the hand controller, and the scope locates the object for you automatically. The scope and mount are otherwise identical to the C11-S. The basic C11 model is also available with enhanced XLT coatings for another $250 or so. Other Celestron models have similar options available.

- The $2,600 CPC 1100 mounts the C11 OTA on an inexpensive alt-azimuth fork mount. Go-to functionality is standard with this model. We consider the CPC 1100 to be Celestron's real entry-level 11" SCT. Anyone who buys the C11-S will almost certainly be disappointed by the inadequate mount. The CPC mount, on the other hand, is sufficient for visual observing, although not for astrophotography.

- The $3,500 NexStar 11 GPS mounts the C11 OTA on a better qual-
  ity alt-azimuth fork mount. We consider the NexStar 11 to be
  Celestron's mainstream 11" SCT, an excellent visual instrument out
  of the box. You can convert the NexStar 11 alt-azimuth mount to
  an equatorial mount by installing an optional equatorial wedge. The
  NexStar 11 isn't the best choice for long-exposure prime focus
  astrophotography, even with the wedge installed, but it is usable for
  casual astrophotography.

- The $4,200 CGE 1100 mounts the C11 OTA on a top-notch equa-
  torial mount. The CGE is Celestron's "serious" 11" SCT model. It is
  well suited for visual observing and serious astrophotography out of
  the box.

Note that the only real difference in these four models is the mount. For
$1,700, you get a Celestron 11" SCT optical tube on a mount that is so
crude and under-sized that using the scope is an exercise in frustration.
Getting the same optical tube on what we consider to be a usable mount
costs an extra $900, and getting it on the mount you really want costs
an extra $2,500. The same is generally true of SCTs in other apertures
and from other manufacturers. If you decide to buy an SCT, don't make
the mistake of undermounting it. A good mount is essential, but it isn't
cheap.

Maksutov-Cassegrain

Until relatively recently, MCTs were a rare sight at star parties. Few
astronomers owned one because few models were available, and those
that were were very expensive. For many years, the only commercially
available MCT was the 3.5" Questar, which ranges in price from $4,000
to $6,800, depending on options. (Questar also makes a 7" MCT, but if
you have to ask the price, you can't afford one.) Several years ago,
Meade jump-started the market for MCTs when it introduced a series of
MCTs ranging in aperture from 4" to 7" and in price from a few hundred dollars to $3,000 or so. In the last couple years, a flood of Chinese-
made MCTs has arrived. These inexpensive instruments range in aperture from 3.5" to 5", are sold by Orion and other retailers, and are surprisingly good optically for their price. MCTs are now relatively
common at star parties. MCTs closely resemble SCTs.

We consider an MCT to be a specialized instrument that is generally
unsuitable as a first or only telescope. MCTs have very long focal
lengths and high focal ratios, typically f/15, compared to the f/6.3 or
f/10 ratios of SCTs and the f/4 to f/6 ratios common in reflectors. That
means that MCTs have very narrow fields of view relative to other types
of scopes, and they are best suited to high-power, narrow-field observing, such as Lunar and planetary work.

Like SCTs, MCTs are physically compact for their aperture, so many astronomers choose an MCT as a grab-'n-go or quick-look scope [Hack #10]. We consider MCTs ill-suited for that purpose, though. Their thick Maksutov corrector plate means that MCTs have long cool-down times. Even small models may take a couple hours to equilibrate, and 7" models typically require several hours of cool down before they can provide their best image quality.

Maksutov-Newtonian

Mak-Newts resemble standard Newtonian reflectors, but have a full-aperture Maksutov corrector plate at the front of the optical tube. Mak-Newts are one of the best-kept secrets in amateur astronomy. The image quality of a top-notch Mak-Newt is superb, indistinguishable from that of an apochromatic refractor. Because Mak-Newts use a very small secondary mirror, they lack the contrast-robbing large central obstruction of traditional Newtonian reflectors and SCTs. The Mak-Newt secondary mirror is affixed directly to the rear of the corrector plate, so there are no vanes to produce diffraction spikes. Like an apo, a Mak-Newt provides pinpoint stars and an extremely high contrast image.

But a Mak-Newt sells for a small fraction of the price of an equivalent apo. For example, a 5" f/6 apochromatic refractor optical tube costs $4,500 to $6,000, while a 5" f/6 Mak-Newt costs about $900. The Mak-Newt takes longer to cool down than does the apo; otherwise, the two are functionally similar. Frankly, we've never understood why anyone would pay the high price for a medium or large apo refractor instead of buying a Mak-Newt of similar aperture, except perhaps that the small secondary mirror of a Mak-Newt can cause some vignetting (progressive darkening of the image as you near the edge) when used for film imaging or used with a very low-power, wide-field eyepiece.

None of the mainstream telescope makers produce a Mak-Newt, so they can be hard to find. Most of the high quality Mak-Newts available are made by two different Russian companies with very similar names, Intes and Intes Micro, which unfortunately have limited U.S. distribution channels. The most reliable U.S. source we know of is ITE (*http://www.iteastronomy.com*), which resells Intes models. Astromart (*http://www.astromart.com*) is also an excellent source.

Their relatively small apertures of 5" to 8" make Mak-Newts poorly suited for visual DSO observing, but for Lunar/planetary observing and astrophotography, Mak-Newts are unsurpassed. A Mak-Newt requires a mount, of course. The most popular choice of equatorial mount for 5" and 6" Mak-Newts is the $1,300 Vixen GP-DX (*http://www.vixenamerica.com*). The 7" and 8" models require a heavier mount, such as the $2,000 Losmandy G11. Figure 1-11 shows a typical Mak-Newt; this one is a 6" f/8 Intes MN-68 mounted on a Vixen GP-DX equatorial mount. Figure 1-12 is a close-up of the Mak-Newt front corrector plate.

Figure 1-11. A typical Maksutov-Newtonian telescope

Schmidt-Newtonian

A Schmidt-Newtonian is similar conceptually to a Mak-Newt, except that the Schmidt-Newt substitutes a Schmidt corrector plate for the Maksutov corrector plate. Optically, there's no reason that an SN scope can't provide excellent images, but most manufacturers have chosen not to produce SN scopes. The one exception is Meade, which a few years ago began producing a series of SN scopes in 6" to 10" apertures. Unfortunately, the Meade SN-series scopes were designed to meet a price point that would appeal to beginning astronomers. Although we have never used a Meade SN-series scope, people whose opinions we respect tell us that the SN-series scopes are acceptable optically but mounted very poorly.

Figure 1-12. The Maksutov-Newtonian corrector plate

Yet another category of catadioptric telescope is commonly available, unfortunately. These hybrid scopes, sold under many brand names, look like standard equatorial Newtonian scopes, but with a shorter tube. Rather than a full-aperture corrector plate, these junk scopes have what is usually described as a "corrector lens" or "built-in Barlow" mounted inside the focuser mechanism. The primary mirror is typically of very short focal ratio—f/3.0 or so—and the built-in corrector lens extends that to an effective f/5 or longer, often f/8 or f/9. The optics of these scopes are terrible, and they are usually mounted on a very poor equatorial mount. Don't waste your money on one of these scopes.

Choosing a Telescope

If you've made it this far, you probably have a pretty good idea of which type of telescope would best suit your own needs and preferences. Before you actually buy a telescope, though, we suggest you consider the following guidelines:

- Avoid "department store" scopes. Actually, there aren't many department stores left these days, but we use the term generically to mean junk telescopes sold by department stores, big-box retailers, cable TV shopping channels, "nature" stores at the mall, and so on. Whatever you do, don't buy a telescope on eBay or a similar online auction site. You *will* get burned. About 99% of the telescopes sold on auction sites are junk.

> We exclude Astromart (*http://www.astromart.com*) from our general condemnation of online auction sites. Astromart is run by astronomers for astronomers, and many knowledgeable amateur astronomers regularly buy and sell gear there. Although we wouldn't advise newbies to buy their first scopes on Astromart, it is the best place we know to buy and sell used telescopes and accessories.

- Avoid any telescope that is promoted based on magnification. In a practice that we think borders on fraud, the boxes of junk scopes are often emblazoned with "475X!" or some other meaningless number. In fact, any telescope can provide essentially any magnification if you use an eyepiece of the proper focal length, but the high-power images provided by junk scopes are so dim and blurred that they're unusable. Not that that matters much, because their mounts are so loose and shaky that you'll probably not be able to keep an object in the eyepiece at high-power anyway.

> We were about to suggest avoiding telescopes with bright, colorful images of astronomical objects on the box, but even some good telescopes use such packaging nowadays. We consider the practice of using full-color images from the Hubble Space Telescope to be deceptive, to say the least, but even reputable vendors have begun doing so. Just don't expect to see anything even remotely similar through the eyepiece.

- Buy from a retailer that specializes in astronomy gear. If you don't have a local astronomy store, check the ads in magazines such as *Astronomy* and *Sky & Telescope*.

Although we don't officially endorse any vendors, we think it's fair to say that we, our observing buddies, and our readers have generally been happy with the prices, products, and service provided by:

- Anacortes Telescope and Wild Bird (*http://www.buytelescopes.com*)
- Astronomics (*http://www.astronomics.com*)
- Hands-On Optics (*http://www.handsonoptics.com*)
- High Point Scientific (*http://highpointscientific.com*)
- Oceanside Photo and Telescope (*http://www.optcorp.com*)
- O'Neil Photo and Optical (*http://www.oneilphoto.on.ca*)
- Orion Telescope and Binocular Center (*http://www.telescope.com*)

Adorama, B&H, and the other New York City camera stores often have excellent prices on astronomy gear, but you must know exactly what you're looking for. Don't expect skilled advice from them. Their shipping charges are often very high, so make sure to get a total price before you order. We prefer to support specialty astronomy retailers. They may charge a few bucks more, but they are run by astronomers for astronomers, and their expert advice is often worth the small extra cost, particularly if you're not entirely sure what you're doing.

- Try out different types of telescopes before you buy. Join your local astronomy club. Attend star parties and public observations where you'll be able to see different types of scopes up close and personal. Arrive early and leave late so that you can watch scopes being set up and torn down.

- Be realistic about your own physical abilities. A smaller scope that you're willing to get out and set up shows you more than a large scope that never leaves the closet.

- Recognize that no one telescope is ideal for all purposes. Buying two moderately priced telescopes may give you more flexibility than spending the same total amount on one more expensive scope. For example, many people buy an inexpensive 8", 10", or 12" Dob or an 8" or 10" SCT as a primary scope, and add a short-tube refractor as a grab-'n-go or quick-look scope.

- If in doubt, buy the next size larger. If you're debating between an 8" and 10" scope, for example, it's usually better to buy the larger scope (assuming you can transport it and so on). Some people buy the next size down, knowing that they'll want to spend money on eyepieces and other accessories. But eyepieces come and go, while the scope will probably remain with you for a long time. Buy the larger scope now, and add accessories later as you can afford to.

- Conversely, don't blow your entire budget on the telescope. You will need some accessories right away, and you'll want many more accessories before long.

- Don't buy a scope based on its perceived capabilities for astrophotography. Nearly every amateur astronomer has aspirations to image the heavens, but unless you're willing to spend at least a few thousand dollars on your first scope, you will be disappointed in the results. Your goal for your first scope should be an instrument best suited to your own preferences for visual observing.

- Do not buy an inexpensive go-to scope. Too much of the cost goes to the mount, motor, and electronics, and too little to the scope itself. Unless you're willing to spend $1,800 or more on an 8" or larger go-to SCT, we think you'll be be disappointed. There are reliable 5" SCT go-to scopes that sell in the $1,000 range, but we think these scopes have too little aperture to serve as a general-purpose scope. Their small aperture effectively restricts them to viewing Solar system objects and the brightest DSOs, which most people will quickly find too limiting. If you have any interest in observing DSOs, start with a scope of at least 8" to 10" of aperture.

- Don't fall prey to analysis paralysis.

Still can't make up your mind? Okay, we'll do it for you, but upon your head be it. (We'll be happy to pick a spouse for you, too, as long as you're willing to live with the results.) Here are the scopes we recommend:

- If you're on a tight budget—or just unsure whether you'll maintain your interest in the hobby—buy the Orion StarBlast. At $170 complete, the financial risk is minimal. Orion markets this as a children's scope, but there are probably more StarBlasts used by serious adult astronomers than by children. At 4.5", the StarBlast provides enough aperture to show pleasing images of Luna, the planets, and the brightest DSOs. You'll probably want a larger scope soon, but you'll never outgrow the StarBlast. We know astronomers who have spent many thousands of dollars on large premium scopes who still keep a StarBlast as a quick-look scope.

- If you're willing to spend $300 to $1,000 initially, buy a mainstream 6", 8", 10", or 12" Dobsonian scope from Orion, Skywatcher (Canada), Celestron, or one of the many other vendors who resell Synta (Chinese) and Guan Sheng (Taiwanese) Dobsonians. Guan Sheng models, sold by Celestron and others, have better mirrors and mechanicals.

 Synta models are not as good optically or mechanically as Guan Sheng models, but the Synta-manufactured Orion IntelliScope Dobs have one unique feature that makes them worth considering. The Computerized Object Locator hand controller, a $99 option, converts the manual dob into a "push-to" automated scope. The scope still doesn't have motorized tracking or go-to capabilities, but you can locate objects automatically just by watching the arrows on the hand controller and manually moving the scope as the arrows indicate.

- If you want a compact, portable scope that tracks and provides go-to functions, choose the Celestron CPC 800 8" SCT, which costs $1,800 with standard coatings or $2,000 with enhanced coatings.

It's obviously very difficult for anyone to make specific recommendations without knowing your circumstances or personal preferences, so we hope you'll take this advice merely as a starting point, and follow the suggestions we made earlier.

HACK #10 Equip Yourself for Urban Observing

Choose the perfect grab-'n-go scope for observing at a moment's notice.

You can, of course, use your primary (or only) scope for urban observing if you are limited to doing so by budget or inclination. But to take advantage of the spur-of-the-moment aspect of urban observing, many serious amateur astronomers keep a special telescope dedicated to that purpose. It's called a *grab-'n-go scope* or *quick-look scope*, which describe its purpose perfectly. A grab-'n-go scope is always set up and always ready to use on a moment's notice. You can carry it out the door and start observing immediately. Figure 1-13 shows our grab-'n-go scope, a 90mm long-tube refractor that we leave set up and ready in our library. (We caught it with its pants down; ordinarily, this scope has a Telrad unit-power finder attached.)

Figure 1-13. A grab-'n-go scope, set up and ready to go

Here are some considerations for choosing the best grab-'n-go scope for urban observing:

Portability

Portability is critical. The ideal grab-'n-go scope should be light enough to pick up with one hand and carry out the door, and should be small enough to be stored unobtrusively in any handy corner or closet. A light, portable scope will be used often. A heavier, less portable scope will sit in the corner gathering dust. The differences can be substantial. For example, our 90mm refractor came with an equatorial mount and weighed more than 30 pounds. We replaced the equatorial mount with an alt-azimuth mount, reducing the total weight of scope and mount to about 10 pounds. That difference may sound minor, but it's not, particularly after a long day at work.

Fast set up and tear down

Urban observing is often done on the spur of the moment. You take the dogs on their evening walk and notice that the sky is particularly clear that night, or driving home from a party, you notice that Mars is ideally placed for a quick look. If your grab-'n-go scope is set up and ready, all you need to do is carry it out the door, set it down, and start observing. Conversely, if you know it'll take 10 or 15 minutes to set up the scope, polar align it, align the finder, and so on, chances are you'll let that observing opportunity go by. The same holds true, in spades, for tear-down time. When you've stayed up past your bedtime watching the dance of Jupiter's moons, the last thing you want to face is another 10 or 15 minutes of tearing the scope down and packing it away.

Fast cool down

When you carry a telescope from a warm house to a cool backyard (or vice versa), the optics need time to equilibrate to the outside temperature (for some reason, astronomers invariably call this process "cool down," even when it's warmer outside than indoors). A scope can't produce its best images until it is fully cooled. Until that happens, the scope produces mushy images. An uncooled scope may produce acceptable images of dim DSOs (Deep-Sky Objects), but for bright, detailed objects—like Luna, the planets, and multiple stars—a properly cooled scope is essential. Because most urban observing targets fall into the latter category, fast cool down is important for a grab-'n-go scope.

Different scopes have different cool-down characteristics. Large scopes generally take longer to cool down than smaller scopes. Refractors cool down faster than reflectors, reflectors cool down faster than SCTs (Schmidt-Cassegrain Telescopes), and SCTs cool down faster than MCTs (Maksutov-Cassegrain Telescopes). Depending on the temperature differential between its storage place and the outdoor air, a small or mid-size refractor may be ready for use immediately, or at most in a few minutes. Larger scopes may take literally hours to cool down sufficiently to provide sharp, high-contrast images of Lunar and planetary detail.

Sufficient aperture

Small apertures contribute to portability, but it's easy to go overboard and end up with a scope that's too small. Many astronomers choose 70mm or even 60mm scopes for urban observing, but we think that's a mistake. Although even a 60mm scope gathers sufficient light for Lunar and planetary observing, light gathering ability is not the only consideration. Aperture also determines maximum usable magnification and the ability to resolve fine detail.

The general rule of thumb is that small, high-quality scopes can use up to 50X to 60X per inch of aperture. Some observers push this as high as 100X per inch, but that requires world-class optics and results in an exit pupil so small that diffraction effects, "floaters" in your eyes, and other issues reduce the amount of detail you can see. Observing Lunar and planetary detail requires using magnifications in at least the 150X to 225X range, so any scope you plan to use for that purpose should support that amount of magnification. A 90mm to 100mm scope spans that range nicely, while a smaller scope may be pushed to and beyond its limits to reach even the lower end of that range.

Aperture also determines the finest detail that can be resolved, with larger scopes able to resolve finer detail than smaller scopes. Small scopes, such as 60mm models, simply cannot resolve sufficiently fine detail to satisfy most serious observers. A 90mm scope resolves 50% finer detail than a 60mm model, and a 100mm scope 67% finer detail.

On that basis, you might think that an even larger scope would allow you to see more detail. That's true in theory, and sometimes in practice. However, there is another consideration. Even an optically perfect large scope is limited in the amount of detail it can resolve by how turbulent the atmosphere is, a characteristic that astronomers call *seeing*. On a perfectly steady night with superb seeing, which in most locations is extremely rare, even large scopes may be aperture-limited in terms of their ability to resolve fine detail. On more typical nights, any scope larger than 6" to 8" may be seeing-limited. On very turbulent nights, which are common in the city, scopes larger than 90mm or so are often seeing-limited.

Low price

Depending on your means, this may not be an issue. Some observers use $3,000 apo refractors as their grab-'n-go scopes. But for most of us, price is definitely an issue. Fortunately, there are many suitable candidates for a good grab-'n-go scope that sell for $150 to $500. At that price, many observers will find it worthwhile to dedicate a scope to urban observing.

Excellent Choices

Here are some scopes we think are excellent choices for an inexpensive grab-'n-go scope:

90mm to 100mm long-tube refractor on an alt-az mount

We think a 90mm to 100mm (3.5" to 4") long-tube refractor on an alt-az mount is the ideal grab-'n-go scope for most urban observers. The aperture is sufficient for viewing Luna, the major planets, and most other objects suitable for urban observing, including the brightest DSOs. A good quality 90mm or 100mm scope can be pushed as high as 200X to 250X, which suffices for viewing Lunar and planetary detail.

Our favorite scope in this category is the Guan Sheng 90mm f/11 refractor, which is the model we use. Orion used to resell this model as their *SkyView Deluxe 90mm Refractor*, but they have since stopped reselling Guan Sheng models and shifted entirely to the less expensive and lower-quality Synta models. The Guan Sheng 90mm refractor is still available from smaller astronomy retailers, usually under their house brand names.

The newer models provide a 2" focuser instead of the 1.25" focuser supplied with our older model. Unfortunately, the Guan Sheng 90mm optical tube is usually bundled with the Guan Sheng EQ (equatorial) mount. It's a decent EQ mount, but for a grab-'n-go scope we much prefer a lighter alt-azimuth mount, such as the Synta AZ-3 sold by Orion. If you can find a Guan Sheng 90mm optical tube for sale by itself, buy it. Otherwise, buy the scope with EQ mount, replace the EQ mount with an alt-azimuth mount like the AZ-3, and resell the EQ mount on Astromart (*http://www.astromart.com*). Or, do what we did and resell the EQ mount to a member of your astronomy club. The mount itself is quite good for a small, inexpensive EQ mount. The problem is that it's too heavy for a grab-'n-go scope.

It's often claimed that a smaller scope is better than a larger scope for urban observing. That's a myth. More aperture always shows you more, even under light-polluted skies. However, it is true that a smaller aperture is less subject to bad seeing conditions, which often predominate in urban environments.

6" Dobsonian

If we couldn't have our 90mm long-tube refractor, our next choice for a grab-'n-go scope would be a good 6" Dobsonian reflector. It's not quite as light or portable as the 90mm refractor, but neither is it heavy or clumsy. One of our regular correspondents lived in a New York City apartment building until recently. He routinely carried his 6" Dob up several flights of stairs to observe on the roof of the building. A 6" Dob is generally f/8, which allows it to provide good images with inexpensive eyepieces. The f/8 focal ratio also allows 6" Dobs to use a small secondary mirror, which contributes to higher contrast and good Lunar/planetary performance.

There are many good inexpensive 6" Dobs available, including models from Orion and Guan Sheng (under several brand names, as usual). Our favorite of this group, albeit the most expensive model, is the Orion 6" IntelliScope. The optics and mechanicals of the IntelliScope aren't quite as good as those of the Guan Sheng Dobs, but the IntelliScope has the inestimable advantage of including digital setting circles (DSCs), which Orion refers to as their Object Locator System. Under light-polluted urban observing conditions, DSCs can allow you to find your target object in seconds rather than spending frustrating minutes looking for it manually.

Depending on circumstances, we might also consider an 8" Dobsonian, which is slightly heavier than a 6" model, but not much larger. The tube of an 8" Dob is of course 2" larger in diameter than that of a 6" Dob, but they are generally the same length because most 8" Dobs have an f/6 focal ratio (a 6" f/8 scope has the same focal length as an 8" f/6 scope). And, while a 6" aperture is a bit small if it is to be your only scope, an 8" scope can serve both as a grab-'n-go scope and a general-purpose scope.

If we could afford only one scope, we might well settle for an 8" Dobsonian. If we could afford multiple scopes, we'd buy a dedicated grab-'n-go scope for urban observing and a 10" or larger scope as our primary instrument.

Orion StarBlast 4.5" Dobsonian

Although we call the StarBlast a Dobsonian, it isn't, really. The StarBlast tube is secured using a Dob-like bearing to a single vertical arm that connects to a baseplate. The baseplate rides on a ground board, again using a Dob-like bearing. So, although the StarBlast is not technically a Dobsonian, it operates much like one, with all of the advantages of stability and smooth motion that implies. Orion markets the StarBlast as a kid's scope, but the truth is that it's a capable observing instrument in its own right. Many serious amateurs keep a StarBlast as a secondary scope, even though their primary scopes are expensive SCTs or giant Dobs.

The StarBlast has some real advantages for use as a grab-'n-go scope, not least its $169 price tag. At 4.5" (114mm), the parabolic StarBlast mirror delivers some serious aperture for a grab-'n-go scope. At only 18" tall and 13 pounds, size and weight are not an issue.

The 450mm focal length of the StarBlast is a two-edged sword. The short focal length allows very wide fields of view, even though the Star-Blast has only a 1.25" focuser, but it also makes it difficult to reach higher powers. For example, using a 24mm Tele Vue Panoptic eyepiece (which costs nearly twice as much as the scope itself), the StarBlast provides a 3.6°+ true field of view. But to reach 150X magnification requires a 3mm eyepiece or eyepiece/Barlow combination, and 225X requires the equivalent of a 2mm eyepiece.

The f/4 focal ratio means that a 28mm eyepiece, which provides a 7mm exit pupil **[Hack #8]**, is about the longest useful focal length, even for young observers. Older folks will probably find 20mm to 24mm to be the longest useful eyepiece focal length. The f/4 focal ratio also means the StarBlast is hard on eyepieces. Even premium eyepieces are being pushed at f/4. Inexpensive eyepieces—including the Explorer II Kellners supplied with the StarBlast—have very soft edge performance.

Finally, StarBlast quality control seems variable. Some StarBlast scopes are excellent, easily reaching 150X or higher with good image quality. Others are mediocre, turning in mushy performance even below 150X. It seems to be the luck of the draw which type you'll receive. Fortunately, Orion has very good return policies. Although it's inconvenient to do so, you can simply keep sending back unacceptable StarBlasts until you get one you're happy with.

Bad Choices

There are also some scopes that we think are poor choices for urban observing, although they are popular for that purpose.

Inexpensive, small go-to scopes

The idea is attractive. Build a computer and some drive motors into a small scope and let the computer find your objects for you automatically. No matter how bright your skies and how hard it is to locate objects manually, you can simply zoom to your target automagically. Or so the marketing literature says.

In practice, it's different. Although go-to scopes work better nowadays than did models from a few years ago, they're still a very poor choice. The fundamental problem with inexpensive go-to scopes is that too much of the price is going to the computer and drives and too little to the optics and mount. If you buy one of these scopes, you end up with an instrument that can put you on target quickly, but won't let you see much when you get there. The optics are too small (and usually of mediocre quality) and the mount is flimsy and shaky. Not to worry, though. The electronics and drives are cheap and fragile, so they'll probably fail soon anyway.

If you really want a go-to scope, be prepared to spend at least $1,800 for what we would consider a minimally acceptable model, which is to say an 8" SCT on a mainstream go-to mount.

Small refractors

As we said earlier in this section, refractors much smaller than 90mm are generally poorly suited as grab-'n-go scopes because of their optical limitations. There are exceptions, of course, but not among inexpensive scopes. If your budget allows you to pay $500 to $2,000 or more for a scope and mount, you may be happy with one of the 80mm StellarVue models, the Tele Vue TV-76, or even the Tele Vue 60. But unless you're on a champagne budget, give small refractors a miss.

Large refractors

If a 90mm or 100mm long-tube refractor is good, then a 120mm or 150mm long-tube refractor must be better, right? Wrong. The problem with larger refractors is weight and portability, both of the scope itself and its mount. If you get far beyond 100mm aperture, you'll find that refractor tubes become long, heavy, and difficult to manage, and their mounts become much too heavy and awkward for grab-'n-go use.

Also, the problem of false color increases with increasing aperture. For example, our 90mm f/11 achromatic refractor produces very little false color, even on very bright objects. To get the same low level of false color with a 150mm (6") achromatic refractor, it would have to be about f/18, which would require an optical tube nine feet long. No one makes such a refractor, of course. Instead, large refractors usually have short focal ratios to allow reasonable tube lengths of a meter or so. A typical 120mm refractor, for example, might have a focal ratio of f/8.3, and a 150mm refractor f/6.7 or thereabouts. The false color from such instruments is intense.

Finally, a longer, heavier optical tube requires a larger, heavier mount for stability. A stable tripod and alt-azimuth mounting head for a 90mm or 100mm refractor is light enough that the scope and mount can be picked up with one hand. Mounts for larger refractors are correspondingly larger and heavier. Even Mills Darden, the 7'6", 1,000 pound lumberjack who was the real-life prototype for Paul Bunyan, would have had a hard time carrying a properly mounted 6" refractor around with one hand.

Short-tube refractors

Short-tube 80mm and 90mm refractors are immensely popular, although we've never understood why. Although they are compact and lightweight, they have at least two severe drawbacks as a grab-'n-go scope for urban observing. First, as short focal-ratio achromats, they produce hideously bad false color on bright objects, including Luna and the planets. Since Luna and the planets are the most common targets of urban observers, we'd rule out short-tube refractors on that basis alone. Second, with the exception of premium models, which are quite expensive, short-tube refractors generally have poor quality objective lenses. Most produce mushy images at anything much over 75X, and none of the inexpensive Chinese models we've seen will support the 200X+ needed for observing Lunar and planetary detail with anything like reasonable image quality.

SCTs and MCTs

Schmidt-Cassegrain Telescopes (SCTs) and Maksutov-Cassegrain Telescopes (MCTs) in the 127mm (5") range are popular with some urban observers, but we think they're less than ideal for that purpose. The main problem is cool-down time, which for an MCT may be an hour or more. The only time we'd choose an SCT or MCT for a grab-'n-go scope is if we could store it in an unheated garage or outbuilding where it would always be acclimated to the ambient outdoor temperature.

In short, there are a few good choices—and many bad choices—for a grab-'n-go scope. Choose carefully according to your budget and needs, and we think you'll be happily surprised with just how much a good grab-'n-go scope will let you see from your own backyard.

CHAPTER TWO

Observing Hacks
Hacks 11–32

Locating and observing astronomical objects requires developing a special set of skills and practices, most of which are not intuitive. It requires a detailed knowledge of the night sky and of specialized astronomical terminology and conventions. There are things you must know and be able to do if you are to be successful.

Just finding the object you want to view can be difficult. The night sky is huge, and many astronomical objects are tiny, dim things. Even after you have found the object and verified its identity, teasing out the maximum possible amount of visible detail is very challenging.

We've watched many beginning observers encounter the same frustrating problems—what we call the "newbie blues"—and we've helped more than a few of them over the hump. All of them, particularly those who have go-to scopes, hope there are shortcuts to learning to observe. There are no shortcuts. A go-to scope is no better substitute for learning the night sky than an automatic transmission is for learning how to drive. Learning to observe is a hard-won skill, but one you can be proud of achieving.

In this chapter, we tell you what you need to learn, know, and do to locate, describe, and observe astronomical objects.

HACK #11 | **See in the Dark**

Have you ever wondered why all cats are gray in the dark?

Our eyes function in two entirely different modes, depending on how much light is available. In daylight or bright artificial light, our eyes function in *day vision* mode. After dark, our eyes shift to *night vision* mode. The physiological changes that occur in our eyes during the shift from day vision to night vision are called *dark adaptation*. Dark adaptation occurs slowly, typically requiring 25 minutes for 80% adaptation and 60 minutes for 100% adaptation. That's why astronomers get upset when someone shows a bright light.

> When we move from dim light to bright light, our eyes undergo physiological changes called *light adaptation*. But while dark adaptation occurs slowly, light adaptation occurs quickly, in two phases. During α adaptation, which requires about 1/20th of a second, the sensitivity of the retina drops by 50% or more. During β adaptation, which requires from one to several seconds, the sensitivity of the retina drops more gradually, and we recover full color vision and visual acuity.

There are many misconceptions about night vision and dark adaptation, even among astronomers. To understand the process of dark adaptation, you need to understand something about the physiology of the human eye. Our eyes have two types of light sensors, called *rods* and *cones*. Rods provide monochromatic vision, but are very sensitive to light. Cones provide full color vision, but are relatively insensitive to light.

Cones and rods are unevenly distributed over the surface of the retina. Cones predominate in the fovea, the center of the retina, where they are densely packed. The fovea contains about 200,000 cones in an area of about one square millimeter, and thus provides acute resolution of fine detail. The entire retina contains about only 7,000,000 cones. That means cones are very sparsely scattered outside the fovea, just enough to show brightness and color with little detail in your peripheral vision. Rods predominate outside the fovea. The entire retina contains about 130,000,000 rods. They are less densely packed—at about 90,000 per mm^2—than cones in the fovea, but much more densely packed than cones outside the fovea. Accordingly, rods provide poor resolution of fine detail relative to the cones in the fovea, but much higher resolution than the sparsely scattered cones outside the fovea.

> The uneven distribution of cones and rods explains, for example, why you have to look directly at this book to read it; if you glance at it from the corner of your eye, you can't resolve sufficient detail to read the words. Conversely, the paucity of light-sensitive rods near the center of your eye explains why it's easier to see dim objects by looking to one side rather than directly at them, a phenomenon astronomers call *averted vision*.

Rods and cones detect light by using dyes to absorb it. As light is absorbed, the dyes bleach and a signal occurs to indicate that light has been sensed.

There is only one type of rod, which is why rods provide monochrome vision. There are three types of cones, one for each of the primary colors of light: red, green, and blue.

Rods

Rods use the dye rhodopsin, and have peak light sensitivity at a wavelength of about 498 nm, in the blue-green part of the spectrum.

L cones

L cones, also called red cones, use the dye erythrolabe and have peak sensitivity at about 564 nm, in the yellow-orange part of the spectrum.

M cones

M cones, also called green cones, use the dye chlorolabe and have peak sensitivity at about 533 nm, in the yellow-green part of the spectrum.

S cones

S cones, also called blue cones, use the dye cyanolabe and have peak sensitivity at about 437 nm, in the violet part of the spectrum.

Nearly all humans have only rods and three types of cones. A tiny percentage of women have a fourth type of cone. Presumably, they see colors they can no more explain to the rest of us than we can explain colors to a blind person.

Vision Modes

Broadly speaking, there are three modes of vision:

Photopic mode

Photopic mode (day vision) occurs at moderate to high lighting levels, 1 milliLambert (mL) or higher, and uses primarily the cones. Photopic vision provides full-color images and high visual acuity, allowing you to resolve fine detail. Photopic light sensitivity peaks in the green part of the spectrum at 555 nm.

Scotopic mode

Scotopic mode (night vision) occurs at low lighting levels, below about 0.001 mL, and uses the rods exclusively. Scotopic vision is monochrome—you see only shades of gray. It provides low visual acuity, making it difficult to resolve fine detail. Scotopic light sensitivity peaks in the blue-green part of the spectrum at 505 nm. (Although blue-green light stimulates rods most efficiently, that blue-green light is still visible only as gray because the rods do not convey color information to your optic nerve and brain.)

Mesopic mode

> *Mesopic mode* occurs at lighting levels in the transition zone between 1 mL and 0.001 mL, or about the brightness range of a moonlit landscape. At these light levels, cones and rods both contribute to vision. Color and finer detail is visible in the more brightly lit or more reflective areas, while objects in shadow are visible only as murky gray.

Technically, mesopic mode isn't a separate mode, but a combination of photopic mode and scotopic mode. Mesopic mode occurs when part of your eye functions in photopic mode and part in scotopic mode. Here are two examples of mesopic mode as it applies to astronomy:

Seeing color in bright nebulae

> If you use a medium to large scope to observe the brightest nebulae, such as the Orion Nebula (M42) and the Ring Nebula (M57), you may see parts of the object in a greenish-gray cast. You can see this color because the amount of light striking your retina is just barely sufficient to trigger some of the green cones. (Some young people can see tinges of blue and red in M42. We want their eyes.) Other objects in the field appear gray because their light is sufficient to trigger only your rods.

Seeing star colors

> With the naked eye, most stars appear white because their light is insufficient to trigger your cones. The red color of a few of the brightest red stars, such as Betelgeuse and Antares, may be visible to the naked eye because they are just bright enough to cross the transition from scotopic to photopic vision. If you use a binocular or telescope, which brightens the image from hundreds to thousands of times, many stars have distinct colors. When you view a colorful star in the same field of view as a faint fuzzy, you are using mesopic vision—photopic (cones) for the star and scotopic (rods) for the faint fuzzy.

Night Vision Fallacies

Here are some common fallacies about night vision and dark adaptation:

Dark adaptation is all-or-nothing. Wrong. Many astronomers believe that any exposure to light damages overall dark adaptation. In fact, not only does each eye dark adapt separately, but each cone or rod also adapts individually and cones do so separately from rods. That means you can keep one eye fully dark adapted even if the other eye is exposed to bright white light. Also, because cones adapt separately from rods, you can use photopic vision for viewing charts, recording observations, etc., without harming the scotopic dark adaptation of your rods.

Pupil diameter is critical to dark adaptation. Not true. The human pupil varies from as small as 2mm in diameter under bright lighting to as large as 8mm under dark conditions, a range of only 4:1 linearly and 16:1 areally. In fact, the usual range is less. It's very rare for a person more than 20 years old to be able to dilate to 8mm; 7mm is the more usual maximum in young adults, 6mm at age 35 to 45, and 5mm is common in people older than 55 or 60. At most, then, the range of brightnesses controlled by pupil diameter is 16:1, and a range of 12:1 or even 6:1 is more usual. In fact, directional sensitivity reduces that factor still more, into the range of 10:1 to 4:1. The range of brightnesses detectible by the human eye is about 10,000,000,000:1, so pupil diameter plays only a miniscule role compared to the sensitivity level of the rods. Also, the pupil constricts and dilates very quickly compared to the time needed for rods to recover their dark adaptation.

> It *is* true that it's important to match the exit pupil of your instruments to the **entrance pupil of your eye [Hack #7]**. For example, if your maximum entrance pupil is 5mm, using an eyepiece or binocular that provides a 7mm exit pupil simply wastes light. If your entrance pupil is 7mm, the exit pupil of a 7X50 binocular almost exactly matches your entrance pupil. Someone whose entrance pupil is only 5mm would be better served by a 7X35 or 10X50 binocular, with a 5mm exit pupil to match the entrance pupil.
>
> It is also true that a constricted pupil impairs night vision. For example, a pupil that dilates to 7mm receives nearly twice as much light as one that dilates to only 5mm, because the first pupil has nearly twice the area ($7^2:5^2=49:25$). One magnitude corresponds to a brightness difference of about 2.51, so a young person whose pupils dilate to 7mm can typically see nearly one full magnitude deeper than an older person whose eyes dilate only to 5mm. In fact, it's worse than that, because the eye lens yellows and darkens as we age. The eye lens of a 15-year-old typically transmits three times as much light—another full magnitude—as the eye lens of someone who is 75.

A dim green light is the best choice for preserving night vision. Nope. This myth probably persists because the military uses dim green lighting in some tactical situations and because military night-vision scopes produce a dim green image. In fact, the military uses dim green light because the photopic (cone) vision needed to provide high visual acuity is most sensitive at green wavelengths. But any green light bright enough to trigger your cones is much more than bright enough to destroy the dark adaptation of your rods, eliminating your night vision.

A bright red light destroys night vision. Red light preserves dark adaptation for a simple reason. The rhodopsin pigment in rods is completely insensitive to light at wavelengths longer than about 620 nm, which is to say deep red. Although the erythrolabe dye present in your L cones has peak sensitivity near 564 nm, its sensitivity extends far into the red part of the spectrum. That means you can use a very bright red light without damaging your scotopic vision at all.

This is an excellent reason to use a red LED flashlight rather than a standard flashlight with a red filter. LED flashlights emit light at one specific wavelength, and red LED flashlights emit at a wavelength to which rhodopsin is insensitive. Red filters, on the other hand, also transmit a fair amount at light at shorter wavelengths, so a bright red-filtered flashlight can impair your night vision.

> Interestingly, rhodopsin sensitivity peaks very close to the 486 nm H-Beta and 496/501 nm O-III lines emitted by many nebulae. Were it not for this coincidence, many faint fuzzies would be even fainter.

HACK
#12

Protect Your Night Vision from Local Lights

Stay dark adapted and keep stray light from the eyepiece (and keep your ears warm).

Night vision is all-important when you observe DSOs. For those fortunate enough to have access to a truly dark observing site, it's not difficult to preserve night vision using standard methods—red LED flashlight, covering your notebook computer screen with red film [Hack #44], and so on. But for many astronomers, the only sites within easy driving distance are, at best, semi-dark. The problem with these sites is often not so much general light pollution as local light pollution—the presence of streetlights and other nearby bright light sources.

For example, our regular "dark" observing site routinely offers mag 5.5+ skies, and on good nights mag 6.0 or better [Hack #13]. In terms of general light pollution, that's a respectable DSO observing site, at least by Eastern U.S. standards. Unfortunately, there are half a dozen mercury-vapor lights within a few hundred yards of the site. Their combined light makes it impossible to become fully dark adapted. In fact, it's bright enough to read a newspaper on the observing pad, literally. Because the site is on private property, it is impossible to install permanent screens against the local light pollution. Portable screens are impractical for various reasons.

Fortunately, there is a cheap, easy solution to such local light pollution problems, as long as you don't mind looking like a complete idiot. All you need is an old towel and a pirate's eye patch. Figure 2-1 shows Robert working at the chart table, looking like an idiot, but with his night vision intact.

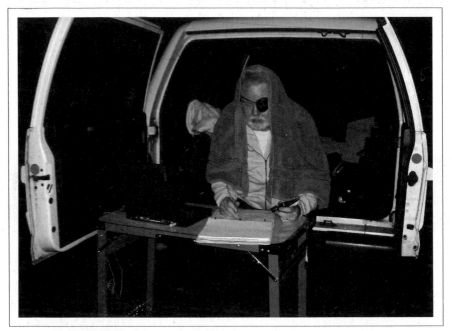

Figure 2-1. Robert, fully equipped with eye patch and towel at the chart table...

Dark adaptation occurs individually for each eye. That means you can keep one eye completely dark adapted by covering it with the eye patch whenever you are not using it to look through the eyepiece. The other eye is never fully dark adapted, but that doesn't matter. You use it for other purposes, such as locating objects with your notebook computer or charts, or recording observations on your log sheet. For that matter, Robert sometimes uses his "regular" eye to locate objects in the finder and Telrad, for which full dark adaptation is less important.

As you prepare to observe an object, position the towel to screen your face and the eyepiece from local light sources and slide the eye patch out of the way. When you finish observing the object, cover your dark-adapted eye with the eye patch before you remove the towel. At particularly bright sites, you may need to use the same procedure when you use the finder or Telrad. Figure 2-2 shows Robert at the eyepiece, using the towel to screen the eyepiece (and his dark-adapted eye) from local light pollution sources.

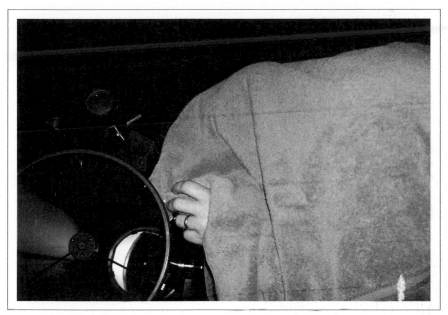

Figure 2-2. ...and at the eyepiece

If you smoke, as Robert does, you create the worst possible local light pollution problem for yourself every time you click your lighter. Burning tobacco usually isn't bright enough to harm your night vision, but the initial flash of the lighter and the glow as you get the tobacco lit is more than enough to harm your night vision severely, even if you close your eyes as you light up. If you use an eye patch, you may not care about losing the night vision in your regular eye. If you do, try Robert's method, shown in Figure 2-3. Again, it looks stupid, but it works. Just make sure not to light anything except your tobacco.

HACK #13   Describe the Brightness of an Object

When speaking of brightness, magnitude and surface brightness are two terms you'll hear a lot. Understand what these terms mean and how they relate to one another.

For astronomy, the primary purpose of a binocular or telescope is to gather light, allowing you to see dimmer objects than are visible to the naked eye. Before you begin observing, it's essential to understand the limits of your viewing apparatus, be it the naked eye, a binocular, or a telescope. Otherwise, you could spend hours looking for objects so dim you'll never be able to see them.

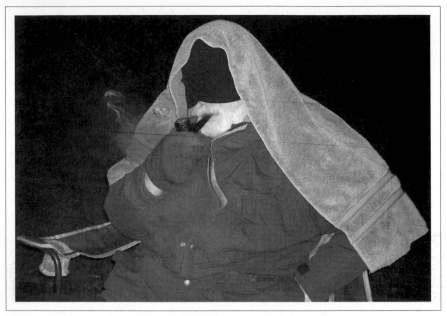

Figure 2-3. Robert, smoking while preserving his night vision

The *magnitude* of an object is a numerical quantification of its brightness. Ancient observers first categorized stars by their brightness, describing the brightest stars as being "of the first magnitude." Slightly dimmer stars were grouped as second magnitude, and so on, down to the dimmest stars visible to the naked eye, which the ancients described as sixth magnitude. Dimmer objects accordingly have numerically larger magnitudes.

The Greek astronomer Hipparchus created the first known star catalog in about 120 BCE. This catalog contained 1,080 stars visible to Hipparchus from his latitude. He organized these stars into constellations, described the position of each star relative to other stars, and rated their brightness from first to sixth magnitude. In about 125 CE, the Egyptian astronomer, cartographer, and geographer Ptolemy updated the Hipparchus catalog in his famous book *Mathematical Syntaxis*, usually called *The Almagest*. Ptolemy added a few northerly stars that Hipparchus had missed, and added more southerly stars that were visible from his Alexandria observatory at 31°15'N but had not been visible to Hipparchus in his observatory at 36°15'N on the island Rhodes. The magnitude system used by Hipparchus and Ptolemy is still in use today, with only minor modifications.

With the advent of photometers and other scientific instruments capable of measuring brightness accurately, astronomers formalized the magnitude system by defining a hundred-fold difference in brightness as five magnitudes.

By definition, then, a first magnitude star is exactly 100 times brighter than a sixth magnitude star, as is, say, a 9th magnitude star versus a 14th magnitude star. The fifth root of 100 is 2.511+, so a star of magnitude 1.0 is about 2.5 times as bright as a star of magnitude 2.0, and a star of magnitude 9.0 is about 2.5 times as bright as a star of magnitude 10.0.

Photometers can provide very precise magnitude values. Star charts commonly list magnitudes to the first or second decimal place. For example, one star might be listed as magnitude 2.75 and another as magnitude 2.74. Although such small differences are impossible to detect visually, they are easily discriminated by CCDs and other sensitive instruments.

> Amateur astronomers normally use tenths of magnitude. For example, when estimating atmospheric transparency and darkness, an astronomer often determines the dimmest star that is visible naked eye near zenith. He might record this "limiting magnitude at zenith" on a given night from a given site as "6.1 LMZ", which means the dimmest star near zenith he could see naked eye with averted vision was magnitude 6.1. (LMZ is used because you can see dimmer stars at zenith than you can at lower elevations, where you are looking through more air, haze, and light pollution.)
>
> The exception is astronomers who observe variable stars, for which magnitude variations in the second decimal place may be important. In the Bad Olde Days, amateur variable star observers had to judge the current magnitudes of their chosen stars by comparing them to the known magnitudes of non-variable stars. Nowadays, many serious amateur variable star observers use CCD equipment that provides exact magnitudes down to the second decimal place.

Types of Magnitude

The term magnitude is used in different ways to describe different aspects of an object's brightness.

Visual magnitude
> *Visual magnitude* describes the brightness of an object as seen by the human eye.

Photographic magnitude
> *Photographic magnitude* describes the magnitude of an object as captured on film or by CCD or other electronic imaging technologies. This term originated in the 19th century. The crude photographic emulsions of the time were sensitive only to blue, violet, and ultraviolet light. Because astronomical objects have widely varying spectra, early astrophotographers soon found that some objects that were bright to the

human eye were dim photographically, and vice versa. Accordingly, they began differentiating visual versus photographic magnitudes.

Nowadays, astronomical imaging equipment can record images from the far infrared to the far ultraviolet, so the photographic magnitude of an object depends entirely on which part of the spectrum you use to define it. For example, some objects emit primarily or exclusively in the infrared portion of the spectrum, which is invisible to the human eye. Such an object has an infinitely high visual magnitude because humans cannot see it at all. Conversely, it may have a very low infrared photographic magnitude because it records on infrared sensitive film or imaging equipment as a very bright object.

Apparent magnitude

Apparent magnitude describes the brightness of an object as seen or imaged from Earth. The apparent magnitude of an object is determined by both the amount of light the object emits and its distance from us. An object with a low apparent magnitude appears bright to us because it emits a lot of light, is close to us, or both. A distant object that emits a massive amount of light may have a lower apparent magnitude (appear brighter to us) than a much closer object that emits less light. Conversely, a very nearby object that emits relatively little light may appear brighter to us than a far distant object that emits 1,000 or 1,000,000 times as much light. Apparent magnitudes are indicated by a lowercase "m". For example, a star with an apparent magnitude of 4.7 is recorded as "4.7m" or "m4.7" when logging observations.

Absolute magnitude

Absolute magnitude describes the inherent brightness of an object, and it is unrelated other than coincidentally to apparent magnitude. Absolute magnitude is defined as the apparent magnitude a star would have if it were 10 parsecs (about 32.6 light years) distant from us. Our own Sun, Sol, for example, has an absolute magnitude of 4.7. That means that if Sol were located 32.6 light years away, its apparent magnitude would be 4.7. Another way of looking at it is that any star that happened to be located 32.6 light years from us would have identical absolute and apparent magnitudes.

The inherently brightest stars we know of have absolute magnitudes of about -8.0. The inherently dimmest stars have absolute magnitudes approaching 16.0. Absolute magnitudes are indicated by an uppercase "M." For example, a star with an absolute magnitude of 4.7, such as our own sun, Sol, is recorded as "4.7M" or "M4.7".

When amateur astronomers mention the magnitude of an object, they almost always mean the apparent visual magnitude.

Magnitude Ranges

Early astronomers used magnitudes ranging from one to six to categorize visible stars. Modern astronomers extended the magnitude range on both ends, and they began applying the concept to celestial objects other than stars.

The first departure came during the 19th century, when astronomers began measuring the precise brightness of stars. They learned that there was a greater range of brightnesses in stars that had formerly been grouped as "first magnitude" than there was between some first magnitude stars and others classically considered to be second magnitude. That discovery led to the assignment of fractional and negative magnitudes.

Magnitudes are rounded in the same way as any other number. For example, any star with a magnitude between 0.5 and 1.49 is called a first-magnitude star, and any star between 2.5 and 3.49 is called a third-magnitude star. The six stars with magnitudes between -0.5 and 0.49 listed in Table 2-1 are called zeroth-magnitude stars. Sirius and Canopus, which are brighter than zeroth magnitude, have no common class name, although we suppose they could be called "negative one magnitude" stars.

Table 2-1 lists the common names, Bayer designations, apparent magnitudes, absolute magnitudes (a "v" indicates the star is of variable magnitude), and distances of the 25 brightest stars in the sky, with our own Sun, Sol, shown for comparison. The ancients described all of these stars as "first magnitude."

The Bayer designation is a "shorthand" terminology used by astronomers to designate bright stars [Hack #14].

Table 2-1. The 25 brightest stars visible from Earth

Common name	Bayer designation	Apparent magnitude	Absolute magnitude	Distance (light years)
The Sun, Sol	n/a	-26.8	4.7	8.33 light minutes
Sirius	α-CMa	-1.45	1.4	8.8
Canopus	α-Car	-0.73	-4.7	196
Rigel Kentaurus	α-Cen AB	-0.27	4.3	4.2
Arcturus	α-Boo	-0.06	-0.2	36
Vega	α-Lyr	0.04	0.5	27
Capella	α-Aur	0.08	-0.6	46
Rigel	β-Ori A	0.15v	-7	815
Procyon	α-CMi	0.35	2.7	11.4

Table 2-1. The 25 brightest stars visible from Earth (continued)

Common name	Bayer designation	Apparent magnitude	Absolute magnitude	Distance (light years)
Achernar	α-Eri	0.53	-2.2	127
Hadar	β-Cen AB	0.66	-3.5	520
Betelgeuse	α-Ori	0.70v	-5	650
Altair	α-Aql	0.8	2.3	16.3
Aldebaran	α-Tau	0.86	-0.7	52
Acrux	α-Cru A	0.9	-3.5	260
Antares	α-Sco	0.98v	-4.7	425
Pollux	β-Gem	1.15	1	35.8
Fomalhaut	α-PsA A	1.16	1.9	22.8
Deneb	α-Cyg	1.25	-7.3	1630
Mimosa	β-Cru	1.28	-4.6	489
Regulus	α-Leo A	1.36	-0.7	98
Castor	α-Gem	1.58	0.8	45.6
Shaula	λ-Sco	1.62v	-3.4	326
Bellatrix	γ-Ori	1.63	-3.3	303
GaCrux	γ-Cru	1.64	-2.5	228
Elnath	β-Tau	1.65	-2	179

The differences are striking. Sirius, the brightest star (other than Sol) visible from Earth has an apparent magnitude of -1.45. With an absolute magnitude of 1.4, Sirius is inherently fairly bright, but the real reason for its brilliance is that it's located less than 10 light years from Earth, which in astronomical terms makes it our next-door neighbor. Canopus, the next-brightest star as visible from Earth, has an apparent magnitude of -0.73, 0.72 magnitude dimmer than Sirius, although at -4.7 its absolute magnitude is 6.1 magnitudes brighter than Sirius. Canopus appears dimmer than Sirius because Canopus is located nearly 200 light years from Earth, or about 22 times more distant than Sirius.

Variable Magnitudes

Although most celestial objects have fixed magnitudes, there are exceptions, which are described in the following sections.

Variable stars. All stars vary somewhat in magnitude, in a cycle that may span from seconds to years. But most stars vary in brightness over a very small range and so are considered for practical purposes to be of fixed magnitude. Some stars, called *variable stars*, vary significantly from their brightest to their dimmest, sometimes by several magnitudes and some-times over very short periods. Depending on their apparent minimum and maximum magnitudes and their degree of variability, variable stars may be anything from prominent at all times to quite dim at their maxima and invis-ible at their minima to prominent at their maxima and quite dim at their minima.

Variable stars are grouped into two types, four classes, and several sub-classes:

Intrinsic variables

> *Intrinsic variable stars* actually increase and decrease their light output over time. There are two classes of intrinsic variable star.

Pulsating variables

> *Pulsating variables* are stars that periodically expand and contract, varying their light output as they do so. There are several sub-classes of pulsating variables, most of which are named for the first star found in that sub-class. *Cepheid variables* have periods of 1 to 70 days and maximum variation of about 2 magnitudes. *RR Lyrae variables* have periods of about four hours to a day and maximum variation of about 2. *RV Tauri variables* have periods ranging from about 30 to 100 days and maximum variation of about 3. *Mira variables*, also known as *long-period variables*, have periods ranging from 80 to 1,000 days and magnitude variations of about 2.5 to 5. *Semi-regular variables* have periods ranging from 30 to 1,000 days and magnitude variations of about 1 to 2, with variations in both period and degree of variation.

Erupting variables

> *Erupting variables*, also called *cataclysmic variables*, are stars in which the core thermonuclear processes periodically run out of con-trol, causing eruptions that increase light output. The most dra-matic example of an erupting variable is a *supernova*, a stupendous and irreversible explosion of a star that can temporarily boost its brightness by 20 or more magnitudes. *Novae* and *recurrent novae* have periods ranging from a day to a year, and magnitude varia-tions from about 7 to 16. *Dwarf novae*, of which there are three sub-types, have periods ranging from 30 to 500 days, and variations from about 2 to 6 magnitudes. *UV Ceti variables*, also known as *flare stars*, are dim, red stars that periodically brighten for several seconds, gaining two or more magnitudes, and then drop back to their minima after a few minutes. *Symbiotic variables* have sporadic periods and magnitude variations to about 3. *R Coronae Borealis variables* have sporadic periods and magnitude variations up to 9.

Extrinsic variables

Extrinsic variable stars produce reasonably constant light output, but show variability caused by other factors. There are two classes of extrinsic variable star.

Eclipsing binary variables

Eclipsing binary variables are multiple star systems in which the orbital plane of the system happens to correspond to our line of sight to the system. As the secondary orbits the primary, it periodically eclipses the primary, reducing the light visible to us on Earth. The period of eclipsing variables ranges from a few minutes to hundreds of years, and the magnitude variations range from 0.1 or less to 2.5 or more. Algol (β-Perseii, called the demon or ghoul star) is a famous example of an eclipsing variable with a relatively short period and significant magnitude variation.

Rotating variables

Rotating variables are stars whose surfaces are patchy, with darker and brighter areas, much like sunspots but on a gigantic scale. The period of a rotating variable corresponds to its own rotation period, and may range from seconds to days. The variability is usually quite small, on the order of 0.1 magnitudes or less.

 For more information about variable stars, visit the American Association of Variable Star Observers (AAVSO): *http:// www.aavso.org.*

Observing variable stars is popular among amateur astronomers for two reasons. First, variable stars can be observed at any time of night on any day of the year from any site, even urban locations. More important, variable star observing is one of the few disciplines in which amateurs can still make a serious contribution to science. There are many variable stars and the resources of professional astronomers are limited, so they depend upon amateurs to gather data for them.

Solar system objects. Solar system objects—Luna, the planets, asteroids, and comets—are inherently of variable magnitude because their orbits vary their positions relative to the Sun and Earth. Comets vary most in magnitude because their orbits may carry them from the outer edges of the solar system to nearly grazing Sol. At dimmest, most comets are invisible in even the largest scopes; at brightest, a spectacular comet may be a naked-eye object during the day. Earth's moon, Luna, also varies dramatically in magnitude, from invisible when new to about -12.6m when full.

Other solar system objects vary less in magnitude, although some show significant magnitude variations. With the exception of Pluto, which has a highly variable orbit, planets and asteroids move in elliptical orbits that approximate circles, so their distance from Sol—and accordingly their illumination—varies relatively little. The chief determinant of planetary and asteroid variability is therefore the separation between Terra and the planet or asteroid in question, which is determined by their relative positions in their orbits. For example, Mars and Earth reach closest approach about every two years, during which Mars is significantly brighter than at other times.

At brightest, Venus has an apparent magnitude of about -4.4, which varies little because it is always quite close to Sol; Jupiter peaks at about -2.7; Mars, about -2.0; and Saturn, about 0.5. Jupiter and Venus both reach negative magnitudes bright enough to be easily visible during full daylight, if you know when and where to look for them. The best time to do that is when one or both of them are near Luna in the daytime sky. For example, Figure 2-4 shows Winston-Salem Astronomical League (WSAL) members Bonnie Richardson, Paul Jones, and Mary Chervenak (left to right) observing the daytime Lunar occultation of Jupiter on 9 November 2004.

Figure 2-4. WSAL members observe a daytime Lunar occultation of Jupiter

An occultation occurs when the orbit of the Moon or another solar system object causes it to block our view of a star, planet, or other object temporarily. Occultations of planets and bright stars are relatively rare events.

With Luna as a reference point—the thin crescent of Luna itself was difficult to locate in the bright daytime sky—all of us were able to view Jupiter naked eye as it was occulted by Luna and then again as it egressed. We were also able to view Venus naked eye because it was only 5° or so distant from Luna, and therefore easy to locate. (Knowing exactly where to look is critical for observing planets during the day; you can scan the sky forever and not see the planet, but once you know exactly where it is it pops out at you, so obvious that you can't believe you couldn't see it before.)

Visibility by Magnitude and Instrument

The range of visible magnitudes depends on many factors, including personal vision, degree of dark adaptation, sky transparency, light pollution, elevation of the observing site, altitude of the star, the cleanliness of your optics, and of course the optical instrument itself. Elevation and altitude can make a major difference because looking through less air allows you to see dimmer objects. The following are approximations and will vary from person to person as well as with differing observing conditions, but they at least give you a reasonable idea of what you might expect to see.

- Naked eye. Under clear, dark conditions, fully dark adapted [Hack #11] and using averted vision [Hack #22], most people can see stars at zenith of magnitude 5.5 or dimmer. Those with younger eyes observing under ideal conditions can often see down to magnitude 6.0 to 6.5 at zenith. Under very clear conditions at high altitude, a young person with superb night vision may be able to see stars as dim as 7.0 to 7.5. We have heard reports claiming sightings of 8.0 magnitude stars naked eye, but frankly we don't believe them.

- Binoculars. A typical standard binocular—depending on aperture, exit pupil, and your entrance pupil—increases light gathering by a factor of 25 to 100 relative to your naked eye, which allows you to see 3.5 to 5 magnitudes deeper using the binocular [Hack #8]. In practice, we've found that under excellent conditions from our regular dark-sky observing site, we can usually see down to magnitude 9.5 or a bit dimmer with a 50mm binocular. Those with younger eyes (or a darker site) may get down to magnitude 10.0 or even 10.5 with a 50mm binocular with superb coatings.

- Amateur telescopes. Although some telescope makers publish "limiting magnitudes" for their instruments, these are approximations at best, assume perfect observing conditions, and even at that are often quite optimistic. Limiting magnitudes for the smallest astronomical reflector and refractor telescopes are generally in the 11.5 to 12.0 range, with 5" and 6" instruments boosting that to 12.5 to 13.0. A typical 8" SCT or Newtonian reflector under good real-world conditions allows you to see stars as dim as magnitude 13.5 or so. A 10" telescope may get you down to magnitude 14.0, a 12.5" scope to 14.5, and a 15" scope to 15.0. A 17.5" or 18" Dobsonian will show you stars as dim as magnitude 15.5, and a 20" model 16.0 or thereabouts. The largest amateur telescopes, giant 30" to 40" Dobsonians, go as deep as magnitude 17.5 or a bit more.

Surface Brightness

The concept of magnitude is applied both to stars, which are point sources of light, and to *extended objects*, which have a visible surface area. Because they are point sources, stars cannot be magnified, although the laws of physics mean that stars actually appear as small disks in any real-world telescope. A star under any magnification therefore always has the same apparent brightness, which is determined only by the aperture of the scope you are using to view it.

Extended objects, such as nebulae and galaxies are different. Because they have surface area (also called *extent*), they can be magnified. As you increase the magnification on an object, the apparent size of the object grows larger, but its apparent brightness grows dimmer. For example, if you double magnification, the apparent linear extent of the object doubles, which is to say that it appears twice as large. Doubling the linear size of the object quadruples its area. That means the light the object emits is spread over four times the area, so the apparent brightness of the object is reduced by a factor of four. In the difference between point-source objects and extended objects lies a truth that has frustrated many a beginning astronomer.

Although non-astronomers find it difficult to believe, there are in fact many extended objects that are larger than the full moon. The Andromeda Galaxy (M31), for example, is roughly 3° long by 1° wide, versus the half-degree circle of the full moon. In other words, M31 has more than a dozen times the surface area of the full moon. There are many other examples of huge extended objects. Just in the constellation Cygnus, for example, are the North America Nebula (NGC 7000) and the Veil Nebula (NGC 6992), both of which have extents greater than that of Luna.

Why, then, don't these objects jump out at us when we are under the night sky? Because they are very, very dim. Or, to put it another way, they have low surface brightness. A beginning astronomer, checking his charts, may note that M31 is listed as magnitude 3.4 (actually, the magnitude of M31 or any extended object depends on which source you use; more on that later in this section). A star of magnitude 3.4 is easily visible from all but the most light-polluted locations, and yet our beginning astronomer can't see M31 and doesn't understand why.

The reason is magnitude versus surface brightness. Magnitudes given for extended objects are *integrated magnitudes*, which are calculated by assuming that all of the light from the extended object has been condensed into a point source. In other words, if M31 were a star instead of a huge extended object, that star would be 3.4m. But M31 is not a star, with its light condensed into an infinitesimally small point. M31 is a gigantic extended object, with its light spread over several square degrees of sky.

The concept of *surface brightness* (abbreviated *SBr*) attempts to rectify the magnitude versus visibility issue for extended objects. To calculate surface brightness, you determine the apparent extent of an object and distribute the total light emitted by that object evenly over the extent. With the light evenly distributed, you can determine for any point on the object's extent how much light is emitted and what the magnitude of a star would be if it were that bright. For example, the planetarium program Cartes du Ciel lists the magnitude of M31 as 3.40, but its surface brightness as 13.50, or more than 10,000 times dimmer. Similarly, the North America Nebula is listed as 4.00m and 12.63SBr and the Veil Nebula as 7.00m and 13.44SBr.

But if M31 were actually as dim as a 13.50m star, it would be impossible to see it with the naked eye, and yet anyone can see M31 naked eye if the site is dark enough. How can that be? There are two answers. Firstly, the human eye can accumulate and integrate light, treating an extended object much the same as it does a point source. When you view M31, your eye accumulates all of the light emitted by the object and presents it to your brain as a soft, dim blur. Secondly, a typical extended object varies greatly in brightness across its extent. Most, like M31, are relatively bright near the center, and increasingly dim near the edges. The brighter areas of some extended objects, including M31, are sufficiently bright to excite the rods (the monochrome-only dim light sensors in your eye) enough to glimpse the object [Hack #11].

Which brings up the reason why the magnitude and surface brightness of extended objects are matters of opinion. Both depend on how you define the extent of an object. Do you use the smaller visual extent, or the much larger photographic extent? Visually, for example, M31 has an extent not much larger than the full moon, depending on how good your dark-adapted vision is and the size of the telescope you use to view it. Photographically, M31 has a much larger extent, on the close order of three square degrees. That's true because long-exposure CCD or film images can reveal dim parts of an object that are beyond the lower threshold of human vision.

As you increase the extent of an extended object, you also decrease the integrated magnitude because you are adding light, albeit in very small amounts. But as you increase the extent defined for an object, you also decrease the surface brightness because the area of the object is growing much faster than the amount of light contributed by the additional area. Accordingly, both the integrated magnitude and the surface brightness are, although this sounds odd, a matter of opinion, because both depend entirely on how you define the extent.

This disconnect has profound implications for actual observing because it determines whether a particular object in a particular scope is easily visible, visible with difficulty, or not visible at all. Consider, for example, the Messier galaxy pair M81 and M82 in Ursa Major. Cartes du Ciel defines M81 as having an extent of 24.9X11.5 arcminutes, an integrated magnitude of 6.9, but a surface brightness of only 13.4. M82 is given an extent of 10.5X5.1 arcminutes, an integrated magnitude of only 8.4, but a surface brightness of 12.5. In other words, M81 is larger and 1.5 magnitudes brighter but has surface brightness almost a full magnitude dimmer.

Actual observations bear this out. Although both galaxies are easy to see under relatively dark skies with even a small scope, M81 is clearly a more difficult object. It is larger, more diffuse, and "feels" dimmer despite its brighter integrated magnitude. Similar issues apply for other extended objects. Our rule of thumb when viewing extended objects is to consider surface brightness and ignore magnitude. We suggest you do the same.

Identify Stars by Name

Learn common star names and how they're pronounced so the other kids won't laugh at you.

From remotest antiquity, every culture has given names to the brightest stars. Such names are called the *common name* or *proper name* of a star. Several hundred of the brightest stars have common names, but most amateur astronomers know and use only a few dozen.

One star may have many common names. Vega, for example, the brightest star in the constellation Lyra, is said to have more than 50 known names.

Hundreds more, no doubt, are lost in the mists of time. Many stars have similar names because they share common root names. For example, the syllable "al" is Arabic for "the" and appears in many common star names, as does "deneb" for "tail."

The same common name is sometimes used for more than one star. When used without qualification, that practice is fortunately limited to less prominent stars. For important stars that share a name, that name is qualified for at least one of the stars. For example, the name Rigel used alone refers unambiguously to the brightest star in Orion. Another bright star named Rigel exists in the constellation Centaurus, but that star is always referred to as Rigel Kentaurus to avoid confusion.

Some common star names—including Sirius, Procyon, Castor, and Pollux—originated with ancient Greek astronomers like Hipparchus and Ptolemy and have come down to us unchanged. The Romans, great engineers but poor scientists, also contributed a few common star names, including Arcturus, Bellatrix, Regulus, and Vindemiatrix. A few common star names, like Polaris and Cor Caroli, are Latin but of relatively recent origin. But the vast majority of common star names come from the Arabic.

The pronunciation and even the spelling of many common star names varies. Original pronunciations and spellings have often been lost or corrupted beyond recognition. For example, the star Almach in Andromeda may be spelled Almaak, Almaach, Almaak, Almak, or even Alamach, with similarly differing pronunciations. Vega is properly pronounced WAY-guh, but if you say it that way people think you're strange. The common pronunciation is VAY-guh, with VEE-guh also sometimes heard.

Table 2-2 lists alphabetically by constellation the common star names that are familiar to most amateur astronomers. The most important ones to know are italicized. When multiple pronunciations are given, the first is preferred or more common, but the other(s) are also used commonly. (The Arabic "al" is commonly pronounced "al," "ahl," or "ul", so we simply leave it as "al." The Arabic "g" is variously pronounced hard (game), soft (gelatin), or with the "zh" sound in Dr. Zhivago.)

Table 2-2. Common names and pronunciations of important stars

Constellation	Principal stars
Andromeda	*Almach* (AL-mawk), *Alpheratz* (al-FUR-uts), *Mirach* (MURR-awk)
Aquila	*Altair* (al-tuh-EER, al-TAIR), Tarazed (TAR-uh-zed)
Aries	*Hamal* (huh-MALL), Mesarthim (mez-are-TEEM), *Sheratan* (SHARE-uh-tan)
Auriga	*Capella* (cuh-PELL-uh), Menkalinen (men-KAL-ih-nen)
Boötes	*Arcturus* (ark-TURE-us)

Table 2-2. Common names and pronunciations of important stars (continued)

Constellation	Principal stars
Canes Venatici	*Cor Caroli* (CORE cuh-ROLL-ee)
Canis Major	Mirzam (MURR-zahm), *Sirius* (SEAR-ee-us), Wezen (WEZ-un)
Canis Minor	Gomeisa (go-MAYZ-uh), *Procyon* (pro-KYE-on, pro-SIGH-on)
Capricornus	Deneb Algedi (den-EB al-ZHED-ee)
Carina	*Canopus* (can-OH-pus)
Cassiopeia	*Caph* (KAF), *Ruchbah* (RUCK-bah, ROOK-bah, RUKE-bah), *Schedar* (SHED-ar)
Centaurus	*Agena* (uh-JEEN-uh, uh-JENN-uh, more properly uh-GEEN-nuh), *Proxima* (PROX-ih-muh)
Cepheus	Alderamin (al-DER-uh-min, al-der-uh-MEEN)
Cetus	Menkar (MEN-car), Mira (MEE-ruh, MEAR-uh, MURR-ah)
Coma Berenices	Diadem (dee-AW-dem)
Crux	*Acrux* (AY-crux)
Cygnus	*Albireo* (al-BEER-ee-oh, al-BURR-ee-oh), *Deneb* (den-EB, duh-NEB)
Draco	Thuban (THEW-ban, THEW-bahn, thuh-BAHN)
Eridanus	Achernar (awk-er-NAHR, AK-er-nahr)
Gemini	Alhena (al-HEN-uh, al-HAY-nuh), *Castor* (CASS-tur), Mebsuta (meb-SUE-tuh), *Pollux* (PALL-ux)
Grus	*Alnair* (al-NAH-ur, al-nah-EER, al-NAYR)
Hercules	Rasalgethi (rah-sool-ZHAYTH-ee, rahz-al-GAITH-ee)
Leo	Algieba (al-ZHEEB-uh, al-JEB-uh), *Denebola* (den-EB-uh-lah), *Regulus* (RAY-gyoo-lus, REG-you-lus), Zosma (ZOSS-muh)
Lepus	Arneb (ARE-neb), Nihal (nee-HALL)
Libra	Zubenelgenubi (zub-BEN-ell-zhuh-NEW-bee), Zubeneschamali (zub-BEN-ess-sha-MALL-ee)
Lyra	*Vega* (VAY-guh, VEE-guh, properly WAY-guh)
Ophiuchus	Rasalhague (rah-SOOL-huh-WAY, RAS-al-hayg)
Orion	*Alnilam* (al-nih-LAHM), *Alnitak* (al-nih-TOK), *Bellatrix* (BELL-uh-trix), *Betelgeuse* (bet-ul-jow-ZAY, BEET-ul-JOOS), *Mintaka* (min-TOK-uh), *Rigel* (RYE-jul), *Saiph* (SIFE, SAFE, sa-EEPH)
Pegasus	*Algenib* (al-ZHEN-ib), *Enif* (ENN-if, EE-nif), *Markab* (MAR-cob), *Scheat* (SHEE-awt)
Perseus	*Algol* (AL-gall), *Mirfak* (MURF-awk)
Pisces Australis	*Fomalhaut* (FOME-ul-HOE, fome-ul-HOWT)
Sagittarius	*Ascella* (uh-SELL-uh), *Kaus Australis* (KOWS ow-STRAWL-us, aw-STRAWL-us), *Kaus Borealis* (BORE-ee-AWL-us), *Kaus Meridionalis* (muh-RID-ee-un-ALL-us), *Nunki* (NUN-kee)

Table 2-2. *Common names and pronunciations of important stars (continued)*

Constellation	Principal stars
Scorpius	Akrab (AWK-rob, also known as Graffias, GRAFF-ee-us), *Antares* (an-TAR-eez, an-TAIR-eez), Dschubba (JOO-buh), Lesath (LESS-uth, LAY-soth), Shaula (SHAH-luh)
Taurus	*Aldebaran* (al-DEB-uh-ron), *Elnath* (EL-nuth)
Triangulum	Metallah (meh-TAHL-uh)
Ursa Major	Alcor (al-CORE), *Alioth* (al-YAHT, AL-ee-oth), *Alkaid* (al-KAH-id or al-KADE), *Dubhe* (DUB-uh, DOOB-uh), *Megrez* (MEG-grez), *Merak* (MER-awk), *Mizar* (mi-ZAHR, MYE-zahr), *Phad* (FAD, also known as *Phecda*, FECK-duh), Talitha (TAH-lith-ah)
Ursa Minor	Kochab (KOE-cab), Pherkad (FUR-cahd), *Polaris* (puh-LAIR-us)
Virgo	Porrima (POR-rim-uh, poh-RIM-uh), *Spica* (SPEE-kuh, SPY-kuh), Vindemiatrix (vin-duh-MEE-uh-trix)

It's important to know common star names because you'll routinely need to know them when you point out objects to another observer (or vice versa), when you align a go-to scope, and for other common activities. It's important to pronounce the names properly to avoid looking like a complete newbie, regardless of how experienced you may be. You needn't learn all of the common star names, nor even all of those listed in Table 2-2. But it does help to learn the few dozen names shown in italic. The best way to learn them is to memorize the named stars in each constellation as you study or work that constellation.

HACK #15 Identify Stars by Catalog Designations

Learn about the stellar catalogs used by amateur astronomers.

Of the thousands of stars visible to the naked eye—and millions visible with optical aid—only a few hundred of the brightest have proper or common names, such as Vega or Sirius. Although proper names are convenient for referring to bright "guidepost" stars, astronomers also need unambiguous designations for dimmer stars.

It is possible, although awkward, to specify a particular star by giving its coordinates **[Hack #17]**. For example, we could say, *the star located at 14h15m40.35s right ascension and +19°11'14.2" declination*, but that gets old fast. As a more convenient alternative, astronomers have developed star catalogs, which assign each star a unique short identifier, such as α-Boötis or HIP 69673. (All three of these designations specify Arcturus, the brightest star in the constellation Boötes.)

Early astronomers began the work of cataloging the stars visible to the naked eye, dividing the stars by constellation and then assigning unique identifiers to each. Modern-day astronomers have continued that practice, but they now treat the sky as a contiguous whole and catalog stars without respect to constellation. Surprisingly, the star naming syntax of some catalogs produced 300 to 400 years ago remains in common use. Here are the stellar catalogs you need to be familiar with.

The Bayer Catalog

The German astronomer Johann Bayer (1572–1625) published *Uranometria*, the first comprehensive star atlas, in Augsburg in 1603. *Uranometria* predates telescopes, so it contains only stars that are visible to the naked eye. *Uranometria* was unique for its time because it mapped far southerly stars, including those in south circumpolar constellations. In *Uranometria*, Bayer introduced his system of labeling the bright stars in each constellation with lowercase Greek letters, a system that is still used today.

Bayer designated the brightest star in each constellation alpha (α), the second brightest beta (β), and so on through the Greek alphabet, shown in Table 2-3, although he made numerous exceptions to the brightness-based hierarchy. For example, Bayer sometimes assigned "brighter" letters to stars that happened to be nearer the head of the traditional mythological constellation figure, and lower letters to those stars near its feet. Bayer appended the genitive form of the name of the constellation [Hack #16] to assign each star a unique label. For example, Vega, the brightest star in the constellation Lyra, has the Bayer designation α-Lyrae.

> Prominent multiple stars and some strings of stars share one Greek letter, with an Arabic numeral as a qualifier. For example, β-Cygni is a stunning double star. The primary, golden Albireo (3.08m) is β1-Cygni. The unnamed blue secondary (5.11m) is β2-Cygni. Similarly, a long string of stars off Bellatrix are assigned the Bayer designations π1 through π6-Orionis.

Table 2-3. Greek letters and their pronunciation

| Greek letter | | | Pronunciation | |
Lower	Upper	Names	Common	Correct
α	A	alpha	AL-fuh	
β	B	beta	BATE-uh	BEET-uh (VEET-uh in modern Greek)
γ	Γ	gamma	GAMM-uh	

Table 2-3. *Greek letters and their pronunciation (continued)*

Greek letter		Names	Pronunciation	
Lower	Upper		Common	Correct
δ	Δ	delta	DELT-uh	
ε	E	epsilon	EPP-sih-lawn	
ζ	Z	zeta	ZATE-uh	ZEET-uh
η	H	eta	ATE-uh	EET-uh
θ	Θ	theta	THATE-uh	THEE-tuh
ι	I	iota	eye-OTE-uh	YOTT-uh
κ	K	kappa	CAP-uh	
λ	Λ	lambda	LAM-duh	
μ	M	mu	MEW	MEE
ν	N	nu	NEW	NEE
ξ	Ξ	xi	ZIE	KSEE (KS as in foX)
o	O	omicron	OH-mih-kron	AW-muh-kron
π	Π	pi	PIE	PEE
ρ	P	rho	ROE	
σ	Σ	sigma	SIG-muh	
τ	T	tau	TOW (OW as in hOW)	TAF (AF as in cALF)
υ	Y	upsilon	UP-sih-lawn	EEP-sih-lawn
φ	Φ	phi	FIE	FEE
χ	X	chi	KYE	KHEE (KH as in Scottish loCH)
ψ	Ψ	psi	SIGH	puh-SEE
ω	Ω	omega	oh-MEG-uh, oh-MEE-guh	aw-MEG-uh

> Two characters represent the lowercase sigma. σ is used at
> the beginning and in the middle of words. ς is used only as
> the final letter in words. ς appears rarely in Bayer designa-
> tions instead of the correct σ.

Bayer had no instruments for measuring magnitudes, and his catalog ranks
stellar brightness incorrectly in many cases. For example, Bayer labeled
Betelgeuse (0.50m) α-Orionis and the brighter Rigel (0.12m) β-Orionis. Similarly, Pollux (1.14m) is β-Geminorum, while the dimmer Castor (1.98m) is
α-Geminorum.

Bayer also intentionally made exceptions for stars he "liked." For example,
Bayer assigned Thuban (3.65m) as α-Draconis, despite the fact that Thuban

is noticeably dimmer than β-, γ-, δ-, ζ-, and even η-Draconis. Bayer presumably honored Thuban for its historical importance rather than its brightness. Five thousand years ago, when the ancient Egyptians were building the first pyramids, Polaris was not the Pole Star. Thuban was.

Despite the fact that stellar magnitudes are now known precisely and Bayer's brightness rankings are known to be wrong, astronomers continue to use Bayer's original designations. It's important to understand Bayer designations because they are used widely in charts, handbooks, articles, and so on. Figure 2-5 shows the constellation Orion with its prominent stars labeled with Bayer designations.

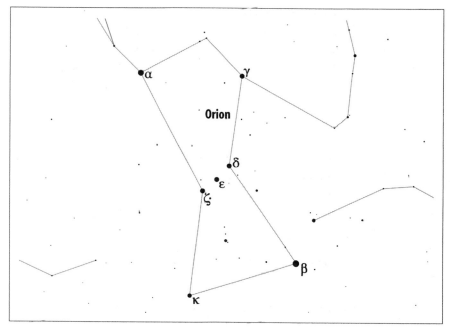

Figure 2-5. Prominent stars in Orion, with Bayer labels

 It's also important to know Greek alphabet symbols because symbols and names are used interchangeably. For example, if you are reading an article about the galaxy NGC 404 in the constellation Andromeda, it may mention that the galaxy is located only a few arcminutes from the bright star Mirach. But the article may refer to that star as beta Andromedae or β-Andromedae. If you don't know that β is the symbol for beta, you won't know which star the article refers to.

The Flamsteed Catalog

The Englishman John Flamsteed (1646–1719) was appointed the first Astronomer Royal of Britain in 1676. He created *Historia Coelestis Britannica*, the first great scientific reference work released under the aegis of the Greenwich Observatory. It contained Flamsteed's detailed observations and a catalog of 2,682 stars with their positions mapped with unprecedented accuracy. Flamsteed never traveled far south, so his catalog contained only stars located at declinations of -50° and higher. Despite this gap, Flamsteed's catalog quickly became the authoritative star atlas because of its accuracy.

> To his extreme annoyance, Flamsteed was published involuntarily. He wanted to delay publication until he considered the work complete, but other scientists including Edmond Halley and Isaac Newton desperately needed Flamsteed's accurate data for their own work, and so eventually forced publication of the work over Flamsteed's objections. The first, unauthorized edition of 400 copies was published in 1712, prepared by Halley and funded by Prince George of Denmark. Flamsteed successfully sued and was able to reclaim and burn these unauthorized copies. The authorized version of *Coelestis Britannica* was finally published in 1725, six years after Flamsteed's death.

Flamsteed's original catalog did not label stars, but merely mapped them by declination and right ascension. The French astronomer Joseph-Jerome de Lalande decided that the absence of star labels made the catalog awkward to use, so he annotated his French edition of *Coelestis Britannica*, adding numeric labels to each of the stars mapped by Flamsteed. Those numbers, universally called *Flamsteed numbers* despite that fact that Flamsteed had nothing to do with them, are still used today.

Figure 2-6 shows part of the constellation Orion, with the prominent stars labeled with their proper names, Bayer designations, and Flamsteed numbers. Notice that the Flamsteed numbers increase from right to left, without regard to where the star is located vertically on this chart. For example, Rigel, at the bottom right of the chart, is assigned the Flamsteed number 19. Bellatrix, at the top center of the chart, is assigned Flamsteed 24. Betelgeuse, at the upper left, is assigned Flamsteed 58.

This is true because Lalande assigned numbers to Flamsteed's stars in order of right ascension, from west to east **[Hack #17]**. Lalande started with the westernmost star in each constellation, assigning it the number 1. He then proceeded east within the constellation, numbering each star without regard to its declination. (The highest Flamsteed number is 140 Tauri.)

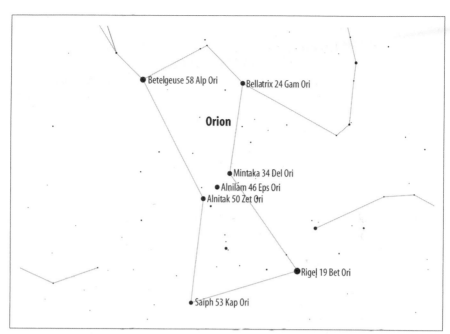

Figure 2-6. Prominent stars in Orion, with proper names, Bayer labels, and Flamsteed numbers

Flamsteed numbers are still commonly used today. Most amateur astronomers use the following rules to decide how to refer to a star.

- If the star has a proper name that is common usage, use it. Every amateur astronomer knows which star you mean when you mention Betelgeuse. (Most will also recognize the α-Orionis Bayer designation; few will recognize the Flamsteed designation 58 Orionis without referring to a chart.)

- If the star has no common proper name, use the Bayer designation if one exists. (Most astronomers recognize the Bayer designation σ-Orionis; almost none recognize the Flamsteed designation 48 Orionis.)

- If the star has no common proper name or Bayer designation, use the Flamsteed number. Some famous stars, for example 61 Cygni, are best known by their Flamsteed numbers.

- If the star has no common name, Bayer designation, or Flamsteed number, use the designation from one of the catalogs described in the following section.

Modern Stellar Catalogs

As useful as the Bayer and Flamsteed catalogs were and still are, they don't go very deep. Bayer maps only naked-eye stars, and Flamsteed maps stars only down to 8th magnitude. Even standard binoculars show stars to 9th magnitude or dimmer, and large amateur telescopes show stars to 17th magnitude or dimmer.

> For comparison, a "beginner" star atlas like Norton's or Cambridge shows stars to 6th magnitude. *Sky Atlas 2000.0*, popular among intermediate and advanced amateur astronomers, shows 81,000 stars down to magnitude 8.5. *Uranometria 2000.0* (the new version, not the classic) shows 280,000 stars down to magnitude 9.75. The *Millennium Star Atlas*, now sadly out of print, shows more than 1,000,000 stars down to magnitude 11.

As the science of astronomy developed, the need for deeper star catalogs quickly became apparent. There was also a need for specialty catalogs that focused on such matters as extremely precise magnitude measurements or detailed spectra. In the mid-19th century, various national and university observatories began developing such catalogs. (Amateur astronomers do not use these catalogs in their native or raw forms; instead, these catalog numbers are incorporated in planetarium software and in printed charts such as *Sky Atlas 2000.0* and *Uranometria 2000.0*). Here are the most important catalogs to know about:

Bonner Durchmusterung (BD) Catalog
> In 1859, the German astronomer F. W. A. Argelander, working at the Bonn Observatory, set out to map all stars down to magnitude 9.5. Cataloging stars by constellation becomes confusing below 8th magnitude, so for the *Bonner Durchmusterung Catalog*, Argelander decided to ignore constellations and map the stars by declination strips 1° wide, each of which incorporated the full 24 hours of right ascension. For each declination strip, Argelander assigned numeric designations to each star in right ascension order, from west to east. For example, Vega is designated BD+38°3238, which indicates that Vega is the 3,238th star in the 38° to 39° declination strip. The original BD catalog mapped stars to 2° south of the celestial equator; it was later extended by the SBD and Cordoba Durchmusterung (CD or CoD) catalogs. The consolidated BD catalogs contain 324,188 stars.

Revised Harvard Photometry (Harvard Revised, or HR)

In 1908, Harvard College issued the *Revised Harvard Photometry Catalog*, which is usually referred to as *Harvard Revised* or *HR*. This catalog lists only 9,110 stars down to magnitude 6.5, but it provides extremely precise magnitude information for each. HD catalogs stars by right ascension, from west to east, beginning at RA 0h00m00s, and numbers each cataloged star sequentially. For example, Caph (β-Cas) is located at RA 0h09m10.70s, and is designated HR 21, while Bellatrix (γ-Orionis), much farther east at RA 5h25m07.90s is designated HD 1790.

Contrary to common belief, the positions of the stars are not fixed in the firmament. The declination and right ascension of all stars changes gradually and in lockstep because of *precession*, the slow wobbling of the earth on its axis. Also, each star is actually moving relative to other stars, although the amount of actual motion is exceedingly small relative to their extreme distance. If a star is close enough to us, it is possible to detect this actual change in position against distant background stars, which is called the *proper motion* for that star.

To take into account precession and proper motion, when astronomers specify stellar coordinates they also include the *epoch* for those coordinates, which specifies the date for which the locations were accurate. Early catalogs, such as HD and HR, used epoch J1900.0, which specified the coordinates of stars as of 1 January 1900. Epoch is usually listed by Julian date, indicated by the J prefix; other date formats are used so seldom that Julian is assumed unless otherwise specified.

Later catalogs, and updated versions of older catalogs, use epoch 1950.0 or 2000.0, which is often indicated in the name of the catalog, as in the names of the printed catalogs *Sky Atlas 2000.0* and *Uranometria 2000.0*. Good planetarium programs and go-to scopes use the *date epoch*, which is the epoch as of the current date. The difference can be surprisingly large. For example, the J2000 coordinates of Vega are RA 18h36m56.19s and Dec +38°46'58.8". The date epoch coordinates on 8 July 2005 are RA 18h37m07.31s and Dec +38°47'16.6". In other words, in only 5.5 years, the position of Vega has changed by 11.12s in RA and 17.8" in declination.

Henry Draper Catalog (HD)

From 1911 until 1915, Annie Jump Cannon at the Harvard College Observatory compiled the *Henry Draper Catalog (HD)*, which was published in the years after WWI. HD was named in honor of the astronomer Henry Draper, whose widow donated the funds needed to compile and publish it.

HD started out as a specialty catalog, with the goal of cataloging the spectral classes of 225,300 stars down to magnitude 10, but it became commonly used as a general catalog. Like HR, HD assigns numeric designations to stars by RA, from west to east, epoch 1900.0, beginning at the vernal equinox. Of course, HD catalogs many more stars, so, for example, Caph (HR 21) is HD 432, and Bellatrix (HR 1790) is HD 35468. In 1949, the *Henry Draper Extension (HDE)* catalog supplemented the original HD catalog, bringing the total number of cataloged stars to 359,083. There is no overlap in the numbering of the HD and HDE catalogs, so stars in the HDE catalog are often listed with the HD designation. HD numbers are used widely today to identify stars that have no Bayer or Flamsteed designation.

Although amateur astronomers seldom use them nowadays, you may encounter *Smithsonian Astrophysical Observatory (SAO)* catalog designations in older publications. In 1966, Harvard University SAO combined and consolidated a dozen older positional catalogs, adding proper motion data for each. The SAO contains 258,997 stars down to magnitude 9. The original SAO used epoch 1950.0; the current version uses epoch 2000.0. SAO divides the sky into 10° declination strips and assigns numeric designations to each star within each declination strip. Astronomers quickly abandoned Benjamin Boss's 1937 *General Catalog of 33,342 Stars* in favor of SAO. In turn, SAO has now been largely abandoned in favor of modern catalogs like the Tycho2, Hipparchos, and HST catalogs described next.

Hubble Space Telescope (HST) Guide Star Catalog (GSC)

The most comprehensive deep-star catalog available is the *Hubble Space Telescope (HST) Guide Star Catalog (GSC)*, for which you may see stars listed under either abbreviation. GSC 1.1 lists positions with 1.7 arcsecond accuracy and magnitudes to within 0.5 accuracy for 18,819,291 objects from magnitude 9 to about magnitude 13.5, with many objects as faint as magnitude 15. Of the listed objects, about 15.2 million are stars and the remainder are mostly dim galaxies. GSC divides the sky into 9,537 regions and then assigns numbers to each object within a region.

The latest iteration of the HST GSC, called GSC 2.2, catalogs an incredible 435,457,355 objects, down to magnitude
19.5. The vast majority of these objects are so dim they are
invisible in all but the largest amateur scopes.

Hipparcos Catalog (HIP)

From 1989 until 1993, the Hipparcos satellite, launched by the Euro-
pean Space Agency, mapped precise locations and parallax data for
118,218 relatively nearby stars down to magnitude 8.5 (complete to
magnitude 7.3), with the goal of determining accurate proper motions
for each. Although it originated as a specialty catalog, the Hipparcos
Catalog (HIP) is now used as a general catalog. HIP assigns numeric
designations to stars by RA from west to east beginning at RA
0h00m00s. For example, Rigel is HIP 24436 and Bellatrix, about 11
minutes more easterly, is HIP 25336. HIP is valued by amateurs for the
accurate stellar distances it provides.

Tycho-2 Catalog (TYC)

The Tycho-2 catalog is also based on data gathered by the Hipparcos
satellite mission and compiled by the Copenhagen University Observatory and the U.S. Naval Observatory. Tycho-2 maps about 2.5 million
stars, with nearly complete coverage of 900,000 stars to magnitude 11
and 1.6 million additional stars to magnitude 12.

For more detailed information about stellar catalogs, visit the
U.S. Naval Observatory Catalog Information and Recommendations page: *http://ad.usno.navy.mil/star/star_cats_rec.shtml.*

Specialty Catalogs

In addition to general or all-sky catalogs, scores of specialty catalogs exist,
some of which are very specialized indeed. Double-star observers, for example, can choose among the *Aitken Double Star Catalog (ADS)*, the Lick
Observatory's *Index Catalog of Visual Double Stars (IDS)*, and the U.S.
Naval Observatory's *Washington Double Star Catalog (WDS)*, to name only
three. Similar catalogs exist for variable stars, nearby stars, carbon stars, and
so on. Those interested in Lunar occultations of stars use the *Zodiacal
Catalog (ZD)*, which includes only stars that Luna can occult.

We don't have room to list (let alone describe) even a small fraction of the
total number of specialty catalogs. If you develop an interest in a specialized
category of stellar observing, you'll soon enough become familiar with the
online and printed catalogs and other resources that target that specialty.

Know Your Constellations

Although stars are important, constellations place those stars in context: constellations are the things you can find in the sky most quickly. And once you've found the constellation that's your point of reference, you can go on to find what you're really looking for.

The first step in learning the night sky is to know the constellations. Before you attempt to identify constellations in the night sky, you should know the names of the constellations you are looking for and how to pronounce those names. There are 88 official constellations. Fortunately, you can learn them in groups because only some of them are visible, according to your latitude and the time of year.

> The best way to learn the constellations is to buy a plani-sphere [Hack #6], which allows you to dial in the date and see an accurate representation of the night sky for that date. Alternatively, the monthly sky charts in *Astronomy* and *Sky & Telescope* magazines provide similar whole-sky views for the current month.

Table 2-4 lists all 88 modern constellations, including the following data:

Name
> The official name of the constellation, as assigned by the International Astronomical Union (IAU), the only body authorized legally to name celestial objects.

Pronunciation
> The pronunciations shown are those commonly used by amateur astronomers. Although many are incorrect, these pronunciations are used so commonly that there is little point in attempting to correct them. For example, Orion is correctly pronounced oh-REE-un rather than oh-RYE-un, and Virgo is WIR-goh rather than VUR-goh, but saying oh-REE-un or WIR-goh will draw strange looks from your observing buddies.

Abbr
> The official International Astronomical Union (IAU) abbreviation for the constellation name.

Genitive
> The Latin genitive case is used to indicate possession of or membership in. In astronomy, the genitive form of the constellation name is used to indicate an object's membership in a constellation. For example, the name sigma Orionis (σ-Ori) indicates a star designated sigma that belongs to the constellation Orion [Hack #15].

Dec

The Dec (Declination) of a constellation determines whether you can see all or part of that constellation from your latitude. If you are north of the equator, subtract 90 from your latitude to determine the most southerly object you can observe. For example, our home is at about 36.2°N latitude. Subtracting 90° from that gives -53.8°, which is the declination of the most southerly object we can see from our home. The star Canopus, alpha Carinae (α-Car), for example, has declination of -52.7°, which means that it never rises more than about 1.1° above our southern horizon.

If you are south of the equator, add 90 to your latitude to determine the most northerly object you can observe. For example, if you are located at 36.2°S latitude (-36.2°), the most northerly object you can observe has a declination of (-36.2° + 90°), or +53.8°. If you are on the equator, at latitude 0°, you can observe any object from declination -90° to +90°, but very northerly and very southerly objects never rise far above your horizon. For example, Polaris (declination +89.25°) never rises higher than 0.75° above your northern horizon.

Season

Constellations are divided by season according to their Right Ascension (RA). A constellation that lies wholly or mostly within the RA range 03:00 to 09:00 is a Winter constellation; 09:00 to 15:00 is Spring; 15:00 to 21:00 is Summer; and 21:00 to 03:00 is Autumn. The season designation is rather arbitrary, but indicates when the constellation is best placed for viewing during evening hours. For example, Orion, a Winter constellation, can be seen as early as mid-Summer, when it rises before dawn, and as late as early Spring, when it sets soon after the Sun.

Table 2-4. The 88 modern constellations

Name	Pronunciation	Abbr	Genitive	Dec (°)	Season
Andromeda	an-DROM-eh-duh	And	Andromedae	+21 to +53	Autumn
Antlia	ANT-lee-uh	Ant	Antliae	-24 to -40	Spring
Apus	A-pus	Aps	Apodis	-67 to -83	Summer
Aquarius	a-KWAIR-ee-us	Aqr	Aquarii	+3 to -24	Autumn
Aquila	AK-will-uh	Aql	Aquilae	+10 to -10	Summer
Ara	AIR-uh	Ara	Arae	-55 to -68	Summer

Table 2-4. *The 88 modern constellations (continued)*

Name	Pronunciation	Abbr	Genitive	Dec (°)	Season
Aries	AIR-eez	Ari	Arietis	+10 to +30	Autumn
Auriga	uh-RYE-guh	Aur	Aurigae	+28 to +55	Winter
Bootes	bow-OAT-eez	Boo	Bootis	+8 to +55	Spring
Caelum	kye-ELL-um	Cae	Caeli	-27 to -49	Winter
Camelopardalis	CAM-eh-low-PAR-duh-lis	Cam	Camelopardalis	+52 to +87	Winter
Cancer	CAN-sur	Cnc	Cancri	+7 to +33	Winter
Canes Venatici	CAWN-es ven-AT-ih-see	CVn	Canum Venaticorum	+28 to +53	Spring
Canis Major	CAWN-is MAY-jur	CMa	Canis Majoris	-11 to -33	Winter
Canis Minor	CAWN-is MYE-nur	CMi	Canis Minoris	0 to +12	Winter
Capricornus	CAP-rih-CORN-us	Cap	Capricorni	-9 to 27	Autumn
Carina	cuh-REE-na	Car	Carinae	-51 to -75	Winter
Cassiopeia	cass-ee-oh-PEE-uh	Cas	Cassiopeiae	+50 to +60	Autumn
Centaurus	sen-TAWR-us	Cen	Centauri	-30 to -65	Spring
Cepheus	SEE-fee-us	Cep	Cephei	+53 to +87	Autumn
Cetus	SEE-tus	Cet	Ceti	+10 to -25	Autumn
Chamaeleon	kuh-MEEL-yun	Cha	Chamaeleontis	-74 to -83	Spring
Circinus	sur-KEE-nus	Cir	Circini	-54 to -70	Spring
Columba	kuh-LUM-ba	Col	Columbae	-27 to -43	Winter
Coma Berenices	KOE-muh BAIR-uh-NEES-us	Com	Comae Berenices	+14 to +34	Spring
Corona Australis	kuh-ROE-nuh aw-STRAWL-is	CrA	Coronae Austrinae	-37 to -45	Summer
Corona Borealis	kuh-ROE-nuh BOR-ee-AL-us	CrB	Coronae Borealis	+26 to +40	Summer
Corvus	KOR-vus	Crv	Corvi	-11 to -25	Spring

Table 2-4. *The 88 modern constellations (continued)*

Name	Pronunciation	Abbr	Genitive	Dec (°)	Season
Crater	KRATE-ur	Crt	Crateris	-6 to -25	Spring
Crux	KRUX	Cru	Crucis	-56 to -65	Spring
Cygnus	SIG-nus	Cyg	Cygnl	+28 to +60	Summer
Delphinus	del-FEE-nus	Del	Delphini	+2 to +21	Summer
Dorado	dor-AW-doe	Dor	Doradus	-49 to -85	Winter
Draco	DRAY-koe	Dra	Draconis	+50 to +80	Summer
Equuleus	eh-KWUH-lee-us	Equ	Equulei	+2 to +13	Autumn
Eridanus	air-uh-DAHN-us	Eri	Eridani	0 to -58	Winter
Fornax	FOR-nacks	For	Fornacis	-24 to -40	Autumn
Gemini	JEM-ih-nye	Gem	Geminorum	+10 to +35	Winter
Grus	GROOSE	Gru	Gruis	-37 to -57	Autumn
Hercules	HUR-cue-leez	Her	Herculis	+4 to +50	Summer
Horologium	hor-oh-LOGE-ee-um	Hor	Horologii	-40 to -67	Winter
Hydra	HYE-druh	Hya	Hydrae	-22 to -65	Spring
Hydrus	HYE-drus	Hyi	Hydri	-58 to -90	Autumn
Indus	IN-dus	Ind	Indi	-45 to -75	Autumn
Lacerta	lay-CERT-uh	Lac	Lacertae	+33 to +57	Autumn
Leo	LEE-oh	Leo	Leonis	-6 to +33	Spring
Leo Minor	LEE-oh MYE-nur	LMi	Leonis Minoris	+23 to +42	Spring
Lepus	LEE-pus	Lep	Leporis	-11 to -27	Winter
Libra	LEE-bruh	Lib	Librae	0 to -30	Summer
Lupus	LOO-pus	Lup	Lupi	-30 to -55	Summer

Table 2-4. The 88 modern constellations (continued)

Name	Pronunciation	Abbr	Genitive	Dec (°)	Season
Lynx	LINKS	Lyn	Lyncis	+34 to +62	Winter
Lyra	LEER-uh	Lyr	Lyrae	+26 to +48	Summer
Mensa	MENS-uh	Men	Mensae	-70 to -85	Winter
Microscopium	mike-roh-SCOPE-ee-um	Mic	Microscopii	-28 to -45	Autumn
Monoceros	MON-oh-SAIR-ose	Mon	Monocerotis	-11 to +12	Winter
Musca	MUSS-kuh	Mus	Muscae	-64 to -74	Spring
Norma	NOR-muh	Nor	Normae	-42 to -60	Summer
Octans	OCT-anz	Oct	Octantis	-75 to -90	Summer
Ophiuchus	oh-FEE-uh-kuss	Oph	Ophiuchi	+14 to -30	Summer
Orion	oh-RYE-un	Ori	Orionis	+8 to +23	Winter
Pavo	PAW-voh	Pav	Pavonis	-57 to -75	Summer
Pegasus	PEG-uh-sus	Peg	Pegasi	+2 to +37	Autumn
Perseus	PERS-ee-us	Per	Persei	+31 to +59	Winter
Phoenix	FEE-nicks	Phe	Phoenicis	-40 to -59	Autumn
Pictor	PIK-tor	Pic	Pictoris	-43 to -64	Winter
Pisces	PYE-seez	Psc	Piscium	-5 to +34	Autumn
Piscis Austrinus	PYE-seez aw-STRINE-us	PsA	Piscis Austrini	-25 to -36	Autumn
Puppis	PUPP-is	Pup	Puppis	-12 to -51	Winter
Pyxis	PICKS-is	Pyx	Pyxidis	-17 to -38	Spring
Reticulum	reh-TICK-you-lum	Ret	Reticuli	-53 to -67	Winter
Sagitta	SADJ-ih-taw	Sge	Sagittae	+17 to +22	Summer

Table 2-4. The 88 modern constellations (continued)

Name	Pronunciation	Abbr	Genitive	Dec (°)	Season
Sagittarius	SADJ-ih-TAIR-ee-us	Sgr	Sagittarii	-12 to -46	Summer
Scorpius	SKOR-pee-us	Sco	Scorpii	-8 to -45	Summer
Sculptor	SKULP-tor	Scl	Sculptoris	-25 to -59	Autumn
Scutum	SKEW-tum	Sct	Scuti	-4 to -16	Summer
Serpens Caput	SUR-penz KAP-put	Ser	Serpentis	-4 to +20	Summer
Serpens Cauda	SUR-penz CAW-duh	Ser	Serpentis	-15 to +6	Summer
Sextans	SEX-tanz	Sex	Sextantis	-11 to +7	Spring
Taurus	TAWR-us	Tau	Tauri	+10 to +30	Winter
Telescopium	tell-uh-SCOPE-ee-um	Tel	Telescopii	-46 to -57	Summer
Triangulum	try-ANG-you-lum	Tri	Trianguli	+26 to +37	Autumn
Triangulum Australe	try-ANG-you-lum aw-STRAWL-eh	TrA	Trianguli Australis	-60 to -70	Summer
Tucana	too-CANN-uh	Tuc	Tucana	-56 to -75	Autumn
Ursa Major	ERS-uh MAY-jur	UMa	Ursae Majoris	+29 to +73	Spring
Ursa Minor	ERS-uh MYE-nur	UMi	Ursae Minoris	+66 to +90	Summer
Vela	VAY-luh	Vel	Velorum	-40 to -57	Spring
Virgo	VUR-goh	Vir	Virginis	-22 to +15	Spring
Volans	vohl-LANZ	Vol	Volantis	-64 to -75	Winter
Vulpecula	vul-PECK-you-luh	Vul	Vulpeculae	+20 to +30	Summer

If you counted, you may have noticed that there are 89 constellations listed. That's because Serpens, the Snake, is the only non-contiguous constellation, split in two by Ophiuchus, the Snake Bearer. The more northerly portion is called Serpens Caput, or Head of the Snake. The more southerly is Serpens Cauda, or Tail of the Snake. Officially, these two chunks comprise a single constellation. Both use the same abbreviation and genitive, but most astronomers treat them as separate constellations.

HACK #17 Understand Celestial Coordinate Systems

Orient yourself to the night sky.

Celestial coordinate systems are used to specify the locations of stars and other astronomical objects. Four celestial coordinate systems exist, but only two of them are used commonly by amateur astronomers. Celestial coordinate systems are analogous to the *geographic coordinate system* of latitude and longitude used to specify locations on the earth's surface, but they are used instead to specify locations on the celestial sphere. The four coordinate systems differ only in what they designate the fundamental plane, called the equator, which bisects the sky into two hemispheres along a great circle.

Horizontal Coordinates

The *horizontal coordinate system*, also called the *altitude-azimuth* or *alt-az* coordinate system, uses the local horizon as the equator, with the two poles straight up (the *zenith*) and straight down (the *nadir*). The location of an object is specified by two values called *altitude* and *azimuth*. Each of these is denominated in degrees (°), minutes ('), and seconds ("). One degree is divided into 60 minutes, and one minute is divided into 60 seconds. Accordingly, one degree contains 3,600 seconds.

> Altitudes and azimuths are sometimes specified in decimal degrees. For example, 63°30' may also be written as 63.5°, because 30' is equal to 0.5°. Using decimal degrees allows you to specify the location of an object with very high precision, for example 63.5342°. Locations specified in degrees/minutes/seconds use decimal fractions of a second to achieve the same fractional precision. For example, a decimal value of 63.5342° can be converted to 63°32'3.172" by multiplying the fractional degree 0.5342 by 60 to yield 32.052', and the fractional minute 0.052 by 60 to yield the fractional second 3.172".

Objects above the horizon have positive altitudes, and those below the horizon have negative altitudes. An object exactly on the horizon has an altitude of 0°. An object directly overhead (at zenith) has an altitude of +90°. An object that is straight down (through the earth, at nadir) has an altitude of -90°.

The second component of an alt-az coordinate location is azimuth, which specifies the angular location of the object relative to north, which is designated 0°. If you face straight north and make a quarter turn (90°) to your right, you are then facing east, or 90°. Another quarter turn to the right takes you to facing south, or 180°. If you make yet another quarter turn to the right, you are facing west, or 270°. A final quarter turn to the right takes you back to north, or 0°. (Just as a 24-hour clock turns from 23:59:59 to 0:00:00, never reaching 24:00:00, azimuth turns from 359°59'59" to 0°0'0", never reaching 360°.)

Because they describe the location of an object relative to the observer's position, alt-az coordinates have the advantage of being intuitive. Everyone understands the concepts of up, down, right, and left. But the inevitable consequence of using the observer's position as the fixed point of reference is that celestial objects move relative to that point of reference as Earth rotates on its axis.

Accordingly, alt-az coordinates are insufficient by themselves to specify the location of a celestial object. You must also specify the location of the observer and the time of the observation. For example, to say that Sirius is located at altitude 32°41' and azimuth 206°59' is as meaningless as saying that a stopped clock reads correctly twice a day. The question is when and, in the case of a celestial object, from what location? But it is correct to say, for example, that for an observer located at latitude 36°09'N and longitude 80°16'W (that's us, folks) at 2130 EST on 9 March 2006 Sirius is located at altitude 32°41' and azimuth 206°59'. Figure 2-7 shows the location of Sirius from that site at that time, with an alt-az coordinate grid superimposed.

But from the same observing site five minutes later, Sirius has moved relative to the observer, as shown in Figure 2-8. It is now located at altitude 32°13' and azimuth 208°17', which is to say 28 arcminutes (about half a degree) lower and 78 arcminutes (just over a degree) farther to the right.

In the past, alt-az coordinates were seldom used because observers had only printed charts, which inherently lack observing location and time references. Nowadays, with go-to scopes, digital setting circles, and planetarium programs that chart object positions in real time, alt-az coordinates are used very commonly. In particular, the overwhelming popularity of Dobsonian telescopes, which use alt-az mounts, has led to the odd situation that many experienced amateur astronomers now understand *only* the alt-az coordinate system.

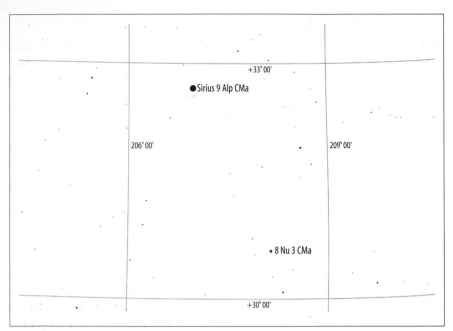

Figure 2-7. The alt-az location of Sirius from 36°09'N and 80°16'W at 2130 EST on 9 March 2006

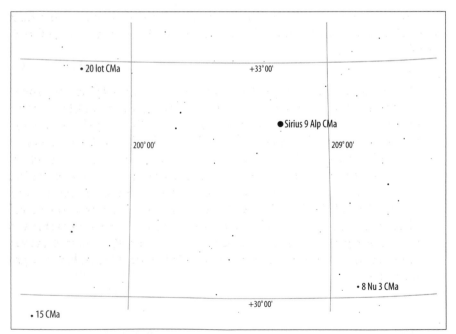

Figure 2-8. The alt-az location of Sirius from 36°09'N and 80°16'W at 2135 EST on 9 March 2006

Equatorial Coordinates

The *equatorial coordinate system*, also called the *EQ coordinate system*, uses the same fundamental plane and the same poles as the geographic coordinate system. The projection of Earth's equator onto the celestial sphere is called the *celestial equator*. The projections of Earth's north and south poles onto the celestial sphere mark the *north celestial pole* and the *south celestial pole*, respectively.

> The bright star Polaris is located less than 0.75° from the north celestial pole, around which the night sky appears to revolve for Northern Hemisphere observers. Southern observers have no such bright star to mark their celestial pole, although the dim star σ-Octantis is reasonably close to the south celestial pole.

The difference between the equatorial and geographic coordinate systems is the point of reference used. The geographic coordinate system uses Earth as a fixed point of reference. As Earth rotates, the coordinate grid rotates with it, fixed to the planet's surface. Equatorial coordinates are fixed instead to the stars, so as Earth rotates the coordinate grid appears to move relative to earth's constantly changing position.

Equatorial coordinates specify the location of an object with two values called *Declination* (abbreviated *Dec*, *DE*, or δ) and *Right Ascension* (abbreviated *R.A.*, *RA*, or α).

Declination

> *Declination* is the equivalent of latitude in the geographic coordinate system. Declination specifies the angular separation of an object north or south from the celestial equator. Declination, like latitude, is denominated in degrees (°), minutes ('), and seconds ("). Declination ranges from +90° for an object located at the north celestial pole, through 0° for an object located on the celestial equator, to -90° for an object located at the south celestial pole. For example, the declination of Sirius is -16°42'58.0" (J2000.0, but we'll get to that later...). Just as the actual distance on the earth's surface represented by one degree of latitude is the same regardless of longitude, the distance on the celestial sphere represented by one degree of declination is the same regardless of right ascension.

The declination of an object determines whether it is ever visible from a specific latitude on Earth. An object at +90° declination, for example, is visible no farther south than Earth's equator, where it will be exactly on the northern horizon. An object at 0° declination is visible from anywhere on Earth, although at either of Earth's poles it will be exactly on the horizon.

To determine the visibility of an object from a northern latitude, subtract 90° from your latitude to determine the most southerly object visible. We are at 36°N latitude, so the declination of the most southerly objects visible to us is 36–90= -54°. An object at -54° declination never rises above the southern horizon at our latitude. An observer at a southern (negative) latitude adds 90° to his latitude to determine the declination of the most northerly objects visible to him. For example, for an observer at 36°S latitude (-36°), the most northerly objects visible are located at declination -36+90= +54°, which never rise above that observer's northern horizon.

Right Ascension

Right ascension is the equivalent of longitude in the geographic coordinate system. Right ascension specifies the angular position of an object proceeding east from the Vernal Equinox. Unlike longitude, right ascension is usually specified by time in hours (h), minutes (m), and seconds (s) rather than by angle in degrees (°), minutes ('), and seconds ("). For example, the J2000.0 right ascension of Sirius is 6h45m08.9s.

Because time and angle are correlated, it is possible to express RA in degrees rather than hours. Earth makes one full rotation (360°) every 24 hours, so one hour (h) of time (or RA) corresponds to 360°/24 or 15° of angle. Similarly, one minute (m) of time corresponds to 15' (15 arcminutes) of angle, and one second (s) of time corresponds to 15" (15 arcseconds) of angle.

Navigators who use the stars to plot their course and position use angular right ascension specified in decimal degrees—called the Sidereal Hour Angle (SHA)—because it makes calculations easier. For example, a navigator would use 101.29° (the approximate decimal equivalent of 6h45m08.9s) for the RA of Sirius.

Right ascension is usually specified in clock time rather than by angle because the RA of an object determines what time that object rises on any specified date. Two objects at the same declination that are separated by one hour of RA rise one hour apart, regardless of what their declination is. For example, an object with RA 14h0m0s rises one hour before an object with RA 15h0m0s, if those objects are at the same declination. (But two objects with the same RA but different declinations do not rise at the same time.) Just as the actual distance on the earth's surface represented by one degree of longitude varies with the latitude, the distance represented by one hour of right ascension on the celestial sphere varies with the declination.

> Right ascension takes its name from the behavior of stars as seen by an observer standing on Earth's equator facing north. From that viewpoint, a star located at declination 0° appears to rise straight up directly on the observer's right, pass directly overhead, and then set directly on the observer's left. Right ascension could therefore equally well be called left descension. For observers located other than at Earth's equator, stars appear to follow an arc as they rise, culminate, and set.

A given point on the celestial sphere appears to move as Earth rotates on its axis. Figure 2-9 shows the same area of the sky in alt-az terms that is shown in Figure 2-7, but this time with equatorial coordinates superimposed rather than alt-az coordinates.

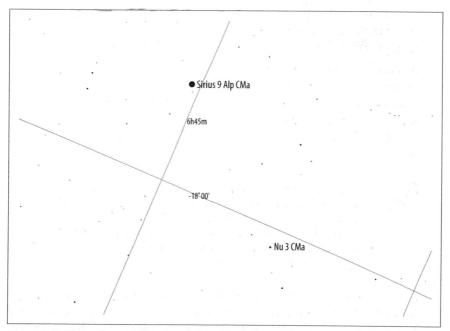

Figure 2-9. The equatorial location of Sirius from 36°09′N and 80°16′W at 2130 EST on 9 March 2006

Figure 2-10 shows the same area of the sky in alt-az terms five minutes later. Again, Sirius has "moved," but this time the coordinate grid has moved along with it. So, although Sirius is now in a different place in the sky, it is in exactly the same place relative to the equatorial coordinate grid.

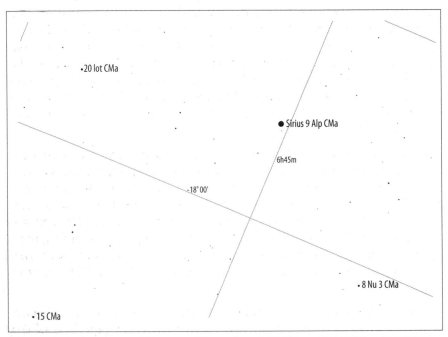

Figure 2-10. The equatorial location of Sirius from 36°09'N and 80°16'W at 2135 EST on 9 March 2006

> Earth rotates west to east, which causes the celestial sphere to appear to make one full rotation east to west each day. Or, more precisely, once every sidereal day, which is about 23 hours and 56 minutes long. (The difference between the 24-hour solar day and the sidereal day exists because as Earth makes one full orbit of the Sun about every 365 days, it also makes one more full "hidden" turn on its axis in the process of completing its 360° orbit.) This four minute difference means that a star rises about four minutes earlier each day than it did the preceding day.

Back to the J2000.0 issue we mentioned earlier. The right ascension and declination of a star change gradually over time because Earth's axis of rotation wobbles rather than pointing directly at the celestial poles. Think of a child's top. When it first begins to spin, it does so smoothly about its axis. But as the top spins down, it begins to wobble, so that the axis begins to describe a circle rather than a point. Earth wobbles in its rotation in identical fashion, which is called *precession*.

As a result of precession, Earth's axis draws a circle in the sky. The precession rate is about 1° every 72 years, which means it takes about 26,000 years for Earth's axis to trace the complete circle. Right now, Polaris is very close to the north celestial pole, but that wasn't the case until recently, and won't be the case in the future. Five thousand years ago, for example, when the Egyptians were building the first pyramids, the pole star was Thuban in the constellation Draco. One thousand years ago, when Robert's Viking ancestors were pillaging Europe—what is "pillaging," anyway?—the pole star was Kochab in Ursa Minor. Twelve thousand years from now, the pole star will be Vega in Lyra.

Precession means that the right ascension and declination of stars changes according to the current position of Earth's axis, so the RA and DE of a star must be specified as of a particular date, which is called the *epoch*. For printed charts, that date has conventionally been set at 50 year milestones, using Julian (J) dates. For example, charts printed early in the 20th century used epoch J1900.0, which is to say the location of objects as of midnight on 1 January 1900. Charts printed during the middle part of that century used J1950.0 positions. Charts printed in the late 20th and early 21st centuries use J2000.0.

Computerized scopes and planetarium software take the concept of epoch a step farther. They use the *date epoch*, which is to say the epoch as of the moment you are using them. For example, Table 2-5 shows the epoch J2000.0 and date epoch as of 9 March 2006 coordinates for Polaris, which is located close to the north celestial pole, and for Mintaka, which is located in Orion's belt, near the celestial equator.

Table 2-5. Epoch J2000.0 positions of Polaris and Mintaka versus date epoch positions

	Polaris		Mintaka	
	RA	Dec	RA	Dec
J2000.0	2h31m48.70s	+89°15'51.0"	5h32m00.40s	-00°17'57.0"
9 March 2006	2h38m58.96s	+89°17'27.6"	5h32m19.38s	-00°17'42.0"

In a little more than six years, the RA of Polaris changes by 7m10.26s, or, in angular terms, nearly 1.8°, while the RA of Mintaka changes by only 18.98s (about 0.079°). But Polaris is located near the north celestial pole, so a 1.8° angular change in RA represents a very small change in terms of distance in the sky, while Mintaka is near the celestial equator, where a small angular change represents a much larger actual distance. (Just as a 0.079° change in longitude at Earth's equator represents about 5.5 miles on the ground, while a 1.8° change in longitude very near Earth's north pole may represent only a few feet on the ground.)

> There is another factor in changing RA and DE, called *proper motion*. In general, stars are so far away from us that their actual motions relative to each other can be ignored safely. But the motion of nearby stars is large enough to be visible in angular terms if measured precisely over a long period using a long baseline. Even so, these proper motions are small enough that amateur astronomers can ignore them for all practical purposes.

Solar system objects—the Sun, Moon, planets, comets, asteroids, and so on—are next-door neighbors in celestial terms, and so they show very large proper motions in both right ascension and declination. That means their coordinates vary noticeably from day to day and, for objects such as the Sun and the Moon that have very large proper motions, even from minute to minute.

Every amateur astronomer needs to be familiar with equatorial coordinates. If you use an equatorial mount, you can locate dim objects by using your setting circles to "dial in" the proper coordinates. Conversely, if you happen to find an object while panning around the sky, you can read the coordinates of that object from your setting circles and use those coordinates to determine which object you were looking at.

> Decades ago, before modern electronics, high-quality equatorial mounts had huge, very precise setting circles equipped with verniers that allowed an observer to locate objects, often with arcminute accuracy. High-quality modern mounts don't have setting circles large enough or precise enough to center an object in the eyepiece using only the object's coordinates, but the setting circles will at least get you close. The setting circles on cheap mounts are so small and so crude that they are nearly useless. They're there to make the cheap mount "look scientific" rather than to be actually used.

But even astronomers who use alt-az mounts, such as Dobsonian scopes, need to understand equatorial coordinates, for a couple of reasons:

- Equatorial coordinates are used universally to specify the location of objects, so it's necessary you understand equatorial coordinates at least well enough to locate objects on your charts using them. For example, you might want to view a comet whose current coordinates are given as

RA 4h37m and DE +19°16'. If you don't understand equatorial coordinates, you will be clueless as to the location of the comet. If you do understand equatorial coordinates, a quick perusal of your charts tells you that the comet is currently located near the bright star Aldebaran in Taurus, between it and the 5th magnitude star 92 Tau. Getting that comet in your eyepiece will take only seconds.

- Astronomers use equatorial directions in casual conversation. For example, if you are looking for a faint fuzzy, an observing buddy might tell you it's located "two degrees northwest" of another object. If you have no idea which way equatorial northwest is, you're no better off for his advice. If you do understand celestial coordinates, you can do an "eyepiece star hop" [Hack #21] to put the object in your eyepiece.

Other Coordinate Systems

In addition to the horizontal coordinate system and the equatorial coordinate system, two other coordinate systems are used in astronomy:

Ecliptic coordinates

The *ecliptic* is the plane in which Earth orbits the Sun. As it happens, the orbital planes of all of the planets except Pluto are very close to the ecliptic, which means they never stray far from the ecliptic. The narrow band on either side of the ecliptic in which the planets are always found is called the zodiac.

Because planets are always near the ecliptic, *ecliptic coordinates* are sometimes used to specify the current location of a planet. The equator for ecliptic coordinates is defined as the circle traced in the sky by the ecliptic plane, whence the name of the coordinate system. *Ecliptic latitude* is the angular separation between a planet and the ecliptic plane. *Ecliptic longitude* is the position of the planet in its orbit, measured along the ecliptic equator from the point where the equator intersects declination 0°.

In practice, ecliptic coordinates are used only by professional astronomers. Amateur astronomers use either equatorial coordinates for the date in question or horizontal coordinates for the date and location.

Galactic coordinates

Galactic coordinates use the (somewhat arbitrarily defined) center plane of our galaxy as the base plane, or equator. Galactic longitude is specified relative to the galactic center in the constellation Sagittarius. Galactic latitude is specified as the angle north or south of this center line. Galactic coordinates are used by cosmologists but not by amateur astronomers.

Although neither of these coordinate systems is commonly used by amateur astronomers, there are times when the location of an object you may wish to observe is specified using one of these systems, so it's as well to understand them.

HACK #18 Print Custom Charts

Make the most of your observing time by taking along the charts you need.

Observing time is precious, particularly for DSO observers [Hack #22], many of whom must drive several hours to a dark-sky site on a new moon weekend and hope for clear skies when they arrive. Many DSO observers have half a dozen or fewer opportunities per year to observe under optimum conditions, so it's critical to make the most of that limited observing time.

The best way to do that is to be prepared before you set up your scope. Know which objects you plan to observe, and know exactly how you're going to find them. By "exactly" we mean not just knowing the general location of the object, but having a detailed plan, including Telrad [Hack #53] and finder circles plotted and, if necessary, a detailed star hop [Hack #21] worked out.

Many observers, including experienced ones, make the mistake of hauling their star atlases and a notebook computer to the observing site, intending to use those resources to locate objects on-the-fly. While you *can* do it that way—and we would certainly never be without star atlases and our notebook computer—it's not the best use of your observing time. With everything pre-planned, we can usually locate one of our target objects in a minute or two at most. If we have to figure out how to locate the object as we go along, it may take as much as half an hour to locate the object through trial and error and then to verify that we've in fact located the object we were looking for rather than some other nearby object. Instead of using your formal charts and planetarium software routinely, reserve them for special needs that arise while you're observing.

In general, for formal observing trips, we try to devote at least two hours of planning time for each hour of expected observing time, which is to say 12 hours of planning time for a typical six-hour observing session. (It gives us something to do on those cloudy nights, and planning an observing session is the next best thing to actually observing....) For *ad hoc* DSO observing sessions, we often fall back on plans from previous sessions that were clouded out.

Although it's possible to develop detailed observing plans on paper, using photocopies of printed charts, it's generally easier and much faster to use your computer to produce custom charts with exactly the amount of detail you need. For example, you can set the limiting stellar magnitude in your software to match the limiting magnitude in your finder, so only stars that are actually visible in your finder are plotted on the custom chart, and you can filter the printed output to include only classes of objects you want to observe, such as Herschel 400 objects only or planetary nebulae only.

> We use the free planetarium program Cartes du Ciel (*http://www.stargazing.net/astropc/*) to generate our custom charts, but any high-end planetarium program will serve as well **[Hack #64]**.

Sometimes, you'll need only an overview chart and one detailed chart to locate a particular object. For example, Figure 2-11 shows the overview chart for locating the Bubble Nebula (NGC 7635) in Cassiopeia. (Actually, we could have gotten by with just a detailed chart, because, like most experienced observers, we can locate M52 from memory.) The triple ring of the Telrad circle is centered on the Bubble Nebula.

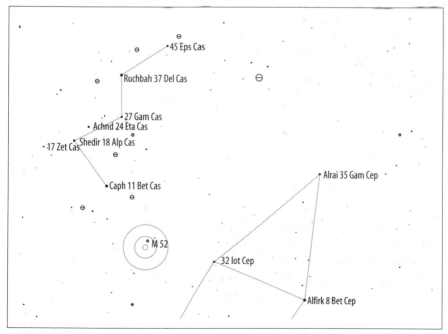

Figure 2-11. An overview chart for locating the Bubble Nebula (NGC 7635) in Cassiopeia

Figure 2-12 shows the detailed chart for locating the Bubble Nebula. The circle is a 5° finder circle, which we would jot down on the printed chart. Otherwise, it's easy to forget whether a circle on a chart is for the finder or for one eyepiece or another. This detailed chart actually serves for finding more than just the Bubble Nebula. While we were in the vicinity, we'd also locate and log the open clusters NGC 7510, King 19, and Markarian 50 **[Hack #26]**. We wouldn't need another detailed chart for these objects because they happen to be in close proximity to each other and near the bright stars 1 Cas and 2 Cas (at the lower left of the finder circle). Both of those stars and all three of the open clusters would be within the field of our wide-field eyepiece.

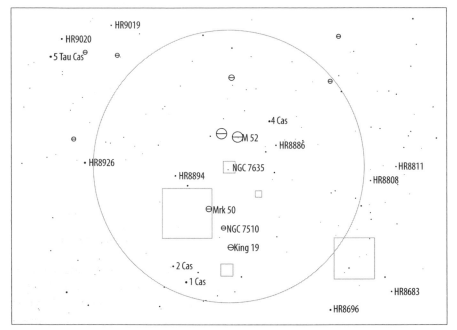

Figure 2-12. A detailed chart for locating the Bubble Nebula (NGC 7635) in Cassiopeia

One detailed chart usually suffices for finding an object. But there are times—for example, when you need to do a long star hop to locate a dim object far from any bright star—when you'll need more than one detailed chart to map the trail from your starting point to the object. Don't hesitate to print as many charts as necessary to get you to your goal, but do remember to number them in sequence so you'll know what order to follow them in.

> Use a laser printer, if you have one. Ink jet printouts often smear and run as the pages become soaked with dew. Alternatively, use transparent acetate sheet protectors to protect the pages from dew.

Custom charts are particularly important when you are observing transient events such as comets, transits and shadow transits of Jupiter's Gallilean moons, and the dance of Saturn's satellites. For example, Figure 2-13 shows the position of Saturn's major satellites as of 7:30 p.m. on 7 February 2005, with a one arcminute circle shown for scale.

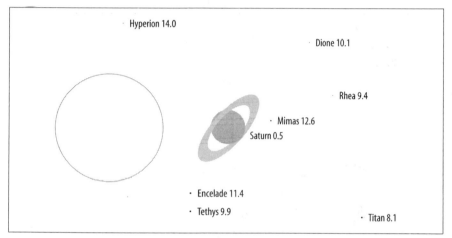

Figure 2-13. The locations of Saturn's major satellites at a particular date and time

Some of Saturn's moons—Titan, Rhea, Dione, and Tethys—are easy to observe even in a small scope because they are relatively bright, well separated from the glare of Saturn, or both. Although we have observed both in our 10" Dob, Encelade and, in particular, Mimas, are much more difficult because they are dimmer and because their orbits keep them so close to Saturn. We have never managed to observe Hyperion, which at magnitude 14.0 is right at the limit of our scope.

All of these moons appear stellar regardless of magnification, so the only way you can verify that you are looking at a moon rather than a dim field star is to have such charts for the particular evening or even the particular hour of the evening when you will observe the object.

In reality, we seldom print ephemeral charts. Instead, we use our notebook computer to provide a real-time chart of the moons as we observe them. But if you don't have a notebook computer (or don't want to risk it outdoors in the dark), detailed ephemeral charts are indispensable.

 Keep Your Charts at the Eyepiece

Refrigerator magnets aren't just for refrigerators any more.

Think back to the last time you tried to find a difficult object. You probably sat at your chart table, trying to memorize the relative positions and magnitudes of several field stars near the object. You then returned to your scope and scanned around a bit, trying to locate that pattern of stars in the eyepiece. Before long, you weren't entirely sure what you were looking for. Back to the chart table. Back to the eyepiece. Back to chart table. Rinse and repeat.

If you have one of the incredibly popular steel-tube Chinese or Taiwanese Dobs, there's a better way. Print out (or copy) the chart, take it with you to the eyepiece, and stick it right on the tube using refrigerator magnets, as shown in Figure 2-14.

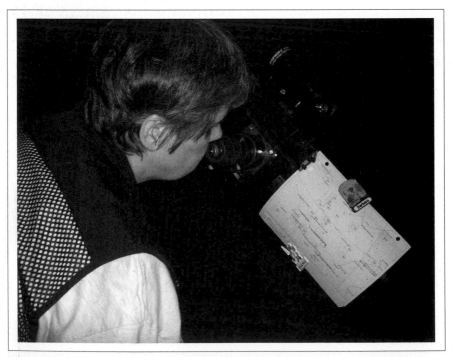

Figure 2-14. Using a chart right at the eyepiece

In fact, nothing limits you to having just one chart. A strong magnet will hold half a dozen or more sheets of paper, ready for immediate use.

HACK #20 Locate Objects Geometrically

You can find things quickly by reckoning from the easy-to-find objects, imagining some lines between them, and going in for the kill with a bulls-eye finder.

The human eye and brain are superbly evolved for detecting patterns. We see lines, angles, and patterns even where none exist, particularly when we look at bright stars against the velvety black background of the night sky. You can take advantage of this by creating imaginary lines and patterns on the celestial sphere, and using those lines and patterns to locate objects geometrically. Geometric navigation allows you to jump directly to objects in seconds rather than spend minutes tracking them down by following a path of dim stars.

A unit-power finder [Hack #53] like the Telrad (shown in Figure 2-15) is an essential aid to geometric navigation. The Telrad works like the heads-up gunsights in WWII fighter planes. When you look through the Telrad, you see a dim, red bulls-eye target against the background sky. The Telrad circles are 0.5°, 2°, and 4°, which means you can use a Telrad to locate almost instantly any object within 4° of a bright star or other easily identifiable object.

Figure 2-15. The Telrad unit-power finder

For example, let's say you want to locate Messier Object 79 (M79), a globular cluster in the constellation Lepus, shown in Figure 2-16. You could track it down the hard way, by star hopping [Hack #21] from the star 9 β-Leporis (Nihal), following a trail of dim stars (not shown in the graphic) in your optical finder until you eventually end up with M79 in your eyepiece. Although doing it that way might give you a sense of accomplishment, it might also take 5 or 10 minutes that could be better spent looking at the object.

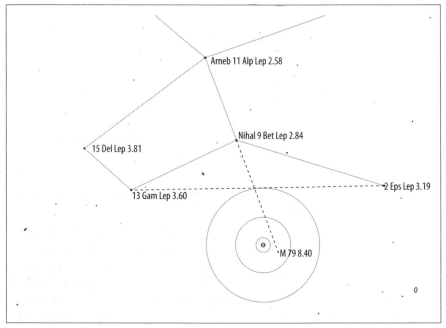

Figure 2-16. Locating Messier 79 with a Telrad

To locate M79 with the Telrad, start with the triangle formed by the three naked-eye stars 9 Bet Lep (Nihal), 13 Gam Lep, and 2 Eps Lep, shown in Figure 2-16. Draw an imaginary baseline from 13 Gam Lep to 2 Eps Lep. Place the outer ring of the Telrad so that it just touches your imaginary baseline about halfway between the two guidepost stars. To confirm the position, use another geometric relationship. Note that the center of the Telrad pattern should be just slightly to the 13 Eps Lep side of the line from 11 Alp Lep (Arneb) to 9 Bet Lep (Nihal). M79 should be visible in your low-power eyepiece. Total time to locate the object? Five seconds or so.

The Telrad is useful for locating any deep-sky object other than those located far from any naked-eye stars, but some objects are easier than others. For example, Figure 2-17 shows how we use the Telrad to locate four galaxies in Ursa Major without using star hops.

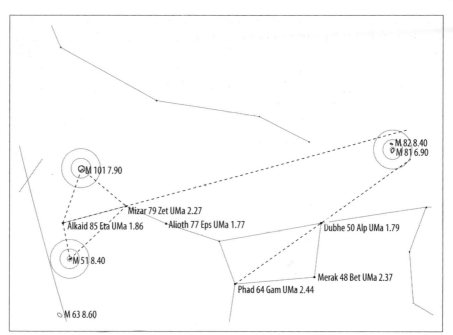

Figure 2-17. Locating Ursa Major galaxies geometrically

Although it is located more than 5° from each of the two nearest bright stars, Alkaid and Mizar in the Big Dipper's handle, M101 is an easy find with the Telrad if you use geometric navigation. Simply create an imaginary equilateral triangle with Alkaid and Mizar as two of the apices. M101 is located at the third apex of the triangle. You can have M101 in your eyepiece in literally five seconds, while the folks with go-to scopes are still waiting for them to grind their way to the target.

M51, the famous Whirlpool Galaxy, is just as easy with the Telrad. Imagine a line from Alkaid to Mizar as the base of a right triangle. From that base, drop an imaginary perpendicular line. Put the Telrad on that line, with the outer circle about 1.6° from Alkaid. Here's where the known sizes of the Telrad circles come in handy. You need 1.6°, and you know that the distance from the outer 4° Telrad circle to the inner 0.5° circle is 1.75°. You can simply eyeball the approximate offset and put the Telrad about where you think it needs to be. Bump the scope just a tiny bit off the perpendicular, away from the Big Dipper's bowl, and M51 is in your eyepiece. Total time required? Five seconds.

The Telrad can be useful even if the object you are trying to locate is more distant from bright stars. For example, the bright galaxy pair M81/82 is located more than 10° from Dubhe. But note the line from Phad to Dubhe.

As it happens, the distance from Phad to Dubhe (about 10.5°) is almost identical to the distance from Dubhe to M81/M82 (about 10.1°), and M81/M82 is very close to that line. It's easy to place the Telrad geometrically to put M81/M82 into the field of view of a 50mm finder, in which they are weakly visible. Once again, it takes a Telrad user five seconds to locate an object that takes much longer to find by star hopping or using a go-to scope.

Even when you can't locate an object directly with a Telrad, it remains useful. For example, if you are trying to locate an object that is far from any bright naked-eye stars and for which no convenient geometric relationship exists, you can still use the Telrad to shorten the required star hop, sometimes dramatically. Rather than do a full star hop from the nearest bright star, which may be quite far away, you can use the Telrad to put your optical finder at a closer starting point for which a geometric relationship does exist. Examine your charts to locate a distinctive pattern of stars that will be bright in your optical finder—5th and 6th magnitude—and then put the Telrad where that pattern should be. Once you are sure you have that pattern in your optical finder, you can use it as a departure point for your star hop to the object.

Although we much prefer the Telrad unit-power finder, many astronomers use the similar Rigel QuikFinder instead **[Hack #53]**.

HACK #21   Learn to Star Hop

When a Telrad is not enough, find and focus on patterns, and then hop pattern by pattern to the target.

Although the Telrad allows you to locate many objects with amazing speed **[Hack #20]**, it's not a panacea. Some parts of the sky are simply devoid of the bright stars that you need to orient your scope with a Telrad. To find an object in those barren parts of the sky, you need to star hop.

Star hopping is the process of locating an object by beginning at a bright "guidepost" star and then using your optical finder to follow a trail of dimmer stars until you arrive at the object. The secret to star hopping is to plan the hops so that each hop provides a distinct pattern of reasonably bright stars in the finder. When the pattern is right, you know that the finder is pointed exactly where you think it is, and you can then move the finder to locate the next pattern of stars.

Before you attempt to star hop, you need to know the field of view of your optical finder and how objects are oriented in it. You can calculate the field of view precisely by drift testing **[Hack #57]** or by using the finder to look at star pairs with known separations **[Hack #56]**. A correct-image finder (right-angle or straight-through) provides an image that is correct top-to-bottom and left-to-right. A traditional straight through finder provides an image that is correct left-to-right, but inverted top-to-bottom. If you are using the latter type of finder, simply invert your star charts to make them correspond to the view in the finder.

You can plan a star hop using either printed charts or planetarium software on your computer. We much prefer using planetarium software because it lets us print out custom charts set to whatever limiting stellar magnitude **[Hack #13]** we want, and it prints finder circles directly on the charts.

There are two ways to plan a star hop using printed charts. The first method, although it is more commonly used, is less desirable:

Use the original chart

> With this method, you use a transparent overlay on the original chart to plan your star hop as you actually do it. There are some disadvantages to using this method, not least that it requires more of your precious observing time. Also, if you use an undriven scope, your last hop is drifting out of view as you plan the next hop.

Some charts come with a transparent overlay with various size circles on it, one of which may correspond closely to your finder's field of view. If your chart does not provide an overlay with a circle of appropriate size, use a laser printer with overhead transparency film to print your own overlay with whatever size circles you need. Ink jet printers are generally unsuitable for this task, as their ink is often water soluble and runs at the first hint of dew.

Use a photocopy of the chart

> Using a photocopied chart has several advantages. First, you can plan your star hops beforehand instead of wasting observing time by doing the star hops while you're observing. Second, you can put the finder circles right on the chart for easy reference. Third, because you're preparing the chart ahead of time, you can take the time to write notes on the chart—the magnitudes of the various stars, characteristics of the object you're looking for, and so on.

The best way to use the copied chart is to prepare a transparent overlay with a finder circle that corresponds to the field of view of your finder at the scale of the chart you are using. (Make sure the photocopy is at 100.0% the size of the original, or the finder circles will be the wrong size. If in doubt, photocopy an accurate ruler or measuring tape and compare the size of it on the photocopy to the actual ruler. You can then reduce or increase the size of your printed overlay circle as necessary.)

Use the pointy end of a compass to punch a small hole at the exact center of the overlay finder circle. Then, as you position the finder circle for each hop, punch a hole in the chart. (Make the hole small enough or large enough that you won't mistake it for a star.) When you finish planning the star hop, set your compass to draw a circle of the correct size and use the compass to draw a finder circle at each hole position.

The following four figures show an example star hop from the first-magnitude star Spica in Virgo to the Sombrero Galaxy, M104. To begin the star hop, center Spica in your optical finder. Pivot the scope along the line from Spica to Porrima, offsetting a bit toward the constellation Corvus. When Spica is near the edge of the finder field, you'll see an arc of four bright (5th and 6th magnitude) stars on the opposite edge of the finder field, as shown in Figure 2-18. (This star hop sequence uses a RACI finder with a 5.5° field of view.)

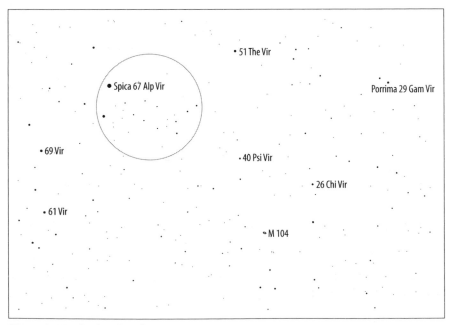

Figure 2-18. The first hop from Spica to M104

These charts show stars only down to 8th magnitude, which is about what you'll see in a 30mm finder from a typical suburban observing site. With a 50mm finder, or at a dark site, you'll see between one and two magnitudes deeper, which means that many, many more stars will be visible in the finder. That's why it's important to select the pattern for each hop from among the brightest stars in the field. Otherwise, it's easy to become confused with the plethora of dimmer stars that are visible.

Continue pivoting the scope on the same line to put that arc of stars on the opposite edge of the finder field, as shown in Figure 2-19. The 5th magnitude star 40 Psi Vir appears in the field of view, as does an arc of three 6th magnitude stars near the edge of the field.

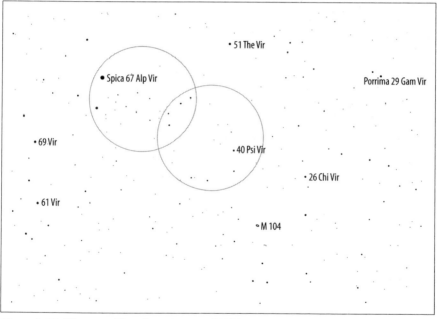

Figure 2-19. The second hop from Spica to M104

Continue pivoting the scope on the same line until 40 Psi Vir is on the trailing edge of the field and the 5th magnitude stars 26 Chi Vir and 21 Vir appear on the leading edge of the field, as shown in Figure 2-20. M104 is now in the finder field of view, although at best it will show weakly in a 50mm or larger finder from a very dark site.

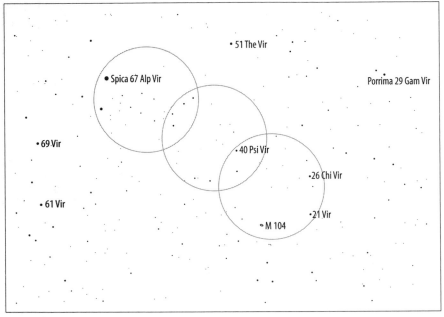

Figure 2-20. The third hop from Spica to M104

Pivot the scope on the line from 26 Chi Vir to 21 Vir until 21 Vir is at the edge of the field. Note the equilateral triangle formed by three 6th magnitude stars, two of which are just outside the finder field in Figure 2-21. When you see that triangular pattern of stars, you've gone just slightly too far. Pivot the scope in the opposite direction until the two stars have just disappeared from the field of view, and M104 is centered in the finder (and the eyepiece). You can confirm the position by the position of the 6th magnitude stars HR4779 (visible here at 6:00) and HR4822 (at 8:00). For further confirmation, note the double star just below the center of the finder field. This 7th magnitude pair is separated by about 5.5 arcminutes, and is quite prominent in the finder.

This star hop is a relatively easy example—in reality, most astronomers would simply use a Telrad to locate M104 geometrically from Spica and Porrima—but it illustrates the process. Rather than 5th and 6th magnitude stars, many star hops use 7th and even 8th magnitude stars. To find an object that's really out in the middle of nowhere, you may have to use 9th and 10th magnitude stars, which are at or beyond the limit of a 50mm finder.

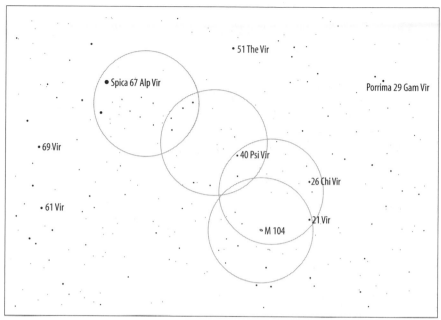

Figure 2-21. The fourth hop from Spica to M104

When that happens, you'll find yourself doing eyepiece star hopping, which is just what it sounds like. You use the optical finder to get as close as possible to the object, and then you begin using the main telescope to locate star patterns dimmer than those visible in the finder. For eyepiece star hopping, it's important to have a finder eyepiece with a wide field of view [Hack #48].

HACK #22 Learn to See DSOs

Seeing a black cat in a coal bin at midnight is easier than seeing some of these dim objects.

One of the major challenges that every beginning astronomer faces is learning to see. That sounds stupid, we know. After all, you've been using your eyes every waking moment all your life. How hard can it be?

Very hard, as it turns out. In daily life, you look at brightly lit, colorful, high-contrast objects that are familiar to you. When you observe DSOs (deep-sky objects—nebulae, galaxies, and so on) you look at unfamiliar, gray, dim objects with almost no contrast. You have to retrain your eyes, your brain, and your way of thinking if you want to see dim astronomical objects.

Imagine a gray scale that runs from 0 (pure black) to 255 (pure white). In daily life, you see objects that span most of that range. In DSO observing, you may be trying to tease detail from an object with an average brightness of perhaps 2 or 3 against a sky background of 1 or 2. Under light-polluted skies, it's even worse because the background sky may be nearly as bright (or brighter) than the objects you are trying to see. That's why you need to get to a dark site to observe galaxies, for example.

Beginners are often flabbergasted at just how dim and low contrast many DSOs are. Even under dark skies, newbies often literally cannot see a "bright" DSO even if an experienced observer has centered it in the eyepiece. For example, one very clear, dark night we were with a group of several experienced observers and one relative newbie. Robert had the bright Messier galaxy M51 framed in the eyepiece of our 10" scope. The experienced observers were discussing the structure visible in M51, including knots, nebulosity, dust lanes, and the connector between NGC 5194 (M51) and its connected companion galaxy, NGC 5195. The newbie thought we were making it all up. He could see only a few stars in the eyepiece, but not even a hint of M51. We weren't making it up. All of that and more was visible to us, but the newbie had not yet learned to see.

So how does an inexperienced observer learn to see what experienced observers call "faint fuzzies" or "lumpy darkness"? (Why one would bother is another question entirely, to which we reply, "because they're there.") Ultimately, the answer is to practice for as long and as often as you can. But there are several techniques that can help you along the way.

Learn Patience and Persistence

The sky varies from night to night—and even minute to minute—as do your eyes. If you are sure an object is in your field of view but you can't see it, keep looking. Even if the sky appears perfectly clear, there may be a slight haze or thin cloud obscuring the object. If the object doesn't reveal itself within a reasonable time, abandon that object temporarily and visit the next object on your list. Return to the missed object five minutes (or five hours) later, and you may be able to see it.

Even experienced observers have "bad nights" when for some reason nothing seems to come together. And we're not talking about nights with poor transparency or other problems beyond your control. We're talking about excellent nights when everyone else is doing very well, but for some reason you're just not able to see much of anything. We've had them. Every experienced observer has had them. When that happens, we just give up on the really faint fuzzies and start observing brighter ones, such as Messier objects. Often, the problem clears up as mysteriously as it developed.

Observe from a Dark Site

It's difficult to overstate the importance of a dark site for DSO observing. General light pollution brightens the background sky, making it difficult or impossible to separate DSOs with low surface brightness from the bright sky background. An object that is distinctly visible with a small scope or even a binocular from a dark site may be completely invisible in a large scope from a light-polluted site. Local lights are also a problem because they prevent you from becoming fully dark adapted.

Observe Objects When They Are High

The ideal time to observe any DSO is when it culminates (reaches its highest elevation). When an object is at low elevation, you're looking through a lot of air and haze, what we call muck. Also, the effects of light pollution are most severe at low altitudes. When the object culminates, it is as high as it gets and you're looking through the least possible amount of air, haze, and light pollution.

If possible, we avoid observing DSOs at less than 30° elevation. If the object is rising, we simply wait until it is high enough to provide a good view. Of course, some objects culminate [Hack #64] at less than 30° at our latitude, so we have no choice. Also, depending on the object and the season, the object may have set below 30° by the time it's fully dark. If so, we'll still observe it, but we'll make a note to observe it at another time of year when it is higher in the sky.

Dark Adapt Fully

For seeing dim DSOs, it's critical that you be fully dark adapted [Hack #11]. It can take as long as an hour to reach full dark adaptation, so an object that's invisible to you soon after full darkness may be clearly visible later in the evening when you are fully dark adapted.

Use the Largest Instrument Available

When it comes to seeing DSOs, APERTURE RULES. A larger aperture gathers more light, and more light is exactly what you need to see dim DSOs. We don't necessarily mean you should run out and buy the largest scope you can afford. There are other issues, not least transportability and convenience. (A scope so large and clumsy that you don't use it shows you less than a smaller scope that you actually use. Duh.) But if you have multiple scopes, use the largest one to do your DSO hunting.

> Nor do we mean that DSOs can't be observed with smaller instruments. Experienced observers have logged all of the Messier objects and even all of the Herschel 400 objects with only a 50mm binocular. Experienced golfers have also shot low rounds using a 2 iron as their only club. Just because you *can* do it that way doesn't mean you *should*.

Use Averted Vision

The center of your eye is very good for seeing detail, but very poor for seeing dim objects. When you observe a dim DSO, do not look directly at it. Instead, look off to one side. Objects that are invisible when you look directly at them become quite distinct with averted vision.

> Using averted vision even on bright objects allows you to see more. For example, an inexperienced observer who locates a galaxy with a bright core often views the galaxy only with direct vision because, after all, it's visible that way. But if you don't use averted vision to view that galaxy, you're missing lots of available detail that falls below the threshold of direct vision.

Keep Both Eyes Open

When you look through the eyepiece of a telescope, it's almost instinctive to close the other eye. Don't do it. Squinting reduces your ability to detect dim objects and detail within those objects. Train yourself to observe with both eyes open, particularly when you look at dim objects. If there is enough local light to make that distracting, cup a hand over the eye you aren't using, wear an eye patch, or use a towel over your head to screen out the local lights [Hack #12].

Absorb Photons

Everyone knows that a camera can capture dimmer objects if you use a longer exposure time, but few people realize that the human eye works the same way. An object that is invisibly dim when you first begin to look at it may become visible after several seconds, as your eye accumulates photons. From our own experience and discussing it with other experienced observers, it seems that the human eye can benefit from "exposure times" as long as a few seconds. Beyond that, there is no additional benefit.

Tap the Tube

As predators, our eyes and brains are hard-wired to be very efficient at detecting motion. A very dim, low-contrast object may lurk just under the limit of visibility as long as it isn't moving. But if it moves, your predator brain kicks in. Aha! Motion! Pounce! But DSOs don't cooperate. They just sit there, unmoving, unless you help them along a bit. The secret is to tap the telescope tube slightly, just enough to make the field of view jiggle. Often, an object you couldn't see will jump out at you as the view jiggles.

Use Different Magnifications and Fields of View

Conventional wisdom about viewing DSOs says you should use only low power, such as 15% or 20% of the aperture of your scope in millimeters. Wrong. The best magnification to use is the one that shows you the most detail in the object. It's true that low power (and a correspondingly large exit pupil) works well for many objects. But it's equally true that many DSOs are best viewed at relatively high power. For example, we often use 125X (50% of our 250mm aperture) to 270X (108% of our aperture) to view small, dim galaxies because that high magnification reveals details that aren't visible at low power.

Magnification and exit pupil are directly related. A magnification equal to the diameter of your primary mirror or objective lens in millimeters yields an exit pupil of 1mm. For example, if your primary mirror is 250mm in diameter, a magnification of 250X yields an exit pupil of 1mm (250/250 =1). In the same scope, 125X magnification yields a 2mm exit pupil (250/125=2).

Field of view is also important, particularly for large extended objects such as open clusters and some nebulae. Ideally, you want a field of view that frames the object in context. For example, if you view a tiny planetary nebula at low power, it looks just like another star in the field. But if you put some magnification on it, it is visible as an extended object within the star field. Conversely, if you view an open cluster with a narrow field of view, you may not be able to see the forest for the trees. The cluster stars look just like field stars. Using a wider field of view allows you to see the cluster as a cluster, set among a sparser group of field stars.

Focus Carefully

When an object is out of focus, its light is spread, becoming dimmer. An object that is on the edge of visibility when in focus may be dimmed sufficiently to be invisible if it is even slightly out of focus. Critical focusing is difficult, particularly under unsteady seeing conditions, where an object that is in focus at one moment may be slightly out of focus the next. It's worth tweaking the focus constantly as you observe an object to keep it in the best possible focus. You'll see more details that way.

Defocus Slightly to Locate Tiny Objects

Tiny objects, such as some planetary nebulae and galaxies, may appear stellar at low magnification. Using more power will reveal them as extended objects, but figuring out where they are in the star field can be difficult. One trick we use to locate such objects is to defocus slightly. With the image slightly out of focus, stars appear different from even small extended objects. Once you locate the extended object, you can apply more power to see the details.

Use Nebula Filters to Increase Contrast

Narrowband filters (such as the Orion Ultrablock or Lumicon UHC) and line filters such as the O-III and Hydrogen Beta can reveal detail in nebulae and comets that's invisible without filtration. In the most extreme cases, an object that is invisible without filtration in a telescope may be clearly visible naked eye with a filter. But even in less extreme cases, a filter can reveal detail that's invisible without the filter [Hack #59].

Don't limit yourself to one filter. Try different filters on each object, whether or not they're supposedly suitable for that object. For example, an O-III filter is usually thought of as a "planetary nebula filter," but there are planetaries for which a narrowband filter works better than an O-III. Also, different filters reveal different aspects of the object. For example, when we use narrowband and O-III filters on the Great Orion Nebula (M42), the narrowband filter shows a larger extent of nebulosity than does the O-III, but the O-III reveals details in the nebulosity that aren't visible with the narrowband filter.

An O-III filter is also handy for locating planetary nebulae, as opposed to just viewing them. Planetaries can be difficult to find because many of them are so small they appear stellar at low magnification. To locate a planetary quickly, once you are sure that it's somewhere in the eyepiece field of view, just move the O-III filter in and out of view between your eye and the eyepiece as you look through the eyepiece. The O-III filter dims most or all of the stars to invisibility, but it passes the light from the planetary nebula. The planetary nebula remains visible as the stars blink on and off.

Keep Looking

Even if you are able to see the object, you won't see all of the available detail in the first five seconds, or even the first five minutes. Keep examining the object, trying to tease out more detail. Once you have seen all the detail you can see in the object as a whole, focus on separate parts of the object, such as a dust lane or one arm of a galaxy, the fringe of a nebula, or the core of a globular cluster. When we view an object as a whole, our eyes and brain eliminate fine detail in favor of providing a gestalt of the object. When we direct our attention to a small part of the object, the object as a whole disappears, and we can see finer detail in that smaller area. It's almost fractal in its effect.

Being able to see fine detail in dim objects is an acquired skill, and one that must be practiced constantly if you are not to lose it. When you are first learning to see, it often seems that it's a matter of one step forward and two steps back. It's frustrating to hear an experienced observer describe an object as "bright" when you can barely see it. Keep at it, though, and you'll soon realize that you're seeing more and more detail in dimmer and dimmer objects. Before long, the newbies will be wondering how you can possibly be seeing details that are invisible to them.

HACK #23 Observe Shallow-Space Objects

Things that go zip, zoom, phizz, DOH-DEE-DOH, and BOOM in the night.

For a long, long time, the stars were referred to as "The Fixed Stars" because it was believed that they were perfect and unchanging. Well, I'll forgive the ancients for thinking they were perfect. Given the lack of light pollution they had and their dependence on stars for navigation, time and date calculations, and religion, they had great knowledge and appreciation for the night sky. But they made one crucial mistake: they ignored things that didn't fit their idea of how things worked. Because it is plain to see, if you look closely enough, that the stars do change. There is a fair bit of action out in the universe.

It is possible to take note of this from your backyard (or, better, from your friends' backyards in the country). The most obvious mobile objects are the Sun and Moon. You may not even think about the Sun being a mobile astronomical object but it is a fine one (always use a safe solar filter when observing the Sun) and, of course, its motion in the sky gives us the day. The Moon, also, you've no doubt seen tooling around the sky, changing phase as it does so. And, whether you've ever explicitly noticed it or not, if you're reading this book, you're most likely aware that the planets move relative to the stars in our sky. The ancients knew about all of these objects moving in the sky and had nice, if incorrect, explanations for all this motion. But there is a lot going on they didn't know about and it makes for an interesting tour of the changing sky. You'll have to look carefully and occasionally take some notes.

Things That Go Zip

Shooting stars have long intrigued people, and you've almost certainly wished upon one (how did that come about, anyway?). These meteors are little bits of fluff—bits of sand, flakes of dust—that hit our atmosphere and burn up in a flash. Every now and then, a bit more substantial chunk enters the atmosphere and burns as bright as the full Moon. These fireballs are impressive and almost completely random. The brightest one I ever saw I didn't actually see. I had my eye at the eyepiece when I noticed that the world was lit up as in movies featuring nuclear war. I pulled my head up but missed the fireball. However, I was rewarded with a smoke trail of the burnt up chunk that drifted away over the next five minutes. This nicely illustrates that, far from being a help, a telescope (or binocular) is actually a hindrance to meteor observing.

Like any meteor, fireballs are more likely at times of meteor showers (see Table 2-6), but they can happen anytime. Also, like any meteor, your chances of seeing one is better after midnight. After midnight, the limb of the Earth you're standing on is facing our direction of travel, making it a little more likely that a meteor will enter the atmosphere above you. If you decide to observe a meteor shower, read up on when the peak should occur. Shower peaks vary. Some are very narrow (the Leonids, for example) while others stretch over days. The Perseids are a favorite as the peak features a good number of meteors, and the peak occurs in the northern hemisphere summer, so it is warm.

To observe a shower, here is what you need: eyes. A notebook (or tape recorder) is also useful as is a nice lounge chair and blankets. Simply stretch out under the sky so that you have as clear a view of as much sky as possible and start counting meteors. You'll also want to know where the *radiant* is. The radiant is the point in the sky from which the meteors appear to originate. The material making up the meteors is debris that has fallen off of comets. The debris now orbits the Sun in the same orbit—but trailing—as the comet. If the orbit intersects Earth's orbit, we'll get a shower as a lot of material will impact the atmosphere where the two orbits meet.

If you were to extend the trail of every meteor that is part of the shower backward, you'd find that all trails intersect at the radiant. This is actually good fun. Get a good all sky chart and sketch the meteors that you observe. After you have 15 or 20, complete the trace of eachtrail backward and see where they meet. For the Perseids, you'll find that those trails meet in the constellation Perseus; showers are named for the constellation in which the radiant is located. Any trails that don't intersect at this point are called *sporadics*. These are simply the random background noise of meteors that happen every night. Note that a good rate during a meteor shower is one meteor every few minutes. What many have in mind when they think of a meteor shower is actually a *meteor storm*. A good example is the 2001 display of the Leonids. The Leonids, in fact, are responsible for a number of meteor storms thanks to very dense pockets of cometary debris that impact Earth on a roughly 33-year cycle.

Table 2-6 lists the major meteor showers that occur regularly each year.

- Maximum lists the range of dates during which the meteor shower reaches its maximum hourly rate.

- Duration is the period during which the meteor rate is more than 50% of the maximum rate. Note that all showers may have very short and random bursts of intense activity during this period.

- *ZHR* (*Zenithal Hourly Rate*) is the number of naked-eye meteors that would be observed from a dark sky (6.5 naked-eye limiting magnitude) if the radiant were at the zenith (directly overhead). Suffice to say, you will probably never see a meteor shower under such conditions. Thus, you will see fewer meteors, on average, than the ZHR. Again, note that bursts with ZHR of >1,000 are possible for a few minutes in most showers.

Table 2-6. Major meteor showers

Shower	Maximum	Duration	ZHR	Source comet
Quadrantids	3–5 January	10–18 hours	120	C/1490 Y1 (?)
Lyrids	20–25 April	1–2 days	20	C/1861 G1
η Aquarids	2–5 May	3–7 days	60	Halley
S δ Aquarids	25–29 July	1–2 weeks	20	unknown
Perseids	11–13 August	1–2 days	100	Swift-Tuttle
Orionids	20–22 October	1–2 days	20	Halley

Table 2-6. Major meteor showers (continued)

Shower	Maximum	Duration	ZHR	Source comet
Taurids	1–8 November	1–2 days	20	Encke
Leonids	16–19 November	0.5–1 day	20	Temple-Tuttle
Geminids	12–14 December	0.5–1 day	120	3200 Phaethon

The Quadrantids are named for the now defunct constellation Quadrans Muralis (which was eliminated in 1933 when the IAU officially defined the 88 modern constellations). The radiant is in northern Boötes. This is, perhaps, the most consistently good meteor shower. However, it is most active when it's most cold. Also, the peak is very narrow. You'll need to be hardy and lucky to catch a good show, but, if you do, you'll be well rewarded.

The Leonids are a very well-known shower thanks to a larger 33-year cycle. The general stream of cometary debris that makes up the Leonid shower contains several extremely dense pockets that can cause ZHRs of 10,000–30,000 that last for a significant amount of time. Most readers will remember fantastic displays of Leonids from 1998–2002. The streams are dense, but narrow, so what you see depends greatly on when you observe. The usual Leonid shower is average but hang on to your hats in the early 2030s.

> Meteor showers are perhaps the simplest celestial event to photograph. You need a 35mm camera and a tripod on which to mount it. Attach the widest field lens you have (35mm is best, but a 50mm lens is also good). Aim the camera at the sky. You can, but don't have to, aim at the radiant. In fact, aiming 20° to 30° away from the radiant gives you a better chance at fireballs. Focus to infinity. Set the exposure to the bulb setting (B on the exposure dial). Use a cable release to open the shutter. Be sure to use film that is ISO 400 or faster, if you're using film. Limit your exposure to a few minutes and end the exposure soon after any meteor passes through the field of view. During some showers, you can record numerous meteors in a single frame. The stars will trail (due to the rotation of the Earth) and meteors will appear as streaks that run against the grain of the star trails. Figure 2-22 was taken during the 2001 Leonid shower and shows a nice fireball passing south of Orion; note the smoke trail that drifted slowly away.

Things That Go Zoom and Phizz

Meteors are the fastest moving celestial objects you're likely to observe (at least until the end times). The dominant source of meteors is the next fastest: comets. Comets are often called dirty snowballs and, although it isn't

Figure 2-22. A fireball passes south of Orion

precisely accurate, it's a good enough name. These are bodies of ice, dust and rock that range in size from a few hundred meters wide to tens of kilometers. Comets have very eccentric orbits. All solar system objects orbit the Sun in an elliptical orbit. Most objects—Earth, for instance—have very low eccentricity. Earth's orbit departs only 0.5% from circular. Comets, on the other hand, may get as close as a few million miles from the Sun at perihelion (nearest approach to Sol) and as far as hundreds of millions of miles at aphelion (farthest departure from Sol). When the comet is approaching the Sun, the body warms, ice melts and evaporates, and trapped gases also escape. The result is a lot of material being ejected from the body of the comet to form a coma and also a tail (or tails—usually a dust tail and an ion tail).

This can be quite impressive. Every 10 or so years, on average, a comet appears that can easily be seen with the naked eye. Most people reading this remember comet Hale-Bopp (1997) and comet Hyakutake (1996). Comet Halley (1986) was a lovely comet and comet West (1976) was stunning. Of course, the ancients knew about these too and were generally really freaked out by them. ("When beggars die there are no comets seen; the heavens themselves blaze forth the death of princes." —William Shakespeare)

If they only come around every 10 years or so, you have no need to get too excited, right? Wrong. Those are just the bright, obvious comets. Every year a number of comets become visible in small telescopes or in a binocular.

You can find a good deal of information about comets from astronomical web sites. Certainly, the big ones get lots of press, but you'll want to follow it from when it's just a little speck of coma to when the party is over and the comet is on its way out to the cold outskirts of the solar system.

Also, comets present an excellent sketching opportunity **[Hack #24]**. Photographs have trouble showing faint things alongside bright things. That is, photographic film and electronic imaging chips have poor range. To register the faint parts of an object, you must overexpose the bright parts. Comets definitely suffer in this regard. There is a wealth of detail around the nucleus (the bright core of the coma) of the comet that is best seen at medium magnification. This almost certainly means excluding the tail (if one is visible at all). However—and this is the cool bit—that wealth of detail changes nightly, sometimes hourly. As jets of gas and steam explode off the surface of the comet, the appearance of the coma, inner coma, and nucleus will change. If you're patient and keep tabs on the comet, you'll get to see a dynamic celestial body being stressed mightily as it falls toward the Sun. (Of course, sometimes a comet is too far away or just too dim to show much of this detail, but you'll never know if you don't look.) Figure 2-23 shows part of my log page for 15 November 1985, including a sketch of comet Halley.

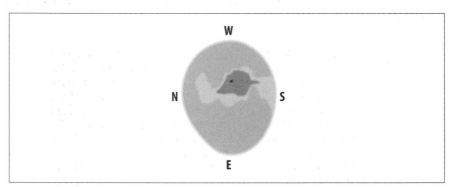

Figure 2-23. Part of my log page with a sketch of comet Halley

Things That Go DOH-DEE-DOH

Unlike meteors and comets, which zip and zoom around Earth's neighborhood, the *minor planets*—also called *asteroids*—follow more leisurely orbits farther out from the Sun (with some notable, and frightening, exceptions). For purposes of this hack, I'm going to lump in Uranus, Neptune, and Pluto with the asteroids. This will probably make all manner of astronomers angry, but, to me, there is little observational difference between Uranus, Neptune, Pluto, and the minor planets. Of the entire group, only Uranus and Neptune show a visible disk and that is only 2 or 3 arcseconds that shows no detail at all in amateur telescopes.

So, why look at them? Well, it's undeniably cool to be able to say you've seen all the planets. And, in fact, Pluto is quite the challenging object, and it's getting more challenging every day as it moves away from perihelion in 1989 and, thus, gets dimmer and dimmer (until it reaches aphelion in 2113—don't wait up). So, it is a good way to hone your skills in locating and identifying objects.

You can find ephemerides for most minor planets and the three outer planets online (*http://www.skyandtelescope.com* is a good site for such things— also see the information at the end of this hack). You'll need celestial coordinates and a good chart for all of these objects (most planetarium software will also chart the brighter minor planets and the three outer planets, as well **[Hack #64]**). There are excellent tips on how to put an object in your eyepiece elsewhere in this book. For now, let's assume you've put the target dead center in your eyepiece. What will you see?

Well, not much. If your target was Uranus or Neptune and you have a low magnification eyepiece in, you'll see what looks like a star. Only at relatively high magnification will the planetary disk of either planet be noticeable. If your target was Pluto or any minor planet, you'll again see what looks like a star. How do you confirm that you have the object you think you have?

> Incidentally, Uranus is, under a dark sky, a naked-eye object and was observed by many early astronomers (Galileo included) before it was recognized as a planet. In fact, there are numerous recorded sightings of Uranus prior to William Herschel recognizing it as a wanderer. (The ancients called planets "wanderers" because planets—and the other solar system objects they couldn't see—change positions relative to the background stars.) Similarly, one minor planet, 4 Vesta, occasionally becomes bright enough to see without optical aid. 4 Vesta can become as bright as magnitude 5.4 and Uranus is never dimmer than magnitude 6.0. You won't mistake either of them for Venus, nor will you see them from downtown Manhattan, but from a reasonably dark sky you have a good chance at seeing them with the naked eye.

Despite looking like a star, your solar system object is going to move. Depending on the object, it will move faster or slower. Or, rather, slower or really, really slow. Generally speaking, the farther away the object is, the slower it will move against the stars (hand it to the ancients, they nailed this). Thus, Pluto crawls along. Uranus and Neptune aren't going to be confused for comets, either. What you'll have to do is sketch the field of view.

Be sure to sketch *all* of the stars visible and carefully note the time, date, and conditions. Either follow the object for a few hours and sketch again or compare the new view to your sketch. Has the object moved? If so, that is your target. If not, come back another night (1–3 days) and sketch the field again. Has one of the "stars" moved? If so, that is the one. If not, you've missed your target. You can also do this exercise with comets. In fact, it's a nice comparison to chart cometary movement with minor planet movement. You should notice a great difference. Cometary motion is sometimes noticeable in just a few minutes. Figure 2-24 shows my 19 July 1987 sketches of the asteroid 6 Hebe (pronounced "hee-bee"). On the left sketch, done just after midnight, I put x-marks next to the objects I suspected of being the asteroid. On the right sketch, done late that evening, the asteroid has revealed itself by its movement, and is marked with an x.

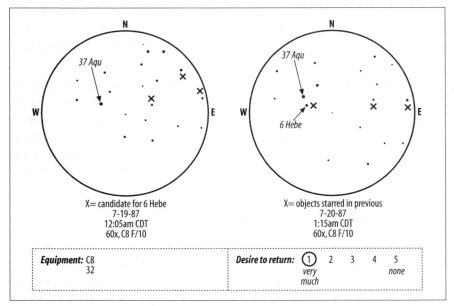

Figure 2-24. *Two sketches of 6 Hebe done about 23 hours apart, showing its motion*

> The Association of Lunar and Planetary Observers or ALPO (*http://www.lpl.arizona.edu/alpo/*) is an outstanding starting point for information on anything having to do with the solar system. So, for minor planets, comets, and meteors this is a good place to start.

—*Dr. Paul B. Jones*

Slow Down, You Move Too Fast, You've Got to Make the Evening Last

Take a second look and some notes. But above all, take your time. Attention to detail can be learned, and here's where you'll learn it.

There are two fundamental elements to visual observation of astronomical objects: the first is to locate the object, and the second is to observe it. Whether you use a telescope, a binocular, or the naked eye, the first is often so difficult for the beginning astronomer that little thought is given to the second. This is completely understandable and quite natural.

Furthermore, you've probably been looking at things all your life and think you know a little about how to do it. Right? Well, then, you must have noticed the pattern of the paper on this page: the swirls, thick and thin patches, color variation, etc. If you have, then you can probably skip the rest of this hack, as that attention to detail is exactly what is required of you in observing astronomical objects. If you haven't, then, hopefully, you'll find here some useful tips on how to improve your observing eye.

First, slow down! There are numerous tips for locating objects elsewhere in this book. For now, assume you've located a target. Now is the time to take a deep breath and relax. Leave the object in the field of view. If you have a scope with a drive, turn it on. If not, be sure you have an eyepiece with as wide a field of view as possible and keep the object close as you take a few minutes to recover from the task of finding the object. Have a drink of coffee or a quick bite.

Look It Over and Under, Up and Down

Now that you've relaxed a bit from the task of finding the object, go back to the eyepiece and have a look. Study the object for a few minutes using the eyepiece with which you found it. This should be your lowest magnification, widest field of view eyepiece [Hack #48]. How many stars can you detect in the object? Immediately around it? If it is a galaxy or a nebula, how does the size and shape change as you go from direct vision to averted vision? Is the view different if you avert your vision to the right or left, up or down? If the object is a globular cluster, can you resolve stars across its face? Just on the outskirts? At all?

The point is, ask yourself as many questions as you can think of about the detail you see. Are there patterns? Any striking shapes or colors? Transfer the image in the eyepiece into your eye. When you're satisfied with the image now in your mind, change eyepieces and go through your questions again. Do this for at least two different magnifications. I generally use three, depending on the object. The wide-field look with the eyepiece (80X) I use to find the object, and then I use a medium-magnification (150X or so for an 8-inch scope) and a high-magnification (250–300X) view. If the object is exceedingly small, I may skip the medium view; if it is large, or has very low surface brightness, I'll skip the high-magnification view. Also, if the object may benefit from a filter—usually a planetary or emission nebula or a bright planet—take a good long look both with and without the filter(s) **[Hack #58]**. Note how the view changes.

> Many people discuss whether a filter "helps" an object. In many cases, this is proper as an object may be either visible only with, or only without, a filter. The Horsehead Nebula, for instance, is essentially invisible without an H-beta filter. Just a few degrees away, M78 is invisible with the same filter. However, most objects are somewhere in between. They look different with different filters. This is especially true of bright objects. You may prefer the look with one filter, or without a filter, but you will often learn something about the object by taking a look with all of your filters. Just be sure to give yourself a few minutes with each view. None of the views are the real object, but simply how that object looks at a particular wavelength or set of wavelengths.

Remember to look at the whole object, get sort of a global picture, and then look at small details. This takes some practice. Viewed through a telescope, astronomical objects rarely have the bold visual impact of professional (or even good amateur) astrophotographs. There are a variety of reasons, but first and foremost is the fact that astronomical photographic exposures are long—minutes to hours—while the human eye can collect light only for a second or two. Thus, the photograph shows parts of objects that are very, very faint or even invisible when viewed through a telescope.

You'll See More with Your Eyes

But there is a tradeoff, and it is one that makes visual observing still worthwhile in an age of photography and electronic imaging. Photographs (film or electronic image) suffer from poor range. What makes it possible for a photograph to show very faint things well makes it very, very difficult for it to show two things in the same image that have very different brightnesses. The eye, on the other hand, has excellent range. Take M42, the Orion Nebula. If you've been involved in amateur astronomy for even a few weeks, you've seen photographs of M42, the pink nebula with wide arms that end in bluish or greenish tinted delicate curtains. But look at the center of the nebula in a photograph. It is a white blob. If you've seen M42 through a telescope—or a binocular—even once, you know there is no blob in the middle of M42. There is a beautiful multiple star, the Trapezium. These four jewels (along with a couple of other much fainter stars you may be able to see) sit in the "middle" of M42. Through a small telescope with a field of view on the order of a degree or so, you can see this nebula looking very much like it does in photographs. There are two key differences:

- The nebula looks gray or green because there isn't enough light to cause the cones to fire [Hack #11].
- You see no blob. That's right, both the nebula and the Trapezium are visible.

In photographs that show the Trapezium, not enough light from the nebula hits the film (or chip) to register an image. In photographs that show the nebula, the Trapezium is very overexposed and appears as a blob, as shown in Figure 2-25. The photographs are impressive, and many new observers are disappointed by the view through a scope; howeve, for my money, the subtlety and delicacy of the visual image is much more beautiful than photographs. You simply have to take your time and see the whole picture along with the details. Figure 2-26 is my sketch of M42 and the Trapezium, done at 290X with my 8" SCT and a 7mm Nagler eyepiece, showing full detail in both objects. (The field of view in the sketch is much smaller than that of the photograph, so the sketch shows only a small portion of the nebula, with the Trapezium group of stars centered.)

Most objects you'll want to observe have the same dual nature: one photographic identity and one visual. While the photographic image is bolder (brighter, more stars, etc.), the visual image reveals numerous features that are absent in the photograph. If you're prepared for the visual image to be less boldly impressive, you've taken the first step in appreciating the subtle image you'll see at the eyepiece.

New Ways of Seeing

Even so, you'll likely need time to learn to see the way experienced amateur astronomers see. Fortunately, there are a few things you can do to speed the learning process along. First, you can practice your observing skills away from the eyepiece. You can use a variety of items as a stand-in for a faint galaxy. Among the items I've heard of being used in this way are eggs, white paper, walls, ceilings, rocks, and carpet. Carpet? Well, you get the idea. Anything that at a glance looks to be featureless but that has subtle texture, color variation, or a hard-to-discern pattern provides a chance to work on your observing skill. I'll often do this while waiting at doctor's offices or for a meeting to start; pick a wall and try to find the texture and the patterns.

Figure 2-25. A photograph of M42 shows detail in the nebula, but the Trapezium is burned out

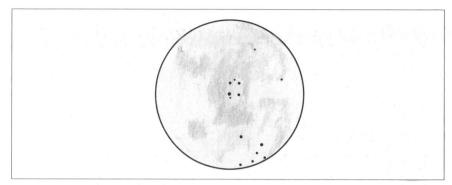

Figure 2-26. A sketch of M42 shows full detail in both the Trapezium and the nebula

You'll be surprised how different a plain, white wall will look after you've studied it for a few minutes—and how much easier these patterns appear after you've done this a few times.

The second thing you can do to increase your observing skill is to sketch objects. At first, I was intimidated by the idea of sketching until I realized that the point wasn't to produce beautiful pictures but to help me see better. If you're worried about producing poor drawings, don't worry, you will. But you have two things going for you: one, you work in the dark and you don't have to show anyone your work; and two, you'll get better quickly. Equip yourself with the following items:

- Sketchbook
- Artist's pencils
- Eraser
- Red light
- A table within reach and a scope with a motor drive also help a lot

You can use any old pen and paper, but I recommend going all out. For only 10 dollars or so, you can buy a proper sketchbook and a set of pencils. I use a set of 24 ranging from very fine to very dark. Using a bound book gives both better paper and a more hardy housing for your work. These tools will let you produce nice drawings that you can refer to in the future. Before you begin, you should prepare a number of circles on each page. These circles will represent the field of view. Don't make them too big or small. You'll need enough room to draw sufficient detail, but making your drawings too big can make it hard to accurately represent the proportions in the object. I make mine with a compass, and they're about three inches in diameter. Figures 2-27, 2-28, and 2-29 show my sketches of Mars, the galaxy M82, and the Trifid Nebula (M20), respectively.

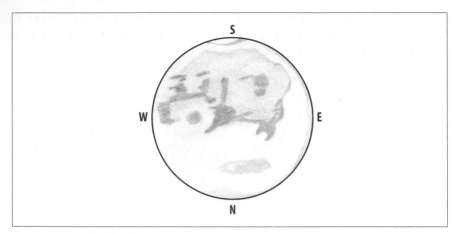

Figure 2-27. A sketch of Mars

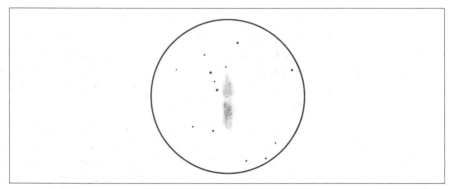

Figure 2-28. A sketch of the galaxy M82

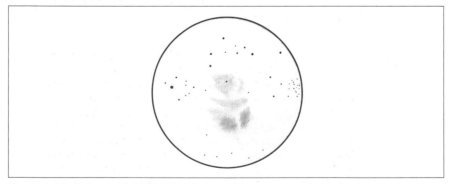

Figure 2-29. A sketch of the Trifid Nebula

After you've set up your equipment (telescope, binocular, etc.), note the date, conditions, and observing site location as you would anytime you log an object. Generally, when I sketch an object, I go through the sequence of eyepieces and filters described previously. I may take a few notes, and I keep in mind the best magnification to show detail and yet properly frame the object. When I feel comfortable with my mental image of the object, I begin by drawing the field stars paying special attention to spacing and relative brightness. Depending on the field, I may draw all of the stars or just a few of the brighter ones. In any case, these stars will serve as a guide for the sketch of the object itself.

> Always label the sketch with the field width of the eyepiece or binocular you used when making the sketch.

Once the field stars are drawn (really, they're just dots, bigger or smaller to represent brightness), I begin drawing the object itself. This is an iterative process: look through the eyepiece, sketch, compare, look, sketch, compare. Actual drawing techniques are described better elsewhere and by others. Remember not to hold the pencil too tightly. Use your finger to smudge the object to soften the image. Don't erase too much (or at all, if possible). Use only a red light to look at the page on which you work.

Again, keep in mind that the real point isn't a beautiful rendering of the object, but rather training your eye to see better. It is the iterative process that is the key to this. Keep going back and forth from eyepiece to paper to eyepiece until your drawing contains all of the subtle details you can see. Look for new details each time and also be sure to confirm old observations (old being anywhere from a minute ago to years ago). Always be on the lookout for a new "angle" on an old object.

> Be sure to keep a lookout for "real" changes in an object. Don't neglect sketching objects you've sketched before. If your observing career spans enough time, your sketches taken together may encompass all, or a large part, of the orbit of a double star. Hubble's Variable Nebula (in Monoceros) changes appearance over time, sometimes over the space of as little as a week; sketches are an excellent way to keep track of this. In fact, one keen amateur astronomer, Jay McNeil, got famous for discovering a new nebula, subsequently named after him, by noting (photographically) a new part of the nebular complex making up M78. There is no reason (other than the rarity of such changes being within reach of amateur equipment) that you couldn't do the same. In any case, by sketching objects over and over, you'll have a better idea of how your sketching is improving.

You can, and probably should, also practice drawing daytime objects. I've practiced on trees, leaves, eggs, and my driveway. These exercises make for good practice at both drawing and seeing. It won't take long before you'll notice that your drawings actually look good. If you wouldn't be ashamed of the sketches you see here, consider that I've only been at this for a couple of years and that I nearly didn't make it out of first grade because I couldn't, and I suppose still can't, fingerpaint.

At this point, it is worth considering observing styles. Observing sessions range from a Messier Marathon **[Hack #30]**, which involves only finding objects with no attention to the object itself, to spending an entire hours-long session on a single object. Your observing style is which you most prefer and most favor. I've done both and you may, too. I generally prefer a session closer to the latter scenario, and I have good friends who prefer sessions closer to the former. What is most important is that you decide what type of session fills your observing needs and keeps bringing you back. There is no right or wrong way to enjoy the night sky. In any case, you'll probably find that engaging in one type of session makes you better able to accomplish and appreciate the other. And, if you wind up with a beautiful piece of art in the form of a sketch, so much the better.

—Dr. Paul B. Jones

 Learn Urban Observing Skills
#25   Observe the universe from your own backyard.

Backyard astronomy isn't what it once was. As a teenager in the mid-60s, Robert was able to observe most of the Messier Objects and scores of other deep-sky objects (DSOs) from his urban backyard using his home-built 6" Newtonian reflector. Nowadays, alas, urban light pollution limits serious DSO observing to rural dark-sky sites.

Light pollution is the bane of DSO observers, and the only way to avoid light pollution is to travel to a dark-sky site. By "dark" we don't mean a suburban park, either. You'll need to travel at least 20 miles from even small towns and 50 miles or more from larger cities to find a truly dark site. But the view is worth the effort. You can see more with a small scope or even a binocular from a dark site than you can see with a large scope from a light polluted site. If you've never observed from a truly dark site, give it a try. You'll be shocked at the difference.

Fortunately, light pollution does little or nothing to hamper other types of observing. If your métier is Lunar and planetary observing, for example, you might as well stay in town. Luna and the planets are just as visible from the most brightly lit urban backyard as they are from the darkest rural site.

There is much to be said for urban observing. You can save yourself a drive, observe more frequently, spend more time observing, and always have a bathroom and a warm refuge available. On weeknights when you don't have time to drive to your dark-sky observing site, you can observe for an hour or two and still be finished in time for the 11 o'clock news. About the only thing you can't do from an urban site is observe galaxies and other dim DSOs.

Urban Observing Targets

Light-polluted urban skies limit your choice of observing targets in two respects. First, the background skyglow makes it impossible to see dim, low-contrast objects such as galaxies. The largest telescope in the world won't let you see such dim objects because the sky background is literally brighter than the object itself. Second, light pollution makes it difficult to locate objects because you can see so few stars. Trying to hunt down faint fuzzies from an urban site is an exercise in frustration. Fortunately, there are many other objects suitable for urban observing.

Luna

Luna (which we dedicated deep-sky observers think of mainly as an annoying source of light pollution) can be a very rewarding target. In fact, some amateur astronomers devote their careers to observing Luna, and seldom observe any other target. Luna is easily visible no matter how bad the light pollution, and under less than perfect viewing conditions. Luna is visible most nights of the month, can take high magnifications, and shows immensely more detail than any other astronomical object.

Although it sounds stupid, we sometimes observe Luna through haze and even light clouds. The atmosphere is often extremely stable on hazy nights, making them quite suitable for Lunar and planetary observing. We are generally limited by seeing (atmospheric turbulence) to about 300X, but on hazy nights we have run as much as 650X on Saturn, 900X on Jupiter, and 1,200X on Luna. No, you can't do that every night, and yes, you have to wait for short steady periods, which may occur only every few minutes and last only seconds, but when the atmosphere steadies down, it's amazing how much detail you can see at these "stupid high powers."

Planets

Although their presence in the night sky changes from season to season, the planets are a rewarding target for many urban observers. Only Pluto is nearly impossible for urban observing. (Not that you're missing much. At its best, Pluto looks like a 14th magnitude star **[Hack #13]**.) The outer gas giants, Uranus and Neptune, show pretty colored discs, but little or no detail is visible. Although some people have logged the brighter Uranian moons from urban locations, we've never had any luck distinguishing these 14th and 15th magnitude objects from the background skyglow.

Mercury is possible near sunrise or sunset at some times of year, although it shows little more than a featureless disk. Venus is often quite high as the Evening Star or Morning Star at some times of year, but its heavy clouds conceal all surface features, leaving its current phase as the only detail of interest. We think Venus, like Mercury, is boring.

That leaves only three planets, but what planets they are. Jupiter, Saturn, and Mars (when it is near closest approach) are magnificent objects, worthy of extended study. We often delight in watching the dance of the Jovian moons, trying to tease out fine detail in Saturn's rings, and looking for the Martian seas and canals that Percival Lowell sketched in detail. (The canals really are visible. Honest. They're an optical illusion, but they are visible.)

Most planetary observers are also Lunar observers, and vice versa. If you focus your urban observing on Luna and the planets, you'll seldom be without something worthwhile to observe.

Double stars

Although we've never seen much point in it, many astronomers enjoy observing double- and multiple-star systems. Okay, we admit we're as impressed as anyone by the brilliant blue and gold Albireo pair or the beautiful ternary β-Monocerotis system, but how long can you look at just two or three stars? Long enough, apparently. We know of one amateur astronomer who's observed more than 1,000 double stars, all from his urban site. The Urban Observing Club of the Astronomical League publishes a good beginner list for urban double-star observing (*http://www.astroleague.org/al/obsclubs/urban/urbanls.html*). This list includes a baker's dozen of the most impressive multiple stars and variable stars.

Bright open clusters and emission/reflection nebulae

Although urban sites are not optimum for observing deep-sky objects (DSOs), many of the brightest open clusters and emission/reflection nebulae are visible easily from all but the most light-polluted sites. The

AL Urban Observing Club publishes an excellent beginner's list with dozens of bright DSOs suitable for urban observers (*http://www. astroleague.org/al/obsclubs/urban/urbanld.html*).

Planetary nebulae

Although planetary nebulae often have high magnitudes, many have relatively high surface brightness. Locating these tiny objects from an urban site is challenging—the smallest planetaries often appear stellar even at medium magnifications—but if you can find them, you can often see significant detail in them.

Although in our experience the so-called "light pollution reduction" filters are of minimal help in reducing the effects of light pollution, they can be quite helpful in locating and viewing planetary nebulae from an urban site. If you have a larger scope, at least 8", an Oxygen-3 (O-III) line filter is probably the best choice for bagging planetaries in-town (just as it is from a darker site). If you have a smaller scope, or if you just prefer to buy a more generally useful filter, a narrowband filter like the Orion Ultrablock or the Lumicon UHC is also quite helpful on planetaries.

Don't buy into the manufacturer hype about broadband "light pollution reduction" or LPR filters like the Orion Skyglow or the Lumicon DeepSky **[Hack #59]**. These filters are usually represented as being useful for reducing "moderate light pollution." If you buy one expecting to improve your urban viewing, you'll probably be disappointed. The visual improvement, if any, will be very small. Before you buy a broadband filter, see if you can borrow one to test for yourself.

Conversely, narrowband filters like the UltraBlock and UHC do indeed cut down background skyglow, but they do so at the expense of making the entire image quite dim. We consider them to be "nebula filters" rather than "light pollution reduction filters," but they are as useful for increasing nebular contrast from an urban site as they are from a dark site.

Globular clusters

Like planetary nebulae, globular clusters often have high magnitudes but relatively high surface brightness. Also like planetaries, globs are sometimes difficult to find. Small globs may appear stellar in a low-power eyepiece, so the best plan is to use the low-power eyepiece to get the scope into the right general vicinity and then use a higher-power eyepiece to verify and view the object. Unlike planetaries, globs do not benefit from a narrowband filter.

Urban Observing Tips

To make the most of your urban observing time, use the following guidelines:

Know what to look for, and when

Impromptu urban observing sessions are by definition not planned. But it's easy to take a lack of planning too far, and end up just standing outside waving your scope around, hoping to see something. You'll make better use of your observing time if you have some idea of what to look for ahead of time. Dedicated urban observers often make lists of the objects they would like to log, and then organize those lists by when the objects are best placed. For example, you might create a consolidated list of 200 objects and then divide them into monthly or bi-monthly lists. That way, if an opportunity for a quick observing session arises in mid-July, you can simply check your 1 July, 15 July, and 1 August lists to determine which objects are high in the sky that night. Well-organized urban observers take the moon phase into account, generating lists of only bright objects during times when Luna is up, and DSOs for times when Luna is not up.

Let Luna be your guide

As with all observing, the state of Luna determines what it's possible to see. During the new moon, or when the moon has already set or not yet risen, you have at least a chance to observe DSOs. The views won't be as good as those from a darker site, but the brighter DSOs can still provide satisfying views on clear, dark nights. If Luna is up, well, observe Luna.

Take the weather into consideration

Weather always determines the success of an observing session, but this is particularly true for urban observing. Cool, dry nights are usually good for observing. When a cold front moves through the area, it pulls air pollution, dust, and haze with it, often providing excellent viewing conditions with very stable air. Periods immediately after a heavy rain are often best of all; the rain does a wonderful job of clearing dust from the air, which improves atmospheric transparency, stabilizes the air, and reduces the effects of light pollution. (It doesn't matter how much light escapes into the air if there's nothing for it to reflect from.)

Atmospheric transparency is always important, but particularly so for urban observing. Use the Clear Sky Clock (*http://cleardarksky.com/csk/*) transparency forecasts to plan your urban observing sessions.

Observe late (or early)

Light pollution is greatest during the period from full dark until midnight or so. After midnight, some (by no means all) businesses have turned off their exterior lights, there are fewer cars on the road, and so on. The atmosphere is also often significantly clearer late at night. Dust has had a chance to settle, water vapor and some of the air pollutants have dispersed, and so on. If it fits your schedule, one of the best times for urban observing is often the hours just before dawn, which are usually the clearest and darkest times available in urban environments.

View near zenith

You can minimize the adverse effects of light pollution and poor transparency by observing objects when they are as high as possible. When your scope is pointed at zenith, you are looking through the minimum possible amount of dirty air and light pollution.

Of course, not all objects culminate anywhere near zenith, and observing near zenith with a Dobsonian or other alt-azimuth-mounted scope is problematic. When we observe with our Dob from our moderately to severely light-polluted suburban skies, we generally restrict ourselves to objects that are at least 60° and no more than 80° elevation. At much less than 60° the light pollution and atmospheric grunge makes it difficult to see objects; at much more than 80° Dobson's Hole **[Hack #42][Hack #43]** becomes a problem.

Use planetarium software to orient yourself

One of the problems with urban observing is that few stars are visible, which may make it difficult to locate your target objects. At times, it can even be difficult to locate familiar constellations visually because you don't have the context of the other constellations. Using planetarium software **[Hack #64]** gives you a real-time map of where the various constellations are located in your night sky. If you are looking for the globular cluster M13 in Hercules on a July evening, for example, your planetarium software may tell you that M13 is located nearly dead west at an elevation of about 70°. Once you know that, it's much easier to locate Hercules. Once you've located Hercules, finding M13 is easy.

Master star hopping with finder and eyepiece

As much as we like unit-power bulls-eye finders like the Telrad, they are often of limited use under light-polluted skies. Using a Telrad effectively requires being able to see more stars than may be visible under urban skies. Fortunately, the same star-hopping skills that are so important under dark skies are also helpful under bright skies. The star hops are often longer because you must start from a brighter star farther from your object, but with patience and skill, it's possible to locate most objects via star hops.

At dark sites, star hopping with your finder allows you to locate nearly any object because the finder shows more than enough stars to allow you to map a course to the object. Under very bright urban skies, even the finder may not suffice. When that happens, use the light gathering power of your telescope with your widest-field eyepiece to take your star hops down to the eyepiece level. It takes longer to locate an object that way, certainly, but it can be done.

Enlist your neighbors' help

As annoying as the skyglow of general light pollution is, local light pollution is even worse. There's not much you can do about streetlights and similar light sources. Well, we do know many amateurs who turn them off temporarily by keeping a laser focused on the sensor, and at least one who claims to turn them off permanently with his .22/250 rifle, but in general you're stuck with them.

You can do something about neighbors' porch lights and flood lights, though. Sometimes, just asking them to turn off their exterior lights suffices. But if you want to convert them from grudging cooperation to enthusiastic participation, invite them over periodically to share the view. Once you've done that and they see for themselves how much difference the absence of local lights makes, at the very least they'll turn off their lights any time you ask them to. They may even invite you to come over and turn them off yourself any time you want to observe.

Reduce the effects of local light pollution

Even if your neighbors are cooperative about their lights, there are many other sources of local light pollution. It's worth making some effort to minimize those problems. Choose the darkest area of your yard, and make improvements to it. Build light screens using a tarp or semi-opaque gardening cloth on a light wooden or PVC pipe framework. Use existing buildings and foliage to block local light sources. Consider planting bushes or a hedge to screen your site from streetlights and other local light sources.

It's also worth the effort to improve your observing equipment and accessories. Make sure your optics are clean. Even a slight film of dirt or small smudges can reduce image contrast significantly under light polluted conditions. Install a dew shield or baffle to screen your objective from stray light. Flock your scope to reduce the effect of off-axis light. Use an eye patch to allow one eye to remain fully dark adapted, and use a towel or dark cloth to cover your head and eyepiece while viewing [Hack #12].

Make sure people know what you're up to. More than one urban observer has been reported to the police as a Peeping Tom. One poor guy was setting up his scope in a public park when several police cars showed up with sirens blaring, responding to reports of a terrorist setting up a missile launcher. The best way to avoid misunderstandings is to tell everyone you see what you're doing and offer them a quick look through your scope.

HACK #26 Sweep Constellations

Stay focused on one or two constellations and bag everything they have to offer while you're in there.

Inexperienced observers often work one list exclusively during a session. For example, a beginner may decide to devote an observing session to bagging as many Messier Objects as possible. The problem with working that way is that you're all over the sky, working many different objects in many different constellations.

Experienced observers, in contrast, often "sweep" constellations, which is to say they devote an entire observing session—or several sessions—to locating DSOs, multiple stars, and other objects in just one or two constellations. The advantage of working this way is that you become intimately familiar with the constellation, if only temporarily, and you are able to locate and observe more objects in less time.

The trick to constellation sweeping is to know which objects to look for in each constellation. The best way to do that is to develop your own customized list for each constellation, incorporating objects from whichever other lists you happen to be working on. As you develop your consolidated list, keep in mind your level of experience, your equipment, and your observing conditions. Many of the "advanced" lists, for example, contain objects that are difficult to find and may be invisible except with a large telescope from a very dark site.

A beginning observer might develop a consolidated list by constellation that includes objects from the following sources:

- Messier Club (*http://www.astroleague.org*)
- Binocular Messier Club (*http://www.astroleague.org*)
- Double Star Club (*http://www.astroleague.org*)
- Urban Observing Club (*http://www.astroleague.org*)

There is considerable overlap in these lists. For example, the open cluster M37 (NGC 2099) in Auriga appears on the lists for the Messier Club, the Binocular Messier Club, and the Urban Observing Club. Knowing that, you can observe M37 so as to get credit for multiple clubs. For example, if you observe M37 with your telescope from a dark site, you get credit only toward the Messier Club. If you observe M37 with your binocular from a dark site, you get credit toward the Messier Club and the Binocular Messier Club. But if you observe M37 with your binocular from your urban back-yard, you get credit for all three clubs for that one observation.

> Just because you *can* get credit for multiple lists doesn't mean that's the way you *should* do it. If you do it that way, you deprive yourself of some very nice views. M37 in a bin-ocular from a bright urban backyard is a pathetic object—at most a faint nebulosity with a few stars. With a binocular from a dark site, M37 is an impressive open star cluster. With a telescope from a dark site, M37 is one of the jewels of the heavens, a scattering of bright diamonds on a pitch-black velvet field.

A consolidated list by constellation for an intermediate observer might add objects from the following sources:

- Deep-Sky Binocular Club (*http://www.astroleague.org*)
- Herschel 400 Club (*http://www.astroleague.org*)
- Caldwell Club (*http://www.astroleague.org*)
- Globular Cluster Club (*http://www.astroleague.org*)
- Royal Astronomical Society of Canada, Finest NGC List (*http://www.rasc.ca*)
- Saguaro Astronomy Club, *110 Best of the NGC* (*http://www.saguaroastro.org*)
- *Night Sky Observer's Guide* (*http://www.willbell.com*)

Again, there is a great deal of overlap in these lists, but that's the reason for developing a consolidated list in the first place. Even if you observe all night long on every clear, dark night, it can take years to complete these lists. But there are other challenges awaiting you.

Advanced observers with large instruments and access to very dark observ-ing sites don't need our help to develop consolidated lists, but as a starting point we recommend the Astronomical League lists for the Herschel II Club, ARP Peculiar Galaxy Club, and Galaxy Groups & Clusters Club. If you are looking for a lifelong challenge, consider a truly daunting project, such as observing the Herschel 2500 objects or all of the NGC and IC objects.

That's the wonderful thing about amateur astronomy—no matter how much you've accomplished, there's always plenty left to do.

HACK #27 Maintain an Observing Notebook

Keep your log sheets, custom charts, object lists, and other reference material organized for easy access.

Being organized makes your observing sessions more productive. One major aid to organization is a proper observing notebook. We use ours to store blank observing log pages, custom-printed star charts, object lists, equipment checklists, and other reference material. We also store our completed observing log pages in our notebooks until we have time to transfer our observing records to our consolidated spreadsheets. Figure 2-30 shows Barbara recording an observation on her log sheet.

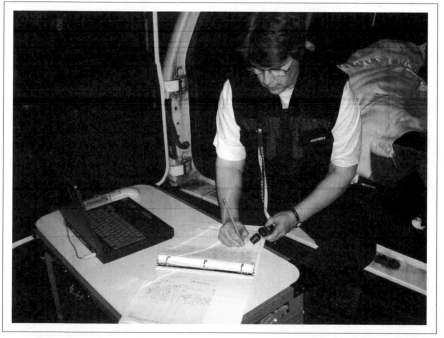

Figure 2-30. Barbara recording an observation in her notebook

Here's some advice based on how we organize our personal observing notebooks:

- Use a standard three-ring binder with clear plastic pockets on the outside covers. The pockets provide convenient places to store custom-printed star charts and other reference material you'll need frequently over the course of the evening. The plastic pockets protect the sheets from dew.

- Buy a ream of pre-punched paper, and use it to print all of your log sheets, charts, and so on.

- Buy a set of tab dividers to organize the contents of your observing notebook. Print labels for the tab dividers using a large, bold font for easy readability under dim red light. Robert's notebook currently has the following tabs, with the current in-progress observing log sheet at the front of the notebook:

  > Charts (tonight's custom charts)
  > H400/Cons (Herschel 400 ledger by constellation)
  > H400/NGC (Herschel 400 ledger by NGC#)
  > H-II/Cons (Herschel II ledger by constellation)
  > H-II/NGC (Herschel II ledger by NGC#)
  > Caldwell/Con (Caldwell ledger by constellation)
  > Urban/DSO (AL Urban Observing ledger for DSO list)
  > Lunar 100 (S&T Lunar 100 ledger)
  > Checklists
  > Reference
  > Blank Logs

- Keep at least a dozen blank observing log pages in your notebook. Even if you don't need them all for a particular session, one of your observing buddies may run short.

- Keep a dozen or so blank pages in your notebook for sketching, taking notes, and so on.

- Purchase a box of reinforcement rings and apply them to all permanent or semi-permanent pages in your notebook. Otherwise, wind and dew make torn pages inevitable, as does thumbing through your notebook in cold weather while wearing gloves.

- Protect permanent and semi-permanent pages by laminating them or covering them with transparent acetate sheet protectors.

- Standard pens often work poorly, if at all, in cold weather. Also, if dewing is severe, ink may run and make a mess of your observing log sheets. Barbara prefers a standard or mechanical pencil. Robert uses a fine-tip Sharpie pen (if it's too cold for a Sharpie to work, it's too cold for Robert to be observing...). Consider tethering your writing instrument to the notebook. It's one less thing to lose, and, as we can attest from personal experience, pens and pencils are always getting lost.

Whatever you do, banish red pens from the vicinity of your observing notebook. One night, we watched one of our regular observing buddies trying to record an observation. His pen wouldn't write, so of course he started scribbling all over his observing logs and charts, trying to get it started. It was some time before he realized that he was using a red pen, whose ink was invisible under his red LED flashlight. Duh **[Hack #11]**.

Creating and Using Observing Logs

A well-designed observing log helps you record and organize your observations. It captures detailed information about each object you observe, as well as information about the observing session itself. We designed our own log sheet to capture all of the information needed for most of the structured observing programs run by the Astronomical League. (You can download a copy of our log sheet in Microsoft Word format from *http://www.astro-tourist.net/files/observing-log.zip* and modify it for your own use.)

Any log sheet you use should capture the following session information:

Date

> We use the local date, but some observers prefer to use the UTC date. It doesn't really matter which you use, as long as you're consistent. In either case, date the page at the time you start it. Observing sessions often run past local midnight. If we have a partially completed log sheet in use at midnight, we simply continue adding entries with times 00:01 or later. When we start a new sheet after midnight, we use the actual current date for that sheet.

Lunar status

> It can sometimes be important to know the state of Luna when you made a particular observation **[Hack #64]**. You can go back and look it up on your planetarium software, of course, but it's handy to have it right on the observing log page. We generally record the rise and set time of Luna in this field, along with its percent illumination.

Page __ of __

> Some observers number their observing log pages sequentially, never resetting to page 1. We've never been that organized, so we number our observing log pages per session, always starting at 1. When we finish a session, we fill in the "of __" part for all pages. That makes it clear later if we have all of the pages or are missing some.

Location

> Log the location of the observing session. We have several regular observing sites, and so we created entries for each to save time when we fill out a log sheet. Make sure to leave a blank field for sessions that take place somewhere other than one of your regular sites.

Observer

> Note who the page belongs to. Again, we've created entries for ourselves, but we have also included a blank field for those times when one of our observing buddies runs short of log pages and needs to borrow some from us.

Observing conditions

> Record the observing conditions for the session, including temperature, wind speed, relative humidity, transparency, seeing, cloudiness, and limiting magnitude at zenith. Obviously, conditions sometimes change during a session, but we generally just record the conditions at the start of the session. Unless the change is dramatic, there's no point in trying to fine-tune your record. We rate transparency and seeing on a scale of 1 to 10, with 1 meaning extremely poor and 10 excellent. Cloud cover we rate in terms of tenths coverage, from 0/10 to 10/10, although at anything over 5/10 there's usually not much point in observing. Limiting magnitude at zenith is the faintest star we can see near zenith with averted vision after we are fully dark adapted, and it is another indication of transparency.

Your log sheet should also capture the following per-object information:

Object

> The name and/or catalog designation [Hack #15] of the object. For DSOs, we recommend using the NGC (New General Catalog) or IC (Index Catalog) number as the primary designator. For example, the Cat's Eye Nebula, a planetary nebula in Draco, is on the Caldwell List as object Caldwell-6 and on the Herschel List as object H.37-IV. When we observe that object, we log it under its official designation as object NGC 6543, noting the common name, Caldwell designation, and Herschel designation as supplementary notes. That way, we can transfer that observation to our consolidated spreadsheet and later sort the list by NGC number and determine which objects we have observed on a variety of lists, including the Caldwell and Herschel lists. We also generally note the constellation in the Notes field.

Time

> Record the time of the observation here, using either local time or UTC time. It doesn't matter which you use as long as you're consistent.

Scope

> Record the instrument you used to make the observation. We use short abbreviations, such as "10" for our 10" Dob, "17.5" for our buddy's 17.5" Dob, "90" for our 90mm refractor, "Bin" for our 7X50 binocular, "eye" for naked-eye observations, and so on.

Ocular

> Record the eyepiece you used, plus any supplemental optics. For example, one of our typical entries in this field might be "14/2X/UB," which means we used our 14mm Pentax XL eyepiece with 2X Barlow and UltraBlock narrowband filter.

Notes

> Record your detailed observation in the Notes field. You can use formal recording nomenclature or your own shorthand (as long as you'll be able to understand it later). We prefer the latter. For example, we might record an observation as "lg sp oc, 30', 25* m7/8, m6/7 rd/wh dbl jst nw ctr, sm neb" to describe a large, sparse open cluster, about 30 arc-minutes in visible extent, with about 25 stars in the magnitude 7 to 8 range and a double star with a magnitude 6 red primary and a magnitude 7 white secondary just northwest of center, with some nebulosity visible. Again, it doesn't matter what method you use, as long as you can decipher your own scratchings later.

HACK #28

Develop an Organized Logging System

Apply accounting techniques to keep track of your observations. Record it once, and you can use it forever.

Accounting records are organized in two ways. A journal records transaction details chronologically as they occur, like your checkbook register. Those transactions are later transferred to a ledger, which organizes the data into categories, such as salary income, rent payments, and so on.

Well-organized astronomers use a similar method for keeping their observing records. The log sheets you fill out during observing sessions are journals because they record transactions chronologically. Periodically, you should transfer those records to a consolidated ledger. A spreadsheet is the ideal tool for creating a ledger because it allows you to sort your observations according to different criteria. For example, if you're "constellation sweeping" in Cassiopeia and want to know which objects you've already observed there, it takes only seconds to sort your ledger by constellation. Similarly, if you're pursuing globular clusters, for example, it's easy to sort your ledger by object type and then NGC number to provide a list in NGC order of globs you've already observed.

Figure 2-31 shows an excerpt from Barbara's observing ledger, sorted first on the "D" column and then on the NGC/IC column to provide a list of the objects by NGC/IC number that she's observed on the Astronomical League Deep-Sky Binocular (D) list. (This is from an old copy of her ledger; she's since completed the DSB list.)

Observing Log
Barbara Fritchman Thompson

Date	Time	Loc	Tmp	Wind	RH	Trans	See	Cloud	LM	Name	NGC/IC	Const	Type	Scope	Ocular	Other	D	H	M	BM	C	U	DS	L	
1-Dec-2004	20:05	Bull	40F	1-3	40	7/10	8/10	0/10	6.0		Cr 399	Vel	OC	7X50	n/a	n/a	X								Coathanger
5-Nov-2004	20:37	Bull	45	1-5	45	8/10	8/10	0/10	6.0		CR463	Cas	OC	7X50	n/a	n/a	X								large cluster of s
5-Nov-2004	19:20	Bull	45	1-5	45	8/10	8/10	0/10	6.0		IC4665	Oph	OC	7X50	n/a	n/a	X								large, bright loos
1-Dec-2004	18:30	Bull	40F	1-3	40	7/10	8/10	0/10	6.0		IC4756	Ser	OC	7X50	n/a	n/a	X								large, loosely sc
1-Dec-2004	18:50	Bull	40F	1-3	40	7/10	8/10	0/10	6.0		Kemble 1	Cam	OC	7X50	n/a	n/a	X								large, cascading
16-Oct-2004	21:205	Bull	56	8-10	40	8/10	8/10	0/10	6.0		Mark 6	Cas	OC	7X50	n/a	n/a	X								faint fuzzy patch
5-Nov-2004	20:55	Bull	45	1-5	45	8/10	8/10	0/10	6.0		Mel 20	Per	OC	7X50	n/a	n/a	X								large cluster of r
9-Sep-2004	22:30	Pitts	70	0	75	8/10	7/10	0/10	6.0		NGC0129	Cas	OC	7X50	n/a	n/a	X								small, rich field c
5-Nov-2004	20:05	Bull	45	1-5	45	8/10	8/10	0/10	6.0		NGC0253	Scl	SG	7X50	n/a	n/a	X								straight on edge
9-Sep-2004	23:30	Pitts	70	0	75	8/10	7/10	0/10	6.0		NGC0457	Cas	OC	7X50	n/a	n/a	X								very faint OC wit
9-Sep-2004	23:35	Pitts	70	0	75	8/10	7/10	0/10	6.0		NGC0663	Cas	OC	7X50	n/a	n/a	X								large cluster of s
16-Oct-2004	20:25	Bull	56	8-10	40	8/10	8/10	0/10	6.0		NGC0869	Per	OC	7X50	n/a	n/a	X								The Double Clus
16-Oct-2004	20:25	Bull	56	8-10	40	8/10	8/10	0/10	6.0		NGC0884	Per	OC	7X50	n/a	n/a	X								The Double Clus
5-Nov-2004	20:15	Bull	45	1-5	45	8/10	8/10	0/10	6.0		NGC1342	Per	OC	7X50	n/a	n/a	X								medium sized c
5-Nov-2004	20:34	Bull	45	1-5	45	8/10	8/10	0/10	6.0		NGC1528	Per	OC	7X50	n/a	n/a	X								very large, loose
5-Nov-2004	20:53	Bull	45	1-5	45	8/10	8/10	0/10	6.0		NGC1582	Per	OC	7X50	n/a	n/a	X								bright, loosely fo
5-Nov-2004	21:23	Bull	45	1-5	45	8/10	8/10	0/10	6.0		NGC1746	Tau	OC	7X50	n/a	n/a	X								large, faint cluste
5-Nov-2004	21:44	Bull	45	1-5	45	8/10	8/10	0/10	6.0		NGC1807	Tau	OC	7X50	n/a	n/a	X								smaller and less
5-Nov-2004	21:44	Bull	45	1-5	45	8/10	8/10	0/10	6.0	H4-7	NGC1817	Tau	OC	7X50	n/a	n/a	X	X							more impressive
1-Dec-2004	19:30	Bull	40F	1-3	40	7/10	8/10	0/10	6.0	H44-5	NGC2403	Cam	SG	7X50	n/a	n/a	X	X							fuzzy patch with
6-Aug-2004	22:50	Bull	65	3-5	53	7/10	4/10	0/10	6.0	H7-7	NGC6520	Sgr	OC	7X50	n/a	n/a	X	X							large grouping of
5-Nov-2004	19:10	Bull	45	1-5	45	8/10	8/10	0/10	6.0	H72-8	NGC6633	Oph	OC	7X50	n/a	n/a	X	X							bright, scatter gr
5-Nov-2004	19:45	Bull	45	1-5	45	8/10	8/10	0/10	6.0		NGC6709	Aql	OC	7X50	n/a	n/a	X								very small, dim s
6-Aug-2004	22:55	Bull	65	3-5	53	7/10	4/10	0/10	6.0		NGC6716	Sgr	OC	7X50	n/a	n/a	X								large open clust
9-Sep-2004	21:05	Pitts	70	0	75	8/10	7/10	0/10	6.0		NGC6819	Cyg	OC	7X50	n/a	n/a	X								rich field of stars
1-Dec-2004	19:55	Bull	40F	1-3	40	7/10	8/10	0/10	6.0	H18-7	NGC6823	Vel	OC	7X50	n/a	n/a	X	X							large grouping of
9-Sep-2004	21:14	Pitts	70	0	75	8/10	7/10	0/10	6.0		NGC6910	Cyg	OC	7X50	n/a	n/a	X								small cluster of s
9-Sep-2004	22:13	Pitts	70	0	75	8/10	7/10	0/10	6.0		NGC6934	Del	OC	7X50	n/a	n/a	X								looks like a smal
5-Nov-2004	19:35	Bull	45	1-5	45	8/10	8/10	0/10	6.0		NGC6934	Del	OC	7X50	n/a	n/a	X								tiny round glob
1-Dec-2004	20:00	Bull	40F	1-3	40	7/10	8/10	0/10	6.0	H8-7	NGC6940	Vel	OC	7X50	n/a	n/a	X	X							large bright rich f
9-Sep-2004	21:22	Pitts	70	0	75	8/10	7/10	0/10	6.0		NGC7063	Cyg	OC	7X50	n/a	n/a	X								open, loose clus
9-Sep-2004	22:45	Pitts	70	0	75	8/10	7/10	0/10	6.0		NGC7160	Cep	OC	7X50	n/a	n/a	X								bright patch of st
1-Dec-2004	20:35	Bull	40F	1-3	40	7/10	8/10	0/10	6.0	H53-7	NGC7209	Lac	OC	7X50	n/a	n/a	X	X							bright rich cluste
9-Sep-2004	23:12	Pitts	70	0	75	8/10	7/10	0/10	6.0		NGC7235	Cep	OC	7X50	n/a	n/a	X								compact cluster
1-Dec-2004	20:35	Bull	40F	1-3	40	7/10	8/10	0/10	6.0	H75-8	NGC7243	Lac	OC	7X50	n/a	n/a	X	X							much more scat
16-Oct-2004	20:56	Bull	56	8-10	40	8/10	8/10	0/10	6.0		NGC7789	Cas	OC	7X50	n/a	n/a	X								Large, beautiful c
9-Sep-2004	23:20	Pitts	70	0	75	8/10	7/10	0/10	6.0		Stock 2	Cas	OC	7X50	n/a	n/a	X								Large OC asteri
1-Dec-2004	18:40	Bull	40F	1-3	40	7/10	8/10	0/10	6.0		Stock 23	Cam	OC	7X50	n/a	n/a	X								looks like a smal
5-Nov-2004	20:49	Bull	45	1-5	45	8/10	8/10	0/10	6.0		TR 2	Per	OC	7X50	n/a	n/a	X								tight faint small c
5-Nov-2004	20:05	Bull	45	1-5	45	8/10	8/10	0/10	6.0		TR 3	Cas	OC	7X50	n/a	n/a	X								large, rich field o
24-Aug-2003	21:20	Urban	75	0	70	8/10	8/10	0/10	4	M13	6205	Her	GC	10	30	n/a								X	Resolves stars f

Figure 2-31. A section of Barbara's observing ledger

Each row of the list records the following information about one observation:

Date through LM

These columns record the date and time of each observation, the location, the weather conditions, transparency, seeing, cloudiness, and limiting magnitude at zenith. Because the date and time are recorded for each object, it's easy to locate the original observing record simply by retrieving the log sheet for the observing session that took place on that date.

Name

The common name of the object, if any. For example, M13 (Messier Object 13, at the bottom of the list) is the common name of NGC 6205. This field can also be used to record secondary designations for the object. For example, the names in the form H4-7 are the Herschel designations for these objects. These are used for the Astronomical League Herschel 400 list, which Barbara is also working on.

NGC/IC

Thousands of galaxies, star clusters, nebulae, and other deep-space objects are assigned an *NGC* (*New General Catalog*) or *IC* (*Index Catalog*) number. Because most objects have an NGC/IC designation, it's convenient to use this designation as the key field. Many objects are known under various common names. For example, NGC 7009 in Aquarius is also known as the Saturn Nebula, Bennett 127, Caldwell Object 55, and Herschel Object IV.1 (4-1). But because all of these other lists are always cross-referenced by NGC/IC number, if you always record the NGC/IC number, you'll always be able to keep track of which objects you've logged and which belong to such lists as the Messier, Herschel 400, Caldwell, and so on. Some objects have no NGC or IC designation. In that case, simply record the designation by which the object is known most commonly, such as Kemble 1 or Stock 23.

Constellation

Most experienced astronomers "constellation sweep" [Hack #26]. In other words, rather than sequentially pursuing all objects on, say, the Herschel 400 list, they focus on one or a few constellations at a time, logging all objects in that constellation regardless of which lists they may be on. If you record the constellation each time you log an object, you can later sort your ledger by constellation to determine which objects in that constellation you've already logged.

Type

This is the type of object logged, for example, an open cluster (OC), globular cluster (GC), galaxy (SG, EG, etc.), planetary nebula (PN), supernova remnant (SNR), emission nebula (EN), multiple star (MS), variable star (VS), and so on. Capturing this information for each object allows you to go back later and sort by object type. For example, you may plan to work planetary nebulae in Sagittarius one evening and want to know which ones you've already logged.

Scope/ocular

Record the instrument and eyepiece you used for each observation.

Other

Record supplementary information, such as whether a Barlow or filter was used.

D, H, M, BM, C, U, DS, L

Barbara is currently working on or has completed the lists for several Astronomical League observing clubs, including the Deep-Sky Binocular (D), Herschel 400 (H), Messier (M), Binocular Messier (BM), Caldwell (C), Urban Observing (U), Double Star (DS), and Lunar Club (L).

Having a field for each of these lists makes it easy for her to sort her ledger by AL club to determine which objects she has already observed on the lists in question. For example, several of the objects she's logged for the Deep-Sky Binocular (D) list also appear on the Herschel 400 (H) list, so she marked those objects as observed for both clubs when she updated her ledger.

Notes

Write your detailed comments about each object. Take as much room as you need here. If you sketched or photographed the object, note the location of that image. You can also use this field for comparative comments, e.g., differences in appearance of an object from the last time you observed it, notes on how different filters affect the appearance, and so on.

Before you create your ledger, give some thought to which data you want to capture. For example

- If you're a dedicated double star observer, you might add columns for separation and position angle, primary and companion magnitudes and colors, and so on.

- If you observe variable stars, you might add columns for magnitude and period.

- If you observe mostly DSOs, you might add columns for magnitude and surface brightness.

- If you're a Lunar observer, you might add columns for the dimensions of the features you log and the feature type (rille, crater, etc.).

Nothing says you have to complete all columns for all objects. If you use all of the columns we've mentioned, many will not apply to all of the objects you observe, but that doesn't matter. What does matter is that you record all of your observations to your ledger. In years to come, your ledger will provide a history of your observing career.

HACK #29 Plan and Prepare for a Messier Marathon

Locate, observe, and log all 110 Messier Objects in one night.

The 18th-century French astronomer Charles Messier (pronounced MEZZ-ee-yay) lived for finding comets. Every clear night, he was out with his telescopes, trying to be the first to identify each new comet. But Messier had a problem. It's very difficult to tell the difference between a dim comet and many other dim astronomical objects. The only way he could know for sure that he'd found a new comet was to observe the object over several nights.

Over that period, a comet moves relative to the background stars, while other objects remain fixed in position.

On the night of 28 August 1758, Messier was looking for comet Halley on its first predicted return when he happened across a dim object in the constellation Taurus that looked very much like a comet. He studied that object intently over the next several nights, only to find that it didn't move against the background stars. Disgusted at the waste of time, Messier decided to create a list of these obnoxious fixed objects that masqueraded as comets so that he and other observers could avoid wasting time looking at them in the future. He carefully logged the position of the object, and recorded it as Messier Object 1, or M1. (That object, now known as the Crab Nebula, is one of the most famous astronomical objects.)

Over the years, Messier added to his list of objects until it eventually included 103 objects, one of which was a duplicate. The irony is that in creating his list of objects to be avoided by comet hunters, Messier unintentionally created a list of the finest objects in the night sky. (The limitations of Messier's equipment meant that he could see only the brightest objects, which are easy and rewarding targets for observers with modern equipment.) Messier would have wanted to be remembered for the many comets he discovered; instead, that aspect of his career is almost entirely forgotten, and he is famous nowadays for his list of objects.

The Messier List now contains 109 or 110 objects—depending on whose argument you find most persuasive. The "extra" objects were all observed by Messier, but not recorded in his list for one reason or another. Nearly all modern amateur astronomers who are interested in deep-sky objects begin with the Messier list. Many amateur astronomers take years to observe all of the objects on the Messier list. Others complete it faster.

For a few nights a year, centered on a new moon date in mid-March to early April, it is at least theoretically possible to view all 110 Messier Objects in one night, starting at dusk and ending at dawn. Those brave (or foolish) enough to undertake this quest call it a Messier Marathon (MM). And a Marathon it is—locating, observing, and logging an average of one object every five minutes or so over a span of about 10 hours.

Planning Your Messier Marathon

Planning is the first key to running a successful Messier Marathon. It's important to think through everything in advance, because you'll have no time to spare when you actually begin running the Marathon.

Choose a date. The first thing to decide is when you'll run your Marathon. In one sense, there's not much to decide, because a full Messier Marathon is practical only at or near a new moon that occurs from mid-March until early April, and the best opportunity to log all 110 objects occurs only when the new moon is in late March. Table 2-7 shows the new moon dates and Messier Marathon weekends for the next 10 years.

> New moon times are specified in days and tenths of days in Universal Time (UT). Depending on your location, the new moon may occur on the prior or next day local time. For example, if you are in the Eastern Standard Time zone (EST), the new moon in the early morning of 6 April 2008 UT occurs in the late evening of 5 April 2008 local time.

Table 2-7. Optimum Messier Marathon dates for 2006–2015

Year	New moon(s)	Primary weekend	Secondary weekend(s)
2006	29.4 March	25 March	1 April
2007	19.1 March	17 March	(none)
2008	7.7 March 6.2 April	8 March	29 March 6 April
2009	26.7 March	28 March	21 March
2010	15.9 March	13 March	20 March
2011	4.9 March 3.6 April	2 April	5 March
2012	22.6 March	24 March	(none)
2013	11.8 March	9 March	16 March
2014	30.8 March	29 March	(none)
2015	20.4 March	21 March	(none)

Not all of these Marathon nights are equally desirable:

- For Marathon dates earlier than mid-March, it is nearly impossible to view all 110 objects because the late-rising objects, particularly M30, are lost in the morning twilight.
- For Marathon dates later than late March, it is nearly impossible to view all 110 objects because the early setting objects, particularly M74, are lost in the evening twilight.

Accordingly, the best years to run a Messier Marathon will be 2006, 2009, and 2014. In 2007, 2010, 2012, and 2014, it may be possible to log all 110 objects, but it will be very difficult because the new moon occurs too early in March and not early enough in April. In 2011, it may be possible to log all 110 objects in early April, but the early-setting evening objects will be extremely difficult. In 2008 and 2013, it will be nearly impossible to log all 110 objects because the new moon occurs much too early in March and much too late in April.

> Latitude is also a factor because evening twilight ends later and morning twilight begins earlier at higher latitudes. The ideal latitude for running a Messier Marathon is about 25°N. If you are at 40°N or higher, it is very difficult or impossible to bag all 110 objects, even if the new moon occurs ideally in late March.

Group Marathons are scheduled on weekends—usually a Saturday night—because most people aren't able to get away during the week. Unfortunately, the best weekend date may be too early or too late in the season. It may also be several days from the new moon, which means moonlight may intrude on your Marathon. Moonlight interferes with finding objects, so if your schedule permits and you don't mind Marathoning by yourself (or you have some like-minded friends) it's best to run your Messier Marathon on the date the new moon actually occurs, or at most one or two days either side of that date.

Weather may also interfere, of course. March and early April have predictably unpredictable weather in many parts of the world. The best chance to complete your Marathon despite weather is to bracket the new moon date. If possible, reserve the site and your lodging for two or three nights before and after the new moon. If the weather forecasts are terrible for the early part of the period, you needn't actually travel to the site early. If the weather is clear and you complete your Marathon early or on time, you needn't stick around if the forecasts are bad for the remainder of your scheduled stay. You may be lucky and get in an evening or two of observing after completing your Marathon. You may even be able to run a second Marathon if you didn't bag all 110 objects on the first go-round.

> If you find yourself clouded out one evening, but the forecast is for clearing weather, do a "calendar-day Messier Marathon." That is, begin your MM at midnight, knock off at dawn, and then come back that evening to log more objects from nightfall until midnight. You can't really claim to have done the traditional Messier Marathon—that 15-hour nap in the middle splits your effort into two half-Marathons—but you will have logged all or most of the Messier Objects in a single day, which is a major accomplishment in itself. And, if the weather remains clear that night, there's no reason you have to stop at midnight. You can continue working, and complete a traditional Messier Marathon at dawn the following morning.

Choose your observing site. A good observing site is critical to the success of a Messier Marathon. Three primary considerations define a good Messier Marathon site:

Unobstructed horizons

Of the three, by far the most important consideration is having unobstructed horizons. To complete the Marathon, you must observe several objects that set early in the evening twilight and several more that rise late in the morning twilight. Your first goal should be to find an observing site with eastern and western horizons as near 0° as possible. Each degree of obstruction costs you about four minutes that you can't afford to lose. For example, if the site has a 15° obstruction on the western horizon, evening objects set one hour earlier, while the sky is still too bright to observe them. The same holds true, in reverse, for obstructions on the eastern horizon. The late-rising morning objects are difficult enough to observe if you have a 0° eastern horizon. Every degree of obstruction on your eastern horizon makes you wait four extra minutes for the object to rise, as the sky grows brighter with each passing minute.

The southern horizon is also important, although not as critical as the eastern and western horizons unless you are observing from a high northerly latitude. Southerly constellations such as Sagittarius and Scorpius culminate well after daylight on Marathon morning and are still quite low at the time you need to observe them. A good Marathon site will therefore have a southern horizon obstruction as close to 0° as possible, and certainly not more than 4° or 5°. The northern horizon is the least critical, although ideally it should be obstructed no more than a few degrees. Don't necessarily rule out an otherwise-ideal site that has an obstructed northern horizon, particularly if you can eliminate the obstruction easily by moving your scope temporarily.

Freedom from light pollution

An ideal Marathon site has pitch-black skies, no light domes, and no local lights. Alas, such sites are rare nowadays, particularly on the US east coast and other heavily built-up areas, so you may have to compromise. Look for a site with skies that are no worse than Bortle 4 (*http:// cleardarksky.com/csk*), and try hard to find a Bortle 3.5 or better site. Avoid sites with intrusive light domes, particularly to the south. Although there's nothing you can do about general light pollution or light domes, don't rule out an otherwise acceptable site because of local lights. You may be able to use screens or other workarounds to block local lights. If you ask nicely and explain why, people are often willing to turn them off for you.

High elevation

High elevation is desirable for several reasons. First, the air blanket between you and the night sky is thinner, which makes it easier to locate and observe dim objects. Second, being at high elevation gets you above air pollution, humidity, low-lying clouds and haze, and other muck, which makes it much more likely the sky will be transparent near the horizons. Third, high elevation minimizes the effects of the light domes produced by nearby cities and other built-up areas. (Even a small town 20 miles or more away can produce a noticeable light dome at a dark site.)

An acceptable observing site is likely to be at least 20 miles from the nearest town of 10,000 or more population, and several miles from smaller population centers. For many of us, that means having to travel to find an acceptable site and to find lodging for the duration of our stay. Rather than simply drive around randomly, hoping to stumble across a good site, contact your local astronomy club and astronomy clubs in the area where you propose to run your Marathon. You may find that other amateur astronomers have already done all the work for you and can suggest the best sites in the area. For that matter, they may already have scheduled a Messier Marathon that you can join.

Once you've chosen a site, particularly if it's on someone else's recommendation, make a preliminary site visit to do a detailed evaluation. If you have a scope with digital setting circles (or can borrow one), take it along and use it to map the horizon from the actual observing pad. Level the scope, point it straight north, and put the finder crosshairs at the top of the highest obstruction in view. Record the obstruction at 0°. Rotate the scope 5° or 10° degrees to the east, and repeat the elevation measurement all the way around until you return to 0°. You'll need this horizon map when you plan your observing sequence and schedule.

> If you don't have access to a scope with DSCs, you can temporarily turn any scope into a makeshift theodolite using a large protractor, a compass, a spirit level, a piece of thread, and a small weight. Use the compass to determine azimuth (don't forget to adjust for local declination), the level to determine baseline altitude, and the protractor with the weighted thread to determine the altitude of obstructions. With care, you can make measurements accurate to a degree, which is more than sufficient. You may need one person to do the sightings while another notes and records the elevations at each point.
>
> Alternatively, you can use a digital camera to produce a panoramic photo that covers the entire horizon. From that photo you can closely estimate the altitude of obstructions around the horizon.

Make sure to get permission to use your chosen site on the dates you plan to be there. If you will be staying at a hotel, make lodging reservations well in advance. Spend some time during your initial site visit touring the immediate area to locate restaurants and similar conveniences. Don't assume they're open all the time. Restaurants and other facilities near remote observing sites may close down for part of the year. Sometimes, a hotel is open year-round, but its restaurant closes during the off-season. Ask. Plan ahead for medical emergencies. Make sure you know where the nearest hospital is (and how to get there), how to contact emergency services, and how to describe the route to your site.

Make sure you know how to get to the site yourself. More than one would-be Marathoner has been embarrassed by his inability to find the site the day of the Marathon. Back roads can be confusing, and the last thing you want is to be driving around trying to find your observing site as the clock ticks down. If the site is difficult to locate, map it out during your preliminary visit. Record distances and turns, or use a GPS to set waypoints. And don't overlook finding your way out. Even if you have detailed directions for finding your way in, it's easy enough to become lost trying to find your way out, particularly just after dawn the morning after.

Develop your own sequence and schedule. It's important to develop a sequence and a schedule for the Messier Marathon. The sequence specifies the order in which you plan to observe the objects. The schedule specifies the time you plan to observe each object. Both are important because during the Marathon you are attempting to observe 110 objects, all of which are moving targets.

- The sequence is important because some objects set early or rise late, leaving you only a short window to observe them. Other objects rise early and set late, and can be observed at any time over a period of many hours. The sequence accommodates this by assigning first priority to observing time-critical objects. The sequence is most important for the dozen or so objects at the start of the Marathon and the final half dozen objects at the end.

- The schedule is important because it helps keep you on track. By assigning a specific time to each object, you know at a glance whether you are behind, ahead of, or on schedule. Without a schedule, you might spend so much time observing a few earlier objects that you run out of time to observe later ones. When you're ahead of schedule, it's nice to know that you can relax a bit without worrying about missing later objects. It's good to know when you're behind schedule, too, because that allows you to push harder while there's still time to catch up.

The best sequence is a personal matter. It takes into account not just the rise, culmination, and setting time of the objects, but your own preferences, the horizons of your site, and so on. The best way to develop a personalized sequence is to start with one of the many published sequences and modify it to suit your own situation. For example, a published sequence may assume 0° horizons all around, and have a later-setting object later in the sequence than one that sets earlier. But at your site, the first object may in fact set earlier because of your horizons. Start with one of the published sequences that appeals to you, and check it against your own situation by using your planetarium software **[Hack #64]** to verify the local times for rise, culmination, and set for each object, taking your horizons into account.

Here are some good sources for standard sequences:

- Harvard C. Pennington. *The Year-Round Messier Marathon Field Guide.* Willmann-Bell, 1997. This is the best book we've found for Marathoners. It covers every detail of planning, preparing for, and running a Messier Marathon, and includes sequences not just for the main Marathon, but for mini-Marathons every month of the year. Pennington provides detailed star charts, with Telrad circles, to help you locate each object quickly. This book is available only from the publisher directly, Willmann-Bell (*http://www.willbell.com*), and it should be a part of any Marathoner's equipment.

- Don Machholz. *The Observing Guide to the Messier Marathon: A Handbook and Atlas.* Cambridge University Press, 2002. This is a renamed second edition of Machholz's original *The Messier Marathon Observer's Guide.* It includes more background information than Pennington's book about Charles Messier and his contemporaries, Messier's quest for comets, the Messier catalog itself, the history of the Messier Marathon, and so on, but we prefer Pennington's book as an actual observing guide.

- The SEDS Messier Marathon Search Sequence page includes links to numerous sequences and other resources (*http://seds.lpl.arizona.edu/messier/xtra/marathon/marathon.html#sequence*).

Treat your sequence as tentative rather than absolute, and be prepared to depart from it as necessary while you are actually running the Marathon. Something as simple as an errant cloud bank may make it necessary to change your observing order, but if you depart from your sequence make sure it's for good reason. On Marathon night, it's easy to convince yourself to change your sequence. Just remember that you put a lot of work and thought into developing the sequence—or at least you should have—so an on-the-spot decision may lead you astray.

Nearly every Marathoner uses a planned sequence, but many fail to attach a schedule to that sequence. We think a schedule is essential to help you stay on track and to avoid missing objects. You can make your schedule as detailed as you wish. We schedule down to the individual object level, but there is an argument of simplicity to be made for scheduling at the level of small groups.

Whichever way you do it, sit down and think through the list. Try to determine how much time you need to spend on each object or group, and then assign it a time or range of times within the sequence that reflects how long you expect to need to find it. It's helpful to have experience in observing all of the objects because different people "trip over" different objects. For example, one person may be able to locate the planetary nebula M76 in seconds, every time. Another equally skilled observer may have practiced finding M76 repeatedly, but may still need five minutes to locate it, every time. It's just a personal quirk, but a few of those can wreak havoc on a schedule if they're not taken into account.

Preparing for Your Messier Marathon

Preparation is the second key to running a successful Messier Marathon. The Messier Marathon is a race against time, so the better prepared you are, the more likely you are to succeed. You can begin preparing at any time before Messier Marathon night, but the best time to begin preparing for the next Messier Marathon is immediately following the last Messier Marathon. We suggest you do the following during the period leading up to your first attempt at a Messier Marathon.

One-year lead-up. During the several months to a year before the Messier Marathon, take the following steps to prepare for Marathon night.

Observe all Messier objects
> The last thing you want on MM night is to be looking for an object you've never seen before. In the months leading up to the Marathon, you should observe all 110 objects, ideally several times each. Get familiar with the appearance of each object and where it is positioned within its constellation and relative to other constellations and guidepost stars. In particular, become intimately familiar with the galaxies in the Coma/Virgo cluster, which Marathoners call the "Coma/Virgo Clutter." Many Marathoners have had their hopes dashed by becoming lost in Coma/Virgo. Practice locating and viewing objects that are setting low in the evening twilight and objects that are rising low in the morning twilight.

Memorize as many objects as you can

Memorize the locations of as many objects as possible to save time on MM night. For example, you should be able to locate all three objects in Orion (the nebulae M42, M43, and M78) from memory, without referring to your charts. That allows you to locate, view, and log three Messier Objects in only a minute or two, saving time for more difficult objects. Memorize other easy objects, such as the three Auriga open clusters (M36, M37, and M38), which you can bag with your binocular in 30 seconds flat. With some effort, you should be able to memorize the locations of at least 30 to 50 objects. Being able to pick off those objects quickly eliminates a great deal of time pressure when you are searching for other objects.

> There is a specialized Messier Marathon called M³, or M-cubed, for the Memory Messier Marathon. You run this Marathon without charts or a computer, without a sequence list or a schedule, in fact without anything except your binocular, telescope, and log sheets.

Secure your MM observing site

Unless your regular observing site is extraordinarily good, both in terms of darkness and horizons, you'll probably need to run your MM from a special site, which may be some distance from your regular haunts. Well before MM night, visit potential sites to determine their suitability in terms of horizons, light pollution, and local lights. Ideally, run at least one observing session at the prospective site. Make sure you have the site reserved for MM night and, if necessary, make lodging reservations well in advance.

Finalize your sequence and schedule

There is no one best MM observing sequence. The best sequence depends on many factors, including the date of your session, the latitude of your observing site, local twilight times, horizon obstructions, and so on. Start with one of the published sequences, but modify it to suit your own conditions. Pay particular attention to the early evening objects and the final morning objects, for which twilight times and horizons are critical. Use your planetarium software to determine what the elevations of critical objects will be on MM night and how those elevations compare with the actual horizons at your chosen site.

Develop a proposed schedule, deciding how much time to devote to each object, taking into account the difficulty of each object based on its position and the status of other objects nearby in the sequence. For example, if you are running a Marathon in early April, the evening objects set very early. You don't want to spend 15 minutes trying to locate M74 setting in the evening twilight if that means you'll miss your opportunity to log several other easier evening-setting objects. Instead, sequence and schedule the easier objects first. Once you have them logged, you can return to M74.

Learn your equipment

MM night is the worst possible time to break in new equipment. Decide what equipment you will use well before the MM, and practice with it extensively. If you have your eye on a new right-angle, correct-image finder, for example, buy it now and get comfortable with it before MM night. If you don't have a Telrad unit-power finder **[Hack #53]**, get one as soon as possible and learn to use it. A Telrad is, if not essential, at least extremely useful for a Messier Marathon.

> Marathon night is likely to be cold except at the most southerly latitudes. Don't underestimate the sheer physical demands of a long observing session in cold weather or the importance of cold-weather gear and preparations. Test your readiness for a long, cold observing session well before Marathon night by actually doing a long, cold-weather observing session. If your gear or preparations are inadequate, the time to find out is before Marathon night **[Hack #4]**.

Prepare your charts

Many amateurs have several sets of charts, ranging from overview charts like the *Orion Deep Map 600* to mag 6 star atlases, and on to more detailed charts like *Sky Atlas 2000.0*, *Uranometria*, and the *Millennium Star Atlas*. None of these are ideal for a Messier Marathon. All provide too little or too much detail, and none focuses on the problem at hand—locating Messier objects quickly.

Rather than use any of these standard charts, we recommend using your planetarium software to produce custom charts for each object, one per page. Set your planetarium software to the coordinates of the observing site. For each object, reset the planetarium time to the approximate time you expect to be locating the object, so that the chart is oriented as the sky will be at the time you observe it. Pre-plan your Telrad placements and star hops, and print those circles right on your custom charts.

Run at least one practice MM

Although it's possible to log all Messier objects in one night only during the March/April MM window, you can do a "mini Messier Marathon" on a night near a new moon at any time of year. The number of Messier objects visible varies with latitude and from month to month, with minima in mid-May to mid-June and early September, but at least 88 objects are possible any dark night from latitudes as high as 40°N. The best way to practice for a MM is to run a MM, so we suggest you run at least one all-night practice MM session in the months leading up to MM night.

One-month lead-up. The new moon weekend a month prior to the MM is a perfect time to do a final trial run, ideally a full practice Marathon. Morning objects are still impossible, but evening objects are higher than they will be on Marathon night. Practice bagging them in the evening twilight. Do at least one full run through the Coma-Virgo Clutter. It will probably be as cold or colder than Marathon night, so now is the time to do any tweaks necessary to your cold-weather gear and preparations.

One-week lead-up. Now is the time for final preparation. Check your equipment. Make sure your scope is collimated and your optics are clean. Replace the batteries in your flashlights, Telrad, mount motors, and so on. Get in at least one final practice observing session, focusing your attention on whichever objects or groups are difficult for you.

One-day lead-up. Relax. You're as prepared as you're ever going to be. There is such a thing as over-preparing. Get your equipment packed up and checked off against your checklists. Charge your notebook, cell phone, and other equipment, and don't forget to charge your spare batteries. Say goodbye to your family and make sure your life insurance policies and will are up to date. It's almost show time.

HACK #30 Run a Messier Marathon

Look over Robert's shoulder as he tries to log all 110 Messier Objects in one night.

No matter how carefully you prepare, the reality of Marathon night is likely to be different from what you expected. As the military strategist Karl von Clausewitz observed, no battle plan survives first contact with the enemy. Few Messier Marathons are any different. In this hack, we'll try to give you a flavor for the reality of running a Messier Marathon.

Robert ran his first (and, to date, only) Messier Marathon on 1/2 April 2003 from an observing site at a private lodge near the Blue Ridge Parkway in southern Virginia. Two other club members, Steve Childers and Paul Jones, participated in this First Annual Winston-Salem Astronomical League Messier Marathon. None of us had done a Messier Marathon previously. Steve used his 10" Dob, with a 27mm Panoptic and 14mm and 10.5mm Pentax XL eyepieces. Paul used his binocular and his 8" SCT with 32mm Tele Vue

Ploessl and 14mm Pentax XL eyepieces. Robert used his binocular and his 10" Dob with 14mm and 40mm Pentax XL eyepieces and a 2X Barlow.

The site was quite dark for the Eastern U.S., about Bortle 3.5 (*http:// cleardarksky.com/csk*). The horizons were excellent from E through NNW, but obstructed from 2° to as much as 34° by the lodge itself and a treeline from N through NE. The main horizon obstruction was nearly dead north, and if you have to have an obstruction during a Messier Marathon, that's where you want it. Weather conditions were excellent.

Based on that experience, here are Robert's comments and advice about running your own Marathon.

Final Preparations (Afternoon–19:30)

Make every effort to arrive at the observing site by late afternoon. Unpack, check your gear, and choose where to set up. If the site is secure, set up your scope now. If local lights are likely to be a problem, set up to avoid them as much as possible. There was one streetlight located a couple hundred feet west of our observing site, but no other local lights. We set up our scopes in line with another phone pole and our vehicles to block the streetlight.

Take a nap until dinner time. If you haven't already done so, set up your scope no later than 18:30 so that it will be cooled down and ready. Plan dinner to end no later than 19:15, including clean up. From 19:15 to 19:30, do final equipment preparation, get your charts out and ready, and so on. Check your finder alignment on Sirius in the south or Capella high in the northwest sky. Put your low-power, wide-field eyepiece in the focuser— Robert used a 40mm Pentax XL that provides a 2° true field in his 10" Dob—and get it focused. Take a deep breath.

Group 1: Early Evening Objects (19:30–20:30)

There's no dipping your toe in the water for a Messier Marathon. You have to hit the ground running to bag all or even most of the early evening group, shown in Table 2-8. Robert used an unconventional sequence for this group, basing it on the order the objects become visible in the growing darkness. Some of these objects are bright and easy, but several are dim and fiendishly difficult to find and see in the evening twilight.

Table 2-8. Early evening objects

Seq #	Time	Object	Seq #	Time	Object
1	19:40	M45	9	19:57	M77
2	19:42	M42	10	19:59	M33
3	19:42	M43	11	20:00	M34

Table 2-8. Early evening objects (continued)

Seq #	Time	Object	Seq #	Time	Object
4	19:46	M52	12	20:03	M32
5	19:46	M103	13	20:03	M110
6	19:48	M31	14	20:06	M79
7	19:50	M78	8	missed	M74
8	missed	M74	15	20:28	M76

Figure 2-32 shows the western horizon at the start of the Marathon. The Pleiades (M45) is the first object to appear as the sky darkens, and it should be an easy naked eye and binocular object within a few minutes after 19:30. By the time you've logged M45, Orion's belt should be visible. Use your finder to center Orion's sword and get M42 and M43 in the eyepiece at low power. By that time, the open clusters M52 and M103 in Cassiopeia should be relatively easy binocular objects. Log them, and then go for M31 with your binocular. If you can't get M31 with your binocular, put your scope on it with moderate power. You may also be able to see M32. If so, log it, but don't waste time trying for M110. The sky isn't dark enough at that point. It's also worth a quick look at this point to see if M33 is visible, but don't waste much time on it. Once you have M31, move back to Orion and place your Telrad or finder to locate M78, just off Orion's belt near the line from Alnitak to Betelgeuse. M78 should be visible in your eyepiece at moderate power.

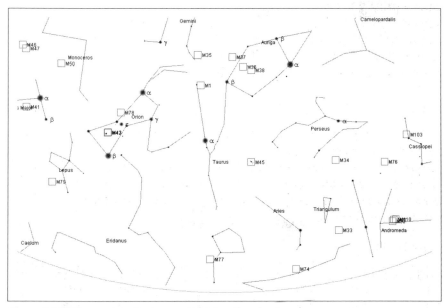

Figure 2-32. The western horizon as the Marathon begins

M74 is by far the hardest of this group, particular for late March or April Marathons. It's a dim galaxy that's very near the horizon while the evening twilight is still bright. Your best chance is to use your optical finder to follow the line from Hamal (α-Ari, 2.0m) to Sharatan (β-Ari, 2.6m) to the 3.6m star η-Psc and then move your scope about 80 arcminutes ENE to center M74. Once you are certain M74 is centered in the eyepiece, change to a moderate- to high-power eyepiece to bring it out against the bright background. (Robert was unable to see M74 even at high power, despite being certain it was in the eyepiece.) Spend at most five minutes looking for M74. If you can't get it, move along to the other objects to make sure you get them before they set.

M77 is the next object. It's a small galaxy with relatively high surface brightness that's easy when it's at high altitude, but difficult on Marathon night because it's only at 6° or so altitude when you begin searching for it. Fortunately, M77 is located in the same eyepiece field as the 4th magnitude star δ-Cet, which makes it relatively easy to locate. Use moderate to high power to bring out M77 against the sky background.

By now, it's nearly 20:00, and the sky has darkened. If you haven't already logged M74, give it one last try, but don't waste too much time on it. If you miss it, you'll be in good company. If you haven't logged M33, give it another try. M33 may be easier with your binocular than with your scope. Then use your binocular to pick up the open cluster M34 in Perseus. Once you have M34, return to your scope, and locate M31 again. By this time, it's dark enough to see M31's companion galaxies, M32 and M110. Log those, and move on to Lepus, where the globular cluster M79 is setting rapidly. Finally, use your Telrad and finder to get the planetary nebula M76 in your eyepiece. Use moderate to high power to verify M76, and log it.

After one hour, you've located, viewed, and logged as many as 15 objects. If your count is lower, don't be discouraged. Most Marathoners fail to bag M74, particular if the Marathon is at a late date, and many miss half a dozen or more of the very difficult Early Evening group. Even if you've missed half a dozen of this group, you still have a good chance to break 100 for the Marathon.

Robert found when he attempted to bag M52 and M103 that they were behind the tree line. That wasn't a major problem. He took his binocular and jogged 25 yards or so to a position where they were unobstructed, logging them and M31 for good measure. After he logged M78 and spent a few minutes missing M74, he bagged M77 and logged M33 and M34 with his binocular.

He then attempted to get M31 in the scope so that he could log M32 and M110. Uh-oh. By that time, the Andromeda group had sunk so low that his vehicle was blocking the view. Oh, well. Nothing for it but to pick up the scope and move it. He carried his scope far enough from his vehicle to give it a clear view of the Andromeda group without looking straight into the now-unobstructed streetlight. After a bad moment failing to see M32 or M110, he put on more power and was able to see M32 and M110. Whew. He carried the scope back to its original screened position, picked up M79, tried and failed again to log M74, and finally picked up M76.

It wasn't until later that Robert realized that he could have moved his vehicle temporarily instead of moving the scope. Duh. The real point here is that you need to remain flexible, particularly if you have horizon issues. Don't be afraid to move your scope, change your sequence, or do whatever it takes to maximize the number of objects you bag.

Group 2: Mid-Evening Objects (20:30–21:00)

By 20:30, the big early push is over, and it's time to start work on the mid-evening group of 21 objects, shown in Table 2-9. Most of this group are open clusters you can bag with your binocular. Begin by locating the supernova remnant M1, which is near the 3rd magnitude star 123 ζ-Tau. With M1 logged, move on to the open clusters M50 in Monoceros; M46, M47, and M93 in Puppis; M41 in Canis Major; M44 and M67 in Cancer; M48 in Hydra; M35 in Gemini; and M36, M37, and M38 in Auriga. Using your binocular from a good dark site, these objects should take you at most a minute or two each to locate and log.

Return to your scope and locate the galaxy pair M65/M66 in Leo, both of which fit in an eyepiece field (along with the galaxy NGC 3628; we wonder how Messier missed that one). Locating these galaxies should take only a minute or two using your Telrad. Once you've logged M65/M66, locate the 5.5m star 52 Leo on the line from Chort to Regulus, and use it as a guidepost to the Messier galaxy trio M95, M96, and M105. With Leo cleared, move along to Canes Venatici to pick up the globular clusters M3 and M53, and the galaxy M64.

Table 2-9. Mid-evening objects

Seq #	Time	Object	Seq #	Time	Object
16	20:35	M1	27	20:46	M36
17	20:36	M50	28	20:46	M38
18	20:37	M46	29	20:50	M65
19	20:37	M47	30	20:50	M66
20	20:38	M41	31	20:55	M95
21	20:40	M93	32	20:55	M96
22	20:41	M44	33	20:55	M105
23	20:42	M67	34	20:58	M3
24	20:43	M48	35	21:02	M53
25	20:45	M35	36	21:05	M64
26	20:46	M37			

About 90 minutes of the Marathon is complete, and you've now logged as many as 36 objects. That's nearly one-third of the total. Robert's count was 35 at this point, and he was feeling a lot better about his prospects for the rest of the night.

Break (21:00–21:30)

We scheduled a half-hour break from 21:00 to 21:30, just to relax a bit after the hectic first 90 minutes and think about what was to come. Paul and Robert were only five minutes or so behind our planned schedule at this point, so we knocked off for a half-hour to drink coffee, discuss what we'd done, and talk about the upcoming group. Steve had had some equipment problems, and so was a bit behind schedule. If you're behind schedule at this point, use the time to catch up, but try to break for at least a few minutes for coffee and to warm up.

Group 3: Late Evening Objects (21:30–Midnight)

The late evening group comprises the 33 objects shown in Table 2-10, most of them galaxies in Ursa Major, Virgo, and Coma Berenices. Clear M51 (the famous Whirlpool Galaxy in Canes Venatici) first, and then log all of the objects in Ursa Major, most of which are relatively easy to find and see.

Table 2-10. Late evening objects

Seq #	Time	Object	Seq #	Time	Object
37	21:30	M51	54	23:05	M58
38	21:35	M101/102	55	23:05	M59

Table 2-10. Late evening objects (continued)

Seq #	Time	Object	Seq #	Time	Object
39	21:40	M106	56	23:05	M60
40	21:45	M40	57	23:07	M89
41	21:50	M81	58	23:07	M90
42	21:50	M82	59	23:16	M91
43	21:55	M97	60	23:16	M88
44	21:55	M108	61	23:23	M87
45	22:00	M109	62	23:25	M84
46	22:05	M102	63	23:25	M86
47	22:08	M63	64	23:30	M98
48	22:12	M94	65	23:30	M99
49	22:15	M68	66	23:32	M100
50	22:20	M83	67	23:40	M85
51	22:37	M104	68	23:48	M13
52	22:53	M61	69	23:50	M92
53	23:00	M49			

With the CVn and UMa objects logged, move on to the Coma-Virgo Clutter, shown in Figure 2-33. The clutter contains scores of bright galaxies in an area of sky you can cover with your hand held at arm's length. If you've practiced the Coma-Virgo Messier galaxies several times and have detailed charts at hand, you should be able to get through the Coma-Virgo Clutter in half an hour or less. Otherwise, you're doomed.

Robert ran the Coma-Virgo Clutter four times; three times during the practice session the preceding night, and once on Marathon night. He used a different sequence for each run, starting once from the Denebola end, twice from the Vindemiatrix end, and once from both ends toward the center. So much for planning. But by Marathon night, he knew the Coma-Virgo Clutter so well he could almost do it in his sleep.

> The trick to negotiating Coma-Virgo is to use a wide-field eyepiece to "galaxy hop" from one cluster of Messier galaxies to the next. Although there are scores of other galaxies visible, the Messier galaxies are generally larger and brighter than the others, and so they can be used as guideposts. By knowing the true field of view of your eyepiece and comparing that field with your detailed charts, you can galaxy hop confidently, knowing that you've correctly identified the brighter galaxies in your field of view [Hack #50][Hack #57].

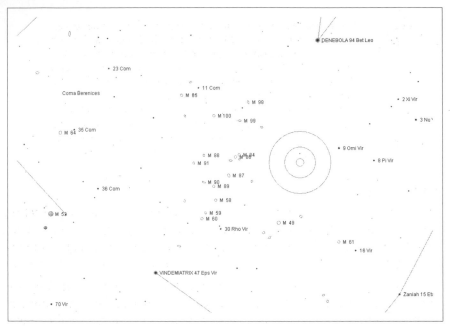

Figure 2-33. The Coma-Virgo Clutter (4° Telrad circle shown for comparison)

It's a mistake to use too much aperture in Coma-Virgo, particularly from a very dark site. With an 8" or 10" instrument, the Messier galaxies stand out. With a large instrument, so many non-Messier galaxies are visible that it's difficult to identify the Messier galaxies in all the clutter.

Once you complete the Coma-Virgo Clutter, move on to the two bright globular clusters in Hercules, M13 and M92, both of which are relatively easy binocular objects. M13 is by far the most impressive globular cluster visible from northern latitudes. Were it not for M13, M92 would be recognized as a magnificent globular in its own right, but the proximity of M13 means M92 gets little attention.

It's now about midnight. In 4.5 hours of work (maybe with a short break) you've logged as many as 69 objects, or nearly two-thirds of the total. At this point, Robert was hanging in there with a total of 68 objects, having logged all of the objects except M74 from the first group. Paul was a perfect 69 for 69, and Steve was sitting at 63. Things were looking good.

Before the Marathon, Paul and Robert—both experienced observers but newbie Marathoners—had discussed what numbers they expected and hoped to make. Both agreed that they'd be satisfied with a total of 75 objects for the night and delighted with 90. Steve, with less experience, said he'd be satisfied with 50 objects and delighted with 70. So, as it turned out, by midnight and with hours yet to go, all of us were within easy reach of achieving our hoped-for totals. We hadn't anticipated that by midnight each of us would already be able to declare our Marathon a personal success.

Nap (Midnight–02:00)

At midnight, it's time for a nap. Seriously. You've been working hard for hours, and are probably chilled. Many of the remaining objects aren't up yet, and those that are will be up for a long time. Take a couple of hours off to rest and warm up. It'll make the rest of the night go a lot easier, at least in theory.

We adjourned to the lodge, lit the fire that we'd pre-laid in the fireplace, made some hot cocoa and microwave popcorn, chatted for a few minutes, and then zonked out. We made sure to set two alarm clocks, afraid that otherwise we'd wake up after sunrise. Paul and Robert woke up on time. They tried to rouse Steve a couple times, but he didn't move. They left him, figuring he was dead, and went back to work on the next group of objects. A while later, Steve staggered out of the lodge and went back to work, already 15 minutes or so behind schedule and muttering something about having friends like us....

Group 4: Early Morning Objects (02:00–04:00)

You have a long run in front of you—36 objects in two hours, nearly one object every three minutes. At this point, despite our naps, all of us were tired and discouraged at the magnitude of the task remaining. At 2:00 in the morning, things look bleak, and standing in a field in the dark with a cold breeze blowing doesn't help matters. You may find yourself wondering, as we did, why you are doing this to yourself [Hack #1].

Robert needed some successes to get the ball rolling, so he abandoned his planned sequence and schedule to log some easy, familiar objects. He used his binocular to bag the large, bright globular clusters M10 and M12 in Ophiuchus, moved on to the binocular globs M4 and M80 in Scorpius, and then used his scope to bag the planetary nebula M57 and the glob M56 in Lyra. With six objects logged in 20 minutes, Robert had reached 74 objects, and things weren't looking nearly as bleak. He was right on schedule, with only 30 objects left to go in the early morning group, shown in Table 2-11.

Table 2-11. Early morning objects

Seq #	Time	Object	Seq #	Time	Object
70	2:05	M10	88	3:26	M11
71	2:06	M12	89	3:27	M16
72	2:08	M4	90	3:28	M17
73	2:10	M80	91	3:30	M18
74	2:18	M57	92	3:37	M26
75	2:20	M56	93	3:40	M8
76	2:25	M5	94	3:41	M23
77	2:30	M19	95	3:42	M24
78	2:35	M62	96	3:42	M25
79	2:42	M6	97	3:43	M22
80	2:42	M7	98	3:45	M28
81	2:44	M107	99	3:52	M20
82	2:47	M9	100	3:53	M21
83	2:50	M14	101	3:56	M54
84	3:04	M27	102	3:57	M69
85	3:07	M29	103	3:58	M70
86	3:15	M71	104	4:03	M55
87	3:21	M39	105	4:09	M75

By shortly after 4:00, we'd all finished logging all 36 objects in this group. We were worn out and cold, so we took a short break to get some coffee and warm up a bit. With that done, we staggered back to our scopes and prepared for the closing stage of the Marathon.

Group 5: Final Objects (04:00–Dawn)

The final five objects, shown in Table 2-12, are, if anything, harder than the early evening objects. The eastern sky is still dark, but it won't be long before it begins to brighten, and the final objects are barely above the horizon. Robert bagged the glob M15 first, just off 2nd magnitude Enif in Pegasus. We were all so punch-drunk by that time that when Robert and Steve mentioned logging M15, Paul said, "What do you mean you logged M15? It isn't up yet." Robert, almost convinced despite himself, replied, "Well, what's that big, bright glob near Enif?" After a few moments, Paul apologized, saying that he'd hopped the wrong direction with his equatorial mount, moving it below the horizon instead of above it. He wasn't the only one confused by then.

Table 2-12. Final objects

Seq #	Time	Object	Seq #	Time	Object
106	4:22	M15	109	4:50	M73
107	4:33	M2	110	missed	M30
108	4:46	M72			

M2, a big, bright globular cluster in Aquarius was next. It was harder to locate than M15, but Robert was eventually able to hop to it from 3rd magnitude Sadalsuud (22 β-Aqr). At that point, he thought he'd probably gotten his last object of the Marathon, and was about to settle for a final score of 106 objects. But he persisted in looking for the small, dim glob M72 in Aquarius and was finally able to locate it and confirm it under high power. By switching back to the 40mm Pentax XL eyepiece, which puts M72 and M73 in the same field of view, Robert managed to locate and confirm the tiny so-called open cluster M73 in Aquarius, which is actually just a group of four dim stars. Hmmm. Robert was now at 108 objects and counting.

By this time, it was 04:50 and the horizon was beginning to brighten. The final object, M30, a medium-size glob in Capricornus with surface brightness of only 11.0 seemed impossible, and so it turned out to be for Steve and Robert. Paul persisted, and was eventually able to locate M30 with his setting circles in the gathering dawn. He put high power on the object to bring it out against the sky, and called Steve and Robert over to confirm that he had M30 in the eyepiece.

The final totals:

Jones—110 objects (missed none)

Thompson—108 objects (missed 74 and 30)

Childers—100 objects (missed 74, 77, 33, 76, 79, 32, 110, 72, 73, and 30)

As dawn broke, we tore down and packed up our equipment and prepared to head home. We decided by acclamation that the Second Annual WSAL Messier Marathon would be held no earlier than March 2103.

HACK #31 Photograph the Stars with Basic Equipment

Shoot star trail images.

If you have been interested enough in astronomy to read up on the subject, you've no doubt run into a number of stunning photographs of celestial objects. Many have been taken by professionals—no one with the astronomy bug can fail to fall in love with Hubble photographs. But, a large number of them are credited to amateurs, like you, with amateur telescopes, like yours, perhaps. It is then inevitable for the new (or not so new) amateur astronomer to decide that she, too, should take beautiful photographs to hang above her fireplace or to publish in one of the many fine astronomy magazines.

There is only one problem. Taking decent astrophotos is a tedious, difficult, time-consuming endeavor. It's also very expensive—superb astrophotos increase the demands in all categories by an order of magnitude and require a pinch of luck.

I don't write this to discourage you. You will want to take astrophotos. Many experienced amateur astronomers scoff at the newbies' urge to take stunning photographs. But what they won't tell you is that they, too, long to take such photos and only through a superhuman effort of will (or the budget balancing efforts of an astrospouse) do they avoid temptation.

Astrophotography, like observing or equipment building, is a simple concept plagued with confusing details. The beautiful photo you admire in *Sky & Telescope* or *Astronomy* represents hundreds of hours and thousands of dollars invested by the photographer. You may wish to dive into astrophotography and, with perseverance, you will someday take good photos. There are a lot of methods and tips you'll need to know and you'll have to find most of them elsewhere, as none of the authors of this book have expertise in the area.

However, we can get you started.

The simplest astrophoto is known as a star trail. As Earth rotates on its axis, the stars appear to move in the sky, rising in the east and setting in the west. In so doing, they describe a circle around the celestial pole. If you photograph a group of stars for several minutes with a stationary camera, you'll capture a portion of this circle and you'll see an arc—a star trail.

In order to have stars appear as points, rather than trails, it is necessary to have the camera and telescope rotate in such fashion as to compensate for Earth's rotation. (An exposure long enough to show stars is long enough to cause trailing unless the camera and scope track.) This requires an equatorial mount and clock drive. Because clock drives aren't perfect, it is also necessary to guide the photograph. That is, you must watch through another telescope on the mount and keep a guide star centered on a crosshair. This sounds difficult enough as it is. Imagine trying to do it in subfreezing weather with a stiff wind. Fortunately, automated guiders are now available. If you have the money, this greatly eases the process of taking astrophotos through a telescope. In this writer's opinion, it also take a good deal of the romance out of the process, as well.

Star trail photographs come in great variety. An ever popular one is to aim a stationary camera at the celestial pole and open the shutter. In the northern hemisphere, you'll capture Polaris making a tight circle about the celestial pole (but, coolly, you will see that Polaris does, in fact, move in the sky) and the rest of the stars making larger circles about the pole. Photographers have used this as a backdrop for any number of interesting photos. I used to enjoy shooting my hometown with the stars wheeling overhead. Use your imagination here, you can come up with many neat foregrounds for star trails. Also, you need not limit yourself to the celestial pole. A shot of a constellation tracing arcs through the sky is usually interesting. Constellations with bright stars, such as Orion or Cygnus, make excellent targets in this regard (see Figure 2-22). Star-trail photography is also an excellent means of photographing bright comets and the only means of photographing meteors. Presented below are a few tips on star-trail photography. Take heart, it is very easy to do.

You'll need the following equipment to shoot star-trail images:

- A 35mm SLR or digital camera capable of taking exposures on the Bulb (B) or Time Exposure (T) setting
- A wide-field lens (a 28, 35, or 50mm lens is excellent for star-trail photography)
- A stable camera tripod
- A cable release with locking mechanism (unless your camera provides the "T" shutter setting)
- Film (or the digital camera chip)

Lens

The lens is very important. You generally want a wide field in star-trail photography. Because the photo isn't guided, fine detail will not be captured. Therefore, zooming in will get you nothing except a larger blur rather than a smaller blur. I recommend a 35mm lens but star-trail photography also works well with a 28mm or 50mm lens.

Exposure

Astrophotos work by collecting light over an extended period of time. Your eye can only collect light for a second or two while film or chip can collect light as long as the shutter is open. This allows objects too dim to see to be recorded vividly in photos. It also makes light pollution— that orange glare lifted up from city streets—a huge problem. An exposure hours long will record skyglow, whether natural or man-made, and the background in the photo will wash out. In order to make nice star-trail photos, you'll have to limit exposures so that the sky remains black in the photo. This will likely require some trial and error on your part, but take the obvious precautions. Don't set up under a streetlight. Find as dark of a location as is practical. Get out in the country and shield your camera from local lights **[Hack #12]** as much as possible. Depending on the sky and the film you choose, exposures may range from a few seconds, which will record only bright stars as very short trails, to several hours, which will record fainter stars and much longer trails.

> Most stars appear white in standard star trail images. Try defocusing the camera *slightly* during part of the exposure, being careful not to move the camera when you refocus. The star trails recorded during that time will be slightly blurred and will show a range of colors, from deep red to bluish white.

Film

Star-trail photography is generally done with slower film than is used for deep-sky, guided, astrophotography. For one, the stars do not stay on the same place on the film, so collecting more light, in this case, doesn't get you a lot. That is, you aren't after really faint stars as much as a good picture of the brighter ones. Two, a fast film will "fog," or washout from skyglow, faster than a slow film. The one exception to using a relatively slow film is in meteor photography where exposures will stay short and capturing faint streaks is the goal.

For the most part, exposure and film must be considered together. If your goal is to take a five-hour star trail around the North Celestial Pole, you'll need a fairly slow film to avoid washing out the sky from skyglow. You may start with ISO 100 film and close the shutter to f/4, 5.6, or even 8. If your goal is to take a 90-second exposure of a bright comet, you'd want a faster film and probably need to go with something like ISO 1600 at f/2.8. Play around and see what you can get away with from your various observing sites.

> Be sure to keep careful records. You'd hate to find a nice film/camera setting/exposure combination that gives great photos but forget what the settings were. A normal log is fine for recording such data, but you can also use a Hollywood system. Before each photo, write down the details of the next photo and photograph that note. That way, the data is right there next to your photo. (This may not be necessary if you use a digital camera; many digital cameras store exposure information inside the image file itself.)

Mount

To take the photo, place the camera on a tripod, aim to the desired point in the sky, lock the camera in place, and trip the shutter (on the "B" setting) with a cable release that holds the shutter open and then walk away. The worst thing that can happen is to bump the tripod while the shutter is open. Stand clear until it is time to close the shutter. Avoid the temptation to "check" the setup once the shutter is open. If you've screwed up somehow, you can't fix it at this point and, if you haven't screwed up, you can definitely still screw it up. The idea is to open the shutter on a camera solidly fixed in one position. A good tripod is a necessity. Obviously, a windy night is not a good night for such photography (or any astrophotography, really).

Processing

If you have a digital SLR, processing is as simple as transferring the file from the camera to your computer. You can then play around with the exposure. Darkening the sky is probably the best thing you can do for your electronic image. If you use film and have the film commercially processed, speak to the lab tech about keeping the background dark on the prints. A darker print may cost you a few faint stars, but the image will be much more pleasing. Also, you should probably ask that they not cut the negatives. Often astrophotos will not have a clear boundary between exposures and occasionally a frame may be cut down the middle by an inattentive tech.

Alternatively, start each roll of film by shooting an ordinary day-time picture. This gives the processing lab an index position, from which they can cut the roll and print other frames correctly.

If you've successfully taken star-trail photos and would like to "move up," you can take piggyback photos. In this technique, you mount the 35mm SLR on an equatorially mounted, clock-driven telescope. You'll have to polar align the scope accurately. You can use slightly narrower fields of view in a piggybacked photo than in a star-trail photo, but a wide-field (28, 35, or 50mm) lens is still appropriate. In this case, you should use relatively fast film (or equivalent setting on a digital camera) and open the shutter to f/2.8 or so. Polar align the scope, center a bright star in a high power eyepiece in the scope, aim the camera, and open the shutter. If you notice the bright star in the eyepiece drifting noticeably, stop the exposure. Because your telescope has a much narrower field of view than the wide-field camera lens, you'll see drifting well before the stars streak on the film. Again, you'll have to experiment with exposures but they will likely be in the range of 1–15 minutes. A successful photo will show stars as points (or circles) and many, many more faint stars than you can see with the unaided eye. You may also see wispy nebulosity, as well.

—*Dr. Paul B. Jones*

HACK #32 Discover and Name a New Planet
Do real science. Let your computer map the universe while you're asleep.

In 1991, astronomers discovered a planet orbiting a distant neutron star. This was the first known *extrasolar planet*, which is to say a planet orbiting a star other than our own Sun. In 1995, astronomers at the Geneva Observatory discovered the first extrasolar planet orbiting a "normal" star, in that case 51 Pegasi. Since that time, a total of 160 extrasolar planets have been discovered—only a few planets per year of perhaps thousands that are within a reasonable distance and waiting to be discovered.

PlanetQuest Collaboratory (*http://www.planetquest.org*) seeks to organize and popularize the search for extrasolar planets, just as the SETI@Home project has done for the search for extraterrestrial intelligence. But while SETI@Home has always been a long shot, the search for extrasolar planets is as close to a sure thing as you can get in science.

As is so often the case in astronomy, observational data are much easier to come by than the computing power needed to process them into usable form. PlanetQuest seeks volunteers to run PQ distributed processing software as a background task on their computers, using idle time to crunch the raw data. With millions of PCs working to process the PQ data, PQ expects to find thousands of new extrasolar planets over the next few years.

Project leaders estimate that the chance of any one person finding an extrasolar planet are one in 3,000 to 5,000, which are pretty good odds. And, get this, if your computer finds a planet, you get to name it.

 If Robert finds an extrasolar planet—which may well happen because he plans to devote a Linux cluster the equivalent of a baby supercomputer to the search—he intends to name it *Laftwet* in honor of his late parents, Lenore Agnes Fulkerson Thompson and William Ewing Thompson, whether or not the Laftwetians approve. If he discovers two extrasolar planets, he'll name one *Barbara*, because he has to sleep sometime.

PQ expects to release beta versions of their distributed processing software late in 2005 and to go live in spring 2006.

Scope Hacks
Hacks 33–43

Hacking is a time-honored custom in amateur astronomy. When Robert started observing in the mid-60s, most people built their own scopes. As a teenager, Robert couldn't afford a commercial scope, so he did what thousands of others did: bought a mirror kit from Edmund Scientific and ground his own 6" mirror, built a finder scope from half of a discarded binocular, assembled an equatorial mount from pipe fittings, and scrounged far and wide for parts to build the mirror cell, focuser, and so on.

Life is easier for amateur astronomers nowadays. In 2000, when Robert decided to jump back into amateur astronomy, someone gave him a catalog from Orion Telescope & Binocular Center (*http://www.telescope.com*). Flipping through it, he spotted a 10" SkyQuest XT10 Dobsonian telescope for only $699. It certainly looked odd to Robert, whose idea of a scope was a tube sitting on a tripod. This one was a tube sitting in what looked like a large box on the ground. OK, so things had changed. But $700 for a complete 10" scope? It must be a piece of junk, right?

Wrong. More Internet searching turned up a lot of answers. The telescope sold by Orion as the SkyQuest XT10 was actually made in Taiwan by a company named Guan Sheng, which mass-produces telescopes of astonishingly good quality at surprisingly low prices. The mechanicals—mirror cell, focuser, and so on—were much better than what Robert had made himself in the 60s, and the optics were, if not quite up to the best premium custom optics, probably better than Robert could have produced himself. Make or buy? The decision was a no-brainer. We ordered an Orion SkyQuest XT10 Dob and have never looked back. (Nowadays, you can buy a very similar 10" Dob for about $500.)

Dobsonian scopes are simple, intuitive to use, have rock-solid mounts, and provide much more aperture for the money than any other type of scope. It's no wonder that Dobs are overwhelmingly popular nowadays. But we think there's another factor.

If you buy a refractor, SCT, or other traditional scope, you've bought a scope. You unpack and assemble it, set it up, and use it. Sure, you can buy more eyepieces and other accessories. Perhaps you can adjust the focuser or tweak the mount a bit, but that's about it. There's not much you can (or should) do to modify the scope itself. If you attend a large star party, you'll see dozens of scopes set up. Every SCT or refractor looks pretty much like every other. There's not much opportunity for personalization or customization.

If you buy a Dobsonian scope, you've bought an ongoing project, or at least the opportunity for one. Although most Dobs work pretty well without any modifications, their inherent simplicity makes it easy to customize them. Dobs appeal to the shade-tree mechanic in all of us. We tweak, modify, upgrade, improve, and tinker with our Dobs. We build new bases, upgrade the bearings, paint the tubes, install cooling fans, replace the mirror cells and focusers, flock the tubes, and on and on. In short, we hack on them.

It's not that a typical Dob actually *needs* all this work, you understand. It's that the simplicity of a Dob *allows* you to make these changes, and each change makes the scope more and more your own personalized instrument. Some Dob owners modify their scopes so heavily that they are no longer rec-ognizable as the commercial telescope that was the starting point.

So, although some of the hacks in this chapter also pertain to other scope types, we focus our efforts on the most important tweaks and upgrades for Dobsonian scopes.

H A C K   Center-Spot Your Mirror
#33   Collimate faster and more precisely.

All telescopes must be collimated if they are to provide the best possible image quality. Collimation consists of aligning the lenses and mirrors in a telescope to share a common optical axis. Because of their physical design, Newtonian reflectors, including Dobsonians, require more frequent collima-tion than other telescope designs. (Fortunately, most Newts are pretty easy to collimate.)

The importance of accurate collimation varies with the focal ratio of the scope. Slow scopes, those with focal ratios of f/8 [Hack #9] or higher, provide reasonably good image quality even if they are only roughly collimated. An f/6 scope, such as a typical 8" Dob, must be collimated with moderate precision to provide good images. An f/5 scope, such as a typical 10" or 12" Dob, must be collimated with extreme precision.

In order to collimate a Newtonian precisely with standard, inexpensive collimation tools [Hack #37], it's necessary that the center of the mirror have a visible mark, called a center spot. Unfortunately, many scope manufacturers don't center-spot their mirrors at the factory. If you have such a mirror, you'll have to apply a center spot yourself.

Don't worry about the center spot affecting the image. The center of the primary mirror, where the center spot is applied, is shaded by the secondary mirror. The shaded center of the primary mirror never "sees" star light, so any marking there has no effect on image quality.

Some astronomers use a felt-tip marker to apply an actual spot to the center of their mirrors. That's a workable solution, but not ideal. It's much better to apply a notebook paper reinforcing ring as a center spot because that little white donut is much more visible when you collimate the scope. The trick is to get the ring applied as closely as possible to the exact center of the mirror. How close to exact center the spot needs to be depends on the focal ratio of the scope. For slow scopes, anything within a couple millimeters is close enough. For f/5 and faster scopes, you need to be extremely precise, within no more than 0.5mm of true center, and 0.25mm is better.

To center-spot your mirror, take the following steps:

1. Remove the mirror cell from your telescope and the primary mirror from its cell.

2. Place the mirror flat, shiny side down, on a sheet of paper large enough to extend beyond the edges of the mirror

3. Use a sharp pencil to trace all the way around the circumference of the mirror, as shown in Figure 3-1. Make sure that the line you trace exactly matches the edge of the mirror.

4. Remove the mirror carefully, and place it face up in a safe location.

5. Carefully cut out the circle you traced, keeping exactly on the pencil line.

6. Fold the paper circle once across its diameter to make a half-circle and then a second time to make a quarter-circle that looks like an ice cream cone.

7. Cut a tiny piece from the pointy end of the ice cream cone, as shown in Figure 3-2, and then unfold the paper circle.

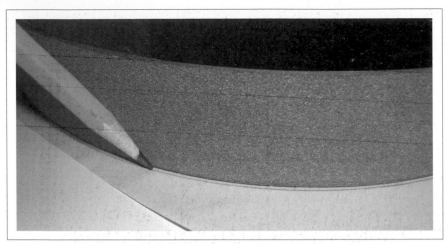

Figure 3-1. Trace the circumference of the mirror onto the paper

Figure 3-2. Cut a center hole just large enough to accept the tip of the pen

8. With the mirror face-up, carefully lay the paper circle on top of it, matching the edge of the circle exactly to the edge of the mirror.

9. Hold a felt-tip pen exactly vertical. Insert the tip through the hole you cut in the paper circle, and make a mark on the surface of the mirror.

10. Remove the paper circle, and apply a notebook reinforcing ring to the mirror, as shown in Figure 3-3. Make sure that the center of its hole is precisely centered on the spot you made with the felt-tip pen.

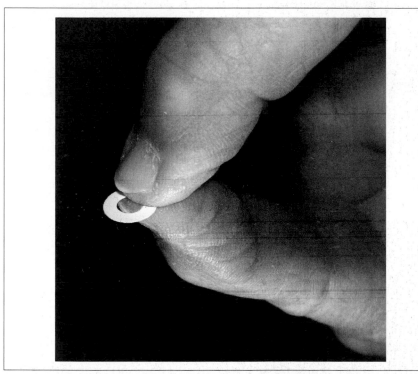

Figure 3-3. Carefully apply the ring, making sure to center it against the mark

Since you have the primary mirror out of the scope anyway, check it to see if it needs to be washed [Hack #34]. After washing the mirror, if necessary, reinstall the mirror in its cell [Hack #35] and the assembled mirror cell into the scope.

HACK #34 Clean Your Primary Mirror
Eliminate crud and protect against corrosion.

The primary mirror of a Newtonian reflector, exposed as it is to outside air, inevitably gathers dust, water spots, film and haze from air pollution and other contaminants. Depending on how often you use your scope, how well you protect it against dust and other contaminants when it is not being used, and the environment at your observing site, your mirror may become dirty enough to need cleaning every few months or it may remain acceptably clean for years on end.

The only way to determine if a primary mirror needs to be cleaned is to remove it from the scope and examine it. Newbies often shine a flashlight down the tube and are invariably horrified at how dirty the mirror appears, even when the telescope is new. No mirror, even one freshly cleaned, can survive the "flashlight test." Every speck of dust and every smudge jumps out at you.

Dust and smudges appear ugly, but the truth is that they have little effect on image quality. Consider this: your secondary mirror sits directly in the path of the incoming light, and so is the equivalent of a chunk of dust or a smudge a couple of inches in diameter, and it has almost zero effect on image contrast and clarity. How little, then, will a few small smudges and some dust affect what you see? Not at all.

The real reason for cleaning a mirror is to remove the surface film of air pollutants that may eventually eat into your mirror's surface, etching it and exposing the reflective aluminum coating to airborne pollutants. Salt air is an even greater danger to your mirror. If your observing site is near the ocean, we recommend cleaning your mirrors frequently.

Conventional advice emphasizes the dangers of cleaning a mirror and recommends doing it as seldom as possible, usually at most every year or two. We think that's overly cautious, although it is true that the mythical average astronomer's mirror probably needs to be cleaned no more frequently than that. But judge for yourself based on your own environment and usage pattern. More importantly, check your mirror periodically. (Just don't judge by shining a flashlight down the tube; no mirror, even one just cleaned, looks good under that harsh lighting.) When your mirror looks as though it needs to be cleaned, clean it.

In truth, a mirror is a pretty rugged item. It is, after all, a chunk of glass an inch or two thick. We've seen bullet-proof glass that was thinner than some primary mirrors. It's true that the mirror surface is finished to an accuracy measured in millionths of an inch and that the reflective aluminum coating is only a few atoms thick. But careful cleaning, no matter how frequently done, has no effect on the surface figure or the reflective coating. The aluminum reflective layer is covered with a thin layer of silicon dioxide—that's quartz—which is very hard and reasonably resistant to scratching.

The real risk of cleaning your mirror isn't scratching the coatings; it's dropping the mirror. When you clean your mirror, take precautions to prevent dropping it. Actually, if you drop a typical primary mirror on your foot, you're more likely to hurt your foot than the mirror. But mirrors have been known to shatter when dropped, and even if the mirror remains in one piece you may scratch the optical surface. Wear rubber gloves for a surer grip, and try to keep the mirror only inches above a padded surface at all times.

We suggest the following method to clean small and medium primary mirrors, those up to about 12.5" in diameter that weigh no more than 20 pounds or so.

1. Prepare your work area. Using the kitchen sink is traditional, but if you clean your mirror there make absolutely sure the sink and surrounding surfaces are completely clean. Even one particle of abrasive cleaner grit can scratch the mirror. Pad the bottom of one side of the sink with a frequently-washed folded towel or use a RubberMaid tub to provide a soft resting surface for the mirror. Fill that side of the sink to a depth of three or four inches with lukewarm water, adding a good squirt of Dawn or a similar clear dishwashing liquid. If your mirror is particularly filthy, add a pint of isopropanol for every gallon of water.

2. Remove the primary mirror cell from your scope, as shown in Figure 3-4, and carry the mirror in its cell to your work area. The mirror cell may be difficult to remove. In extreme cases, you may even need to use a hammer to free the cell from the tube. (Use a rubber mallet or a scrap piece of lumber to protect the tube and cell.) Because the primary mirror cell may come loose suddenly, it's a good idea to have a pillow or other soft landing spot under the mirror cell in case you drop it. Also, if you are removing the mirror cell from a Dobsonian scope, remember that the tube will nosedive as soon as you pull the mirror cell. We generally support the front end of the tube with a chair, stool, or similar object, or have a helper support the front of the tube until the primary mirror cell is removed.

3. Remove the primary mirror from its cell. If it is not already marked, it's a good idea to make index marks on the side of the mirror and on the mirror cell so that you can later reinstall the mirror in the original orientation. When you handle the mirror itself, wear rubber gloves, which both prevent getting fingerprints on the mirror and give you a much better grip on the mirror, particularly once it's wet and slippery. Avoid touching the surface of the mirror; handle it only by the edges as much as possible.

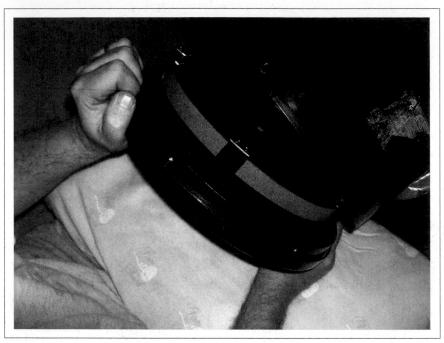

Figure 3-4. Removing the primary mirror cell

4. Run the tap water until it is lukewarm. Hold the primary mirror at an angle, as shown in Figure 3-5, and rinse the surface thoroughly with tap water. The goals are, first, to rinse off any grit or other abrasive particles, and, second, to remove as much as possible of the water spotting, film, and other grunge without touching the surface of the mirror. Don't hurry this process. We generally spend five minutes or more rinsing our 10" primary mirror. Make sure the water stream touches every part of the surface. The stream of water from the tap is a gentle, safe way to dislodge grunge without touching the mirror.

If you are concerned about thermal shock, make sure your mirror has equilibrated to room temperature and rinse it with room temperature water. Many mirrors use Pyrex glass, which is very resistant to thermal shock. But even plate-glass mirrors are not at risk if the temperature difference is small. We wouldn't bring a mirror in from a cold outdoor storage place and run hot water over it before it had a chance to warm up to room temperature, but we can't imagine that running lukewarm water over a room-temperature mirror could cause a problem.

Figure 3-5. Rinsing the primary mirror

5. After the mirror is thoroughly rinsed, lower it gently (face-up) into the sudsy water on the other side of the sink. Swish it around gently in the sudsy water for a minute or so, and then allow it to rest on the towel or other soft surface at the bottom of the sink. Then allow the mirror to soak for at least 10 minutes or so. If it's particularly dirty, it won't hurt to let it soak for an hour.

6. Swish the mirror one last time for a minute or so, and then carefully lift it from the sudsy water and rinse it thoroughly under tap water set to about the current temperature of the sudsy water. Examine the surface of the mirror carefully. Unless the mirror was truly filthy, it's probably now as clean as it needs to be.

7. If any streaks or smudges remain on the mirror, decide whether to continue cleaning the mirror or declare the job done. Minor streaks and smudges have almost no effect on image quality, so it's usually best and always safer to accept whatever minor problems remain. If in doubt, play it safe by skipping the following step. Otherwise, proceed to the next step.

The center of the mirror, that portion shaded by the second-ary mirror, does not contribute to the image you see in the eyepiece. Any spots or other contamination near the center of the primary mirror (including, of course, the center spot) can safely be ignored.

8. Resubmerge the mirror in the sudsy water. Holding the mirror at a slight angle under the surface of the water, draw a pad of sterile surgical cotton or a cotton ball across the surface of the mirror from one side to the other. Do not press down on the cotton. You're not scrubbing the mirror, but only using minor pressure—usually the weight of the cotton itself is sufficient—to remove any remaining film, spots, or other contaminants. Use each piece of cotton only once; replace the cotton for each swipe.

9. Once the mirror is clean and thoroughly rinsed with tap water, the next step is the final rinse. Hold the mirror nearly vertical, resting on a soft pad, and flood it with at least a gallon of distilled water to which you have added a drop or two of surfactant to reduce the surface tension of the water and prevent beading. We use Kodak Photo-Flo surfactant, which many astronomers prefer, but a drop or two of clear dishwashing liquid such as Dawn works just as well and leaves no residual film.

10. Hold the mirror vertically and allow the final rinse to drain completely. If the mirror is clean, the rinse water will "sheet" off the mirror, leaving few or no droplets on the mirror surface. You can remove whatever droplets remain by using the corner of a paper towel to absorb them without touching the mirror's surface, as shown in **Figure 3-6**.

11. When the mirror surface appears pristine and dry, place the mirror face up on a folded towel or other soft surface and allow it to dry thoroughly. If you're concerned about dust, cover the surface of the mirror with sheets of facial tissue. Once the mirror is completely dry, put it back in the mirror cell and reinstall it in the telescope.

After you reinstall the mirror, you'll need to recollimate your scope **[Hack #38]** and realign the finder **[Hack #55]**.

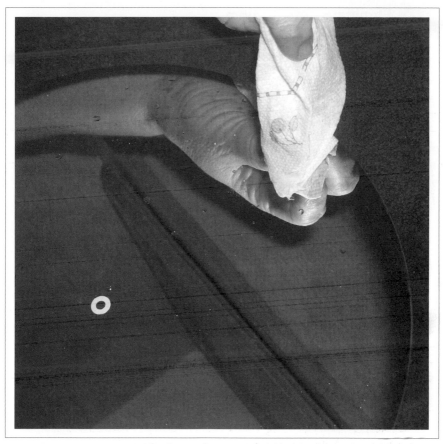

Figure 3-6. Removing water droplets with a paper towel

For large, heavy mirrors, the best cleaning method is a matter of debate. Mirrors in amateur instruments can be as large as 40" in diameter and weigh up to several hundred pounds. Obviously, it's impractical to clean a mirror even half that size using the method described previously. Handling a large, heavy, wet piece of glass that costs thousands of dollars is a recipe for disaster. But, there are alternatives.

Some astronomers clean their large mirrors in situ using water, alcohol, and cotton. Others clean their large mirrors using collodion, which is a highly flammable solution of nitrocellulose or guncotton in diethyl ether. Collodion is applied directly to the mirror surface, where it dries quickly to a thin, tough film. The film is then peeled away from the mirror, carrying dust, smudges, and other contaminants with it. A mirror cleaned properly with collodion is as clean as it is possible for it to be, but collodion is expensive and we consider it too dangerous for routine use.

Eliminate Astigmatism

HACK #35

Pinching can turn good optics bad and your pinpoint stars into fuzzy lines.

Astigmatism is an optical aberration that may be present in your eyes, in your telescope optics, or both. In a telescope that suffers from pure astigmatism, slightly defocusing a star image in one direction results in the circular star image becoming a short, straight line. Defocusing in the other direction yields another straight line, but oriented at 90° to the first line. Such pure on-axis astigmatism is rare; most telescopes that suffer from astigmatism also suffer from other aberrations and miscollimation, all of which can cause bewilderingly distorted images.

> Unless you are an optics expert, looking at a defocused star won't tell you much except perhaps that your telescope is not providing very good images. An optically perfect telescope that is perfectly collimated displays stars as Airy discs, a bright center spot surrounded by symmetric rings of decreasing brightness. There are so many possible causes of imperfection—from unstable atmosphere to air currents in the telescope tube to miscollimation to mediocre optics—that it's very difficult to determine not just what's causing a problem, but what the problem actually is. Accordingly, unless your scope already star-tests perfectly, we recommend that you follow the steps described in this hack. They can't hurt, and they may help.
>
> If you want to learn about all of the technical details involved in testing optics, get a copy of *Star Testing Astronomical Telescopes*, available directly from Willmann-Bell (*http://www.willbell.com*).

There are many possible causes of astigmatism, including your own eyes, a low quality eyepiece, a poorly figured primary mirror, a poorly supported primary mirror, a secondary mirror that is slightly spherical rather than flat, a mirror made from poorly annealed glass, and so on. But by far the most common cause of astigmatism in amateur telescopes is pinched optics. Fortunately, that's an easy problem to solve.

The worst offenders are the ubiquitous Chinese and Taiwanese Newtonian reflectors and Dobs, which are sold under numerous brand names, including Orion, Skywatcher, Celestron, and many others. Many of these scopes use mirror clips to retain the primary mirror. The problem is that mirror clips aren't intended to lock the primary mirror into place, nor even to keep it from moving, but only to keep it from falling out of the mirror cell if the scope tube is pointed downward.

But the Chinese and Taiwanese scope makers, apparently concerned about damage during the long trip across the Pacific Ocean, often torque down the mirror clips much too tightly. We've encountered some so tight we suspect they used an air wrench to tighten them. Although it seems unlikely that these soft rubber clips could deform a mirror no matter how great the pressure, even gentle pressure against the mirror is enough to distort it by a few millionths of an inch. That's more than enough to wreck image quality.

To eliminate this most likely cause of astigmatism, pull the primary mirror cell from your scope and examine the mirror clips. If you've never touched them, you'll probably find they're screwed down tight and in full contact with the surface of the mirror. In fact, chances are good that the rubber mirror clips have actually adhered to your primary mirror.

To fix the problem, remove the screws that secure each clip, being careful not to let the screwdriver or screws contact the primary mirror. Peel the clips off the mirror if necessary, and then reinstall the clips. Tighten the screws only far enough to retain the clips while leaving a slight gap between the surface of the mirror and the rubber. The mirror clips are not intended to clamp the mirror in place. Their only purpose is to keep the mirror from falling out of the cell. When the clips are adjusted properly, you should be able to slide a business card between the mirror and the rubber clip, as shown in Figure 3-7.

Figure 3-7. Leave a small gap between the mirror clip and the mirror surface

After you reinstall the mirror clips, use a small dab of nail polish or Loc-Tite on the head of each screw to make sure it stays put, as shown in Figure 3-8. The last thing you want is a loose screw falling onto your primary mirror and scratching it.

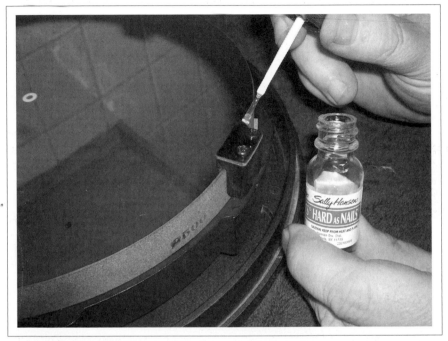

Figure 3-8. Use nail polish or Loc-Tite to secure the mirror clip screws

The secondary mirror is another possible source of astigmatism in a Newtonian reflector, albeit much less likely than tight primary clips. Actually, the secondary mirror itself is seldom the problem. Even in inexpensive scopes, the secondary mirror is generally close to being optically flat. The problem is the secondary mirror holder, which may bind the mirror, torquing it slightly out of flat. To eliminate this possible cause of astigmatism, remove the secondary holder and examine it to determine how the secondary mirror is secured.

- Some secondary holders secure the mirror with a small dab of room-temperature vulcanizing (RTV) adhesive. That's not an ideal solution, but it generally works as long as you keep the dab small.

- Other secondary holders retain the secondary mirror with a frame that functions much as primary mirror clips do, simply by physically preventing the mirror from falling out of the holder. If your secondary holder is of this type, open the frame to verify that it puts no direct pressure on the secondary mirror. Inexpensive scopes often use an injection molded plastic frame to retain the mirror. Sometimes, a stray bit of plastic that remains from the molding process presses on the mirror itself. If so, use a sharp knife or file to remove the excess plastic so that the mirror can float freely within the secondary holder.

Because of their physical designs, other types of telescope are less likely to be astigmatic than are Newts. Still, it's relatively common for a Chinese or Taiwanese refractor to suffer from pinched optics because its lens cell clamps the objective lens too tightly. This problem seems particularly common in 80mm and 90mm short-tube refractors.

If you have such a refractor, check the lens cell to make sure the objective lens is not clamped too tightly. The objective lens should have enough slack to rattle slightly in the cell when you shake it. Lens cells vary, so we won't attempt to provide detailed instructions. Some lens cells are easily adjustable. Others are adjustable but require special tools. Still others aren't adjustable at all. Check the manual to determine if it is possible to adjust the cell.

HACK #36 Eliminate Diffraction Spikes and Increase Contrast

Build a $0 aperture mask to turn your $500 Dob into a $2,000 apo refractor.

Well, not really. But a simple, cheap hack can let your inexpensive Dob provide most of the benefits of an expensive apo refractor. Read on.

There's an old saying among astronomers: Aperture Rules. And it's true. A larger aperture gathers more light and provides higher resolution of fine detail. Assuming the larger scope has at least decent optics, you can simply see more with a larger scope than with a smaller one, period. That's the reason Newtonian reflectors, particularly Dobsonians, are so popular. They provide a lot of decent-quality aperture for not much money.

But if you attend a large star party, you'll see something puzzling. There'll be a lot of people using Dobs up to 20" or larger. There'll also be a lot of people using 8", 10", and 12" SCTs. But mixed in with these mid-size and large scopes, you'll see a fair number of people using 3" to 5" premium apochromatic refractors, such as those made by Tele Vue, Takahashi, and TMB. And those refractors, small as they are, aren't cheap. Even a small apochromatic refractor can easily cost $2,000 or more without a mount. The 5" models sell for $4,500 and up. So why would anyone pay so much for such a small scope? In a word, image quality. (Well, OK, two words.)

The big Newtonian reflectors and SCTs provide a lot of aperture, but they are obstructed scopes. They have a secondary mirror in the light path and, in the case of Newts, a spider assembly and secondary holder as well. Figure 3-9 shows a view looking down the tube of our Newtonian reflector. The 10" primary mirror is visible at the back of the tube. The secondary holder, held by its three spider vanes, is visible at the front of the tube.

Figure 3-9. The secondary holder and spider vanes of a Newtonian reflector

All of that gubbage in the light path has two undesirable effects:

Lower contrast
> Any *central obstruction* (CO) reduces image contrast relative to an unob-structed scope like a refractor. The amount of contrast reduction depends on the percentage of central obstruction. Typical Newts have a CO ranging from 15% to 25% linearly (2.25% to 6.25% by area). Slower Newts, such as f/8 **[Hack #9]** models, tend to have central obstruc-tions in the lower part of that range. Faster Newts, such as f/4.5 and f/5 models, typically have central obstructions at or near the top of that range. Typical SCTs and MCTs have a CO ranging from 33% to 40% linearly (10.9% to 16% areally).

At 4% or less areal CO, the effect of the central obstruction on contrast is unnoticeable. At 6% to 8%, the reduction in contrast becomes noticeable, particularly on low-contrast targets, such as planetary surface detail. At 8% to 10% areal CO, contrast reduction becomes severe; at 12% areal CO, many observers consider the contrast reduction unacceptable. (This is why SCTs, despite their flexibility and popularity, are not the best choice for planetary observing.)

Diffraction spikes

Diffraction from the circular secondary mirror in Newts and SCTs causes overall contrast reduction, but Newts have a further problem. Those straight spider vanes also cause diffraction, but rather than scattering the diffracted light over the whole image, they concentrate it into sharp spikes that are visible around bright objects, such as planets and bright stars. Many observers find diffraction spikes distracting, and they can actually interfere with some observing activities, such as splitting double stars.

Refractors have no central obstruction, and so do not suffer from these diffraction effects. The upshot is that refractors provide sharp, high-contrast views that are more esthetically pleasing to many observers than the views through a Newt or SCT. But the superiority of a refractor view is an illusion. A larger obstructed scope provides more image detail than a small refractor, but the image is not as pleasing. Still, a lot of astronomers are willing to pay large sums for that pleasing image.

Steve Childers, Paul Jones, and Robert did a field test one night in Steve's front yard. Steve set up his 10" Dob, Paul his 8" SCT, and Robert his 90mm refractor. We then turned all three scopes toward Jupiter. The planet was well placed, and the atmospheric stability (seeing) was excellent. We wanted to see how these three scopes, of widely differing types and apertures, performed on Jupiter. To keep the comparison fair, we ran similar magnifications—about 225X for the Dob, 230X for the SCT, and 210X for the refractor.

We won't keep you in suspense. We all agreed that the 90mm refractor provided the most esthetically pleasing views. They were sharp, crisp, and of exceedingly high contrast. But in terms of actual detail visible, the 10" Dob beat the 8" SCT and simply blew away the 90mm refractor. Aperture Rules.

If you have a medium or large Dob or other obstructed scope, you can gain most of the benefits of that $2,000 apo refractor at a cost of $0 and a few minutes' work. All you need is a sheet of cardboard, a ruler, a compass, a pencil, and a sharp knife. You'll use these to build an *off-axis sub-aperture mask*, usually called an *aperture mask* for short.

An aperture mask blocks the scope's aperture entirely except for a small off-center circular hole, cut so the secondary holder and the spider vanes do not intrude. Light enters the scope only through that off-axis hole. Because the secondary holder and spider vanes are masked off, they contribute no diffraction to the image. In effect, the large masked mirror functions like a small mirror without any obstruction.

To create an aperture mask for your scope, follow these steps:

1. Measure the inside diameter of the the tube and the diameter of the secondary holder. Measure the primary mirror, or simply use the size specified. In our case, the tube was 300mm inside diameter, the secondary holder 63mm, and the primary mirror 250mm.

2. Use your compass to draw three concentric circles of those sizes on a sheet of stiff, thin cardboard.

3. Use the ruler to draw a line from the center of the circle to one edge.

4. Measure the distance between the inner circle (the secondary edge) and the middle circle (the primary mirror edge). Call that distance X (X is half the difference between the diameter of the primary and the diameter of the secondary). In our case, X was about 93mm.

5. Calculate 0.5X and place a pencil mark at the mid-point of the radius line between the two inner circles.

6. Set your compass to 0.5X (less a couple millimeters for slop), put the center pin of the compass at the pencil mark you made in the preceding step, and score a circle. This circle should touch neither the inner circle (the secondary edge) nor the middle circle (the primary mirror edge). We could have drawn the aperture circle up to 93mm in diameter, but that would assume a perfect fit and no slop. Instead, we decided to use an 85mm circle to allow for some misalignment and slop while still being sure that neither the secondary holder nor the edge of the primary mirror would intrude into the aperture mask hole.

Intuitively, it seems as though it should matter whether you have a three-vane or four-vane spider. In fact, it doesn't. The limiting factor on the diameter of the aperture mask circle is the distance between the edge of the primary and the edge of the secondary in either case.

7. Carefully cut the cardboard around the outer circle to form the body of the aperture mask. Make radial 1/4" long cuts every couple inches around the circumference of the circle to make the aperture mask easy to seat and remove.

8. Cut out the inner off-center small circle that will be the aperture mask hole. Try to make a smooth, regular, circular cut, but don't worry too much about minor mistakes. (Even a square or oval hole works, although a perfect circle is optimum optically.)

To install the aperture mask, simply slide it into the front of the tube, arranging it so that none of the spider vanes are visible through the aperture. As it happened, we installed our aperture mask, shown in Figure 3-10, with the aperture hole near the focuser. We could have placed the mask between any two of the vanes. Unless your mirror is hideously bad, it doesn't matter which pair of vanes you choose.

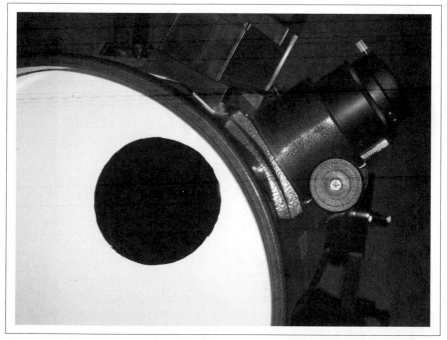

Figure 3-10. The aperture mask in place on our 10" Dob

So how does it work? Pretty well, actually, if you like the idea of turning a 250mm scope into an 85mm scope. The 85mm aperture-masked scope is a lot dimmer than a 250mm scope, and it can't resolve detail as finely. But it does provide the kind of image you'd expect looking through a $2,000 85mm apochromatic refractor. Sharp, snappy, high contrast, and visually pleasing, with no false color or other aberrations.

An inexpensive Dob with an aperture mask isn't the true equivalent of the expensive apo, of course. If it were, no one would buy apo refractors. If you think of the apo as a fine Swiss watch, the Dob is more like a crude Soviet tank. If you value fine workmanship, the apo refractor is simply in a different class.

The other difference is focal length. A typical 85mm or 90mm apo refractor has a focal length of 450mm to 600mm. Our 85mm aperture-masked Dob still has the native focal length of the underlying scope, 1,255mm. At high magnification, such as for Lunar and planetary viewing, the different focal lengths don't matter. But for rich field viewing, such as browsing Milky Way star fields, the shorter focal lengths of the apo refractors give them a huge advantage. Our aperture-masked Dob has a maximum possible true field of about 2°, versus twice that or more for the apo refractors.

Is it worth building an aperture mask yourself? Sure, why not? It costs next to nothing, takes only a few minutes to do, and you may fall in love with the apo-like images. But before you decide to make the aperture mask a permanent part of your observing kit, we suggest you do some side-by-side comparisons. View the Moon, planets, double stars, and other objects with and without the mask in place. We think you'll decide, as we have, that the advantages of larger aperture greatly outweigh the disadvantages.

HACK #37 Build a Film Can Collimating Tool
Align your scope on the cheap.

Collimation is the process of aligning a telescope so that all of the mirrors and lenses share a common optical axis. There are numerous collimation tools available commercially, including sight tubes, Cheshire eyepieces, laser collimators, autocollimators, and so on. Two of the most popular tools are a combination sight-tube/Cheshire and a laser collimator, such as the Orion models, shown in Figure 3-11.

None of these collimating tools allows you to collimate a scope perfectly. Their purpose is to get the collimation close enough that you can do final tweaks on a defocused star to achieve perfect collimation **[Hack #40]**. Star-collimation allows you to adjust alignment almost perfectly, but it's nearly impossible to star-collimate a scope unless it is already reasonably well collimated.

How close you need to get to perfect collimation before you can star-collimate depends on the focal ratio **[Hack #9]** of the scope. An f/5 or faster scope must be very close to perfect before it's possible to star-collimate it. An f/8 or slower scope need only be moderately well collimated.

Figure 3-11. A sight-tube/Cheshire (left) and a laser collimator

For fast scopes, we recommend using a combination sight-tube/Cheshire to do the preliminary collimation. But for slower scopes, there's no need to spend the $35 or so that a sight-tube/Cheshire costs. Instead, you can make your own sight tube for $0 and a few minutes' work. All you need is an empty 35mm film can, a sharp knife, a center punch or nail, and a small drill bit (we used a 1/16" bit, but anything close to that is fine).

Coincidentally, a film can is almost exactly the same diameter as a 1.25" eyepiece, and the lid is just large enough to prevent the can from sliding down into the focuser. A black plastic Kodak film can with a gray plastic top is ideal. To begin, use a sharp knife or scissors to cut the bottom off the can, as shown in **Figure 3-12**. Work carefully, and try to avoid bending the film can out of round. Discard the bottom of the can.

The next step is to create a perfectly centered peephole in the lid, which will allow you to place your eye exactly on the optical axis of the scope. Fortunately, the lid has a small raised nub that marks its exact center. Place the film can lid upside down on a flat surface, and use a center punch, nail, or heavy needle to mark the exact center of the lid, as shown in **Figure 3-13**.

Once you have a center-punched dent to prevent the drill bit from "walking," use the 1/16" drill bit to cut a clean, circular hole in the center of the film can lid. Snap the lid back onto the film can, and you're done. To use your film-can collimator, simply insert it into the focuser as you would an eyepiece, put your eye to the peephole, and verify that all of the optical components appear as concentric circles in your field of view **[Hack #38]**.

Figure 3-12. Remove the base of the film can using a sharp knife

Figure 3-13. Center punch the lid of the film can

 The exact procedures required to collimate a scope vary for different scopes. Refer to the manual for your scope for detailed instructions.

Tune Your Newtonian Reflector for Maximum Performance

Align your optics to provide the best possible image quality.

Collimation is the process of aligning all mirrors and lenses in a telescope so that they share a common optical axis. A well-collimated scope provides the best images its mirror or objective lens is capable of providing. A poorly collimated scope has significantly degraded image quality—how degraded it is depends upon how poor the collimation is.

To understand how important proper collimation is, consider two telescopes of the same aperture and focal ratio. One scope has a typical mass-market Chinese mirror, accurate to perhaps 1/4 wavelength or about 0.80 Strehl. (Strehl ratio is a statistical measure of overall mirror quality based on interferometry testing; a perfect mirror—impossible in the real world—has a Strehl ratio of 1.0.) The second scope has a premium mirror, made by a master optician such as Carl Zambuto or R. F. Royce. That premium mirror may cost 5 or 10 times as much as the mass-market mirror, and it is accurate to perhaps 1/20 or 1/40 wavelength, say 0.98 Strehl. There is no comparison between these mirrors. The first is mediocre. The second is world-class.

So we set up the two scopes and point them at Jupiter or Saturn. The inexpensive scope is perfectly collimated. The premium scope is just *slightly* out of collimation. How do the images compare? The cheap mirror beats the premium mirror, and not just by a little bit. The cheap mirror wins hands-down. The moral here is that if you want a premium scope but can't afford one, don't despair. Learn to collimate properly instead, and the images in your inexpensive scope will be at least as good as those in a typical premium scope. (Until you run into someone with a premium scope who also knows how to collimate. Oh, well.)

Proper collimation is particularly important for telescopes with fast focal ratios [Hack #9]. An f/10 SCT, for example, provides only very slightly degraded images, even if it is only roughly collimated. Conversely, an f/6 reflector must be collimated with reasonable care to provide anything near its best image quality. Around f/5.6 to f/5, precise collimation becomes critical. Large Dobs often have focal ratios of f/4.5 down to perhaps f/4.2, and we know of one large Dob with an f/3.5 mirror. At focal ratios that fast, precise collimation is not just critical but is an ongoing concern during observing sessions. For example, owners of very fast Dobs often collimate several times during an observing session. As time passes, the temperature drop is sufficient to take the Dob out of collimation. Even changing the elevation of the Dob from, say, 30° to 60° may require tweaking the primary collimation to bring the scope back into alignment. Needless to say, owners of very fast scopes soon become accomplished experts at collimating.

> There are actually two types of collimation: optical collimation and physical collimation. In a scope that is optically but not physically collimated, the optics are all aligned properly, but the optical axis does not coincide with the physical axis of the tube. For example, the optical axis may be off-center or tilted relative to the physical axis of the tube.
>
> Precise optical collimation is necessary for any type of scope if it is to provide good images. Precise physical collimation is unnecessary for a manual telescope but is important for a go-to scope or one that uses digital setting circles. This is true because the computerized object location function in such scopes assumes that the scope is both optically and physically collimated. If the optical and physical axes differ, the object locating functions of the scope will not function properly. If you have such a scope, refer to the manual for details about how to collimate it properly.

Collimating a Newtonian reflector has the reputation of being a black art, but in truth there's nothing very complicated about it. The secret, if there is one, is to do things in the proper order. Most beginners make the mistake of focusing their efforts exclusively on the primary mirror. That's understandable, we suppose. After all, the primary mirror is a big, impressive chunk of glass, and the primary collimation screws are readily accessible. But you can collimate the primary mirror until you're blue in the face, and it won't help unless you've first collimated the secondary mirror. Here's the proper sequence of steps:

1. Adjust the position of the secondary mirror holder until the secondary mirror is centered under the focuser.

2. Adjust the rotation and tilt of the secondary mirror until the primary mirror is exactly centered in the reflection from the secondary mirror.

3. Finally, adjust the tilt of the primary mirror to bring its optical axis into alignment with the common optical axis of the focuser and secondary.

The first step in collimating a Newtonian reflector is to center the secondary mirror under the focuser. To begin, insert your sight tube or film-can collimating tool [Hack #37] in the focuser, as shown in Figure 3-14, and peer through it to view the secondary mirror.

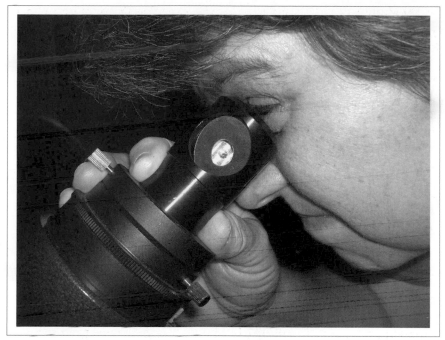

Figure 3-14. To begin collimating the secondary, insert the sight tube in the focuser

The secondary mirror is actually elliptical, but it is set at a 45° angle (to reflect at 90°), so from the focuser it appears to be a circle. Slide the sight tube in or out of the focuser (or use the focuser itself) to adjust the position of the sight tube until the circular edge of the secondary mirror is exactly concentric with, and just inside, the circle formed by the bottom of the sight tube, as shown in **Figure 3-15**.

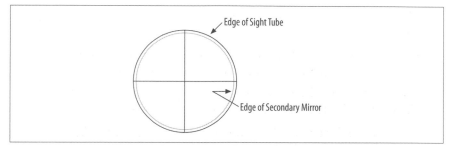

Figure 3-15. Adjust seating depth until the edges of the secondary and sight tube are concentric

The focusers on most mass-market scopes, including inexpensive Dobs, typically have relatively loose tolerances and quite a bit of slack in them. For example, you may be able to wiggle the sight tube significantly within the focuser, changing the apparent position of the secondary mirror. Similarly, the focuser may abruptly shift position noticeably when you reverse focusing direction.

Short of replacing the stock focuser with an expensive after-market Crayford focuser, the best solutions are either to use Teflon tape to remove some of the slack or simply to try to strike a happy medium. That is, set the focuser at a midrange position, where you would typically use it when actually observing, and tighten the setscrew against the sight tube, just as you would if you were using an eyepiece.

The "hall of mirrors" effect caused by the multiple reflections between the secondary and primary mirrors can be confusing when you use a sight tube. To eliminate this problem until you have the secondary mirror collimated properly, ask a helper to hold a sheet of paper between the secondary and primary mirrors, as shown in **Figure 3-16**. Blocking the primary mirror this way eliminates confusing multiple reflections. Depending on your scope, it may also be helpful to put a sheet of white paper on the inside of the tube, opposite the focuser, to make the edge of the secondary mirror easier to see.

Figure 3-16. Hold a sheet of paper between the secondary and primary to block multiple reflections

If the secondary mirror appears circular and is centered in the sight tube, you're ready to proceed to the next step. If the secondary mirror is offset up or down the tube, you'll need to slide the secondary mirror holder in or out until the secondary mirror is centered. Most secondary holders have a single central retaining screw, shown in **Figure 3-17**, that must be loosened to adjust the position of the secondary holder. (Yes, our scope is dusty; that's what happens to a working scope.)

Figure 3-17. Loosen the secondary retaining screw to adjust the position of the secondary

Be very careful when loosening the secondary retaining screw. If it reaches the end of its travel, the portion of the secondary holder that holds the mirror may drop free and fall onto your primary mirror, damaging the primary mirror, the secondary mirror, or both. That's a good reason to remove the primary mirror cell before you collimate the secondary. Alternatively, keep the scope tube horizontal so that if the secondary holder does fall it won't slide down the tube and bounce off the primary mirror.

In particular, some early Dobsonians made by Synta and sold by Orion and others had secondary retaining screws that were much too short to allow a full range of adjustment on the secondary. The first time you collimate your secondary, verify that the screw is long enough to support the secondary holder safely. If it is not, take the original screw to a hardware store and get a longer version with the same thread.

To move the secondary mirror closer to or farther from the primary, loosen the retaining screw and adjust the overall seating depth of the three adjustment screws visible in Figure 3-17. Drive all three screws farther in to move the secondary mirror closer to the primary, or back all three screws farther out to move the secondary toward the front of the tube and away from the primary.

The next step is to adjust the rotation and tilt of the secondary mirror so that the primary mirror appears to be centered in the reflection from the secondary. If you have a laser collimator, use it to perform this step. To do so, insert the laser collimator into the focuser and turn it on. Look down the tube to see the spot of the laser on the primary mirror. Use the three adjustment screws on the secondary holder to adjust the rotation and tilt of the secondary mirror until the laser beam strikes the primary mirror in the middle of the center spot [Hack #33], as shown in Figure 3-18.

Figure 3-18. The view looking down the tube while using a laser to collimate the secondary mirror

The primary mirror is the circular object at the center of the image, surrounded by the gray of the scope's tube, with the focuser visible as a blurred image at the upper right and the secondary holder even more blurred at the lower left. The primary mirror reflects a crescent of the upper part of the tube (at left), the ceiling of the room where this image was taken, the spider vanes that hold the secondary, the secondary mirror itself, and part of Robert's arm and hand. The two bright spots reflected at the upper right of the primary mirror are the bottom of the laser collimator—reflected in the secondary mirror—and the laser spot on the secondary mirror itself. None of that matters, but it is visually confusing. What does matter is the laser spot on the primary, which is the bright spot at the center of the primary, is centered in the notebook reinforcing ring. Having that laser spot centered in the reinforcing ring means that the secondary is collimated properly, pointing to the center of the primary mirror.

If you don't have a laser collimator, you can use the sight tube for this step. To do so, adjust the tilt and rotation of the secondary mirror (or have someone do it for you) as you view the reflections through the sight tube. When the secondary mirror is collimated properly, you will see the secondary mirror edge concentric with the edge of the sight tube, the primary mirror edge concentric with the secondary mirror edge, and the crosshairs of the sight tube centered on the center spot of the primary mirror.

Once you have the secondary properly collimated, the hard part is done. Fortunately, you probably won't have to repeat this process unless you remove the secondary mirror for cleaning, install an upgraded focuser, or take some other action that affects secondary collimation. Most secondary holders maintain collimation well. We seldom have to adjust secondary collimation from one year to the next.

The final step in collimating your scope is to collimate the primary, which consists of adjusting the primary mirror tilt until the optical axis of the primary is the same as the optical axis of the secondary and focuser. Most mirror cells use three adjustment screws (see Figure 3-19) and three locking screws. Turning an adjustment screw in or out slightly changes the tilt of the primary. Once the primary is collimated, you tighten all three locking screws to clamp the mirror cell in place. It's much easier to collimate the primary if you have a helper. One person sits or lies at the rear of the scope and tweaks the adjustment screws while the other person watches the effect from the front of the scope.

The fastest, easiest, and most precise way to collimate the primary is to use a Barlowed laser [Hack #39]. If you don't have a laser collimator, you can use a Cheshire or sight-tube/Cheshire to collimate the primary, albeit somewhat less precisely.

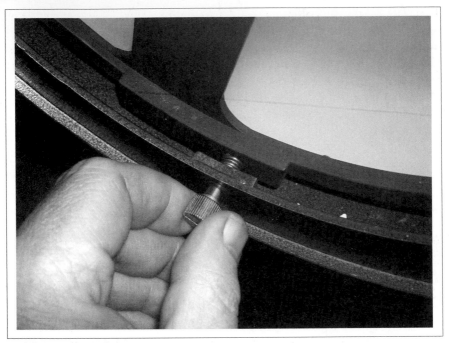

Figure 3-19. A primary mirror cell adjustment screw

To do so, illuminate the fine-ground 45° reflective surface of the Cheshire (visible at the center of Figure 3-14). If you are collimating after dark, point your red flashlight at the Cheshire opening. Tweak the primary adjustment screws until the center mark on the primary mirror coincides with the donut shape visible in the Cheshire.

Collimating the primary without a helper can be confusing because you constantly have to move between the rear of the scope, where the adjustment screws are, and the front of the scope, where you can view the effect of the adjustments you've made. If you frequently collimate the primary by yourself, it's worthwhile to create a graphic map on a 3×5 card to indicate the effects of tightening each of the three primary collimating screws. For example, if you are using the Barlowed laser method to collimate your primary **[Hack #39]**, you might note that tightening primary collimation screw #1 (label the screws on your map) moves the shadow toward 3 o'clock; tightening screw #2 moves the shadow toward 7 o'clock; and screw #3 moves the shadow toward 11 o'clock.

That's all there is to collimating a Newtonian reflector. It takes a few minutes to collimate the secondary, and perhaps a minute or two to collimate the primary. Surprisingly, many Newtonian reflector owners seldom collimate their scopes, and most never collimate them. But the 5 or 10 minutes you spend to get your scope properly collimated pays off in much better image quality and more visible detail, particularly on planets and other low-contrast objects.

You're not quite finished, though. When you finish using your sight-tube/ Cheshire and laser collimator, the scope is very close to being perfectly collimated. Close, but not close enough. The final step to achieving perfect collimation is to use a defocused star to tweak the collimation to perfection [Hack #40].

HACK #39 Collimate Your Primary Mirror Quickly and Accurately

Use a Barlowed laser to show the error of the primary only, taking the secondary out of the equation.

In the January 2003 issue of *Sky & Telescope* magazine, Nils Olof Carlin describes a simple and accurate method of collimating the primary mirror of a Newtonian reflector. Carlin's method uses a laser collimator and Barlow lens in combination, as shown in Figure 3-20.

Figure 3-20. Insert the Barlow between the laser collimator and the focuser

The new method is preferable in several ways to the standard one in which the laser collimator is used alone. The most striking improvement is that the new method is insensitive to collimation errors of the other elements in the optical path, namely the secondary and the focuser. These should have been collimated before the tilt of the primary is adjusted, of course, but decoupling the primary from the other elements vastly simplifies the final part of the work, namely, adjusting the primary mirror.

Carlin's method, like the standard one using a simple laser collimator, involves inspecting the light coming back from the primary. In both cases, the reflected light returns to a target, and the tilt of the primary is adjusted until the return beam is centered on this target. With the usual method, the target is the face of the laser, and the narrow return beam must be centered on the hole in the face. In Carlin's method, the beam is a diverging one, and the target is a disk of white cardboard, pierced by a hole at its center, which may be freestanding or attached to the Barlow. The tilt of the primary is adjusted until the shadow of the center spot you have affixed to the primary mirror falls on the hole in the target.

Both methods become a bit clumsy when the target—the laser face or the bottom of the Barlow—sits so high in the focuser tube that it is not easily seen from the front of the scope. Carlin suggests using an auxiliary mirror to allow a better view of the target. We have tried this and find it inconvenient. What has worked much better for us is a target which extends a bit into the body of the scope. This is easily visible from the front of the tube and, being located at fingertip distance, allows the shadow of the center spot to be moved with sub-millimeter accuracy. (Not that such accuracy is needed: the position of neither the center spot on the primary nor the center of the target is likely to be all that accurate.)

There are many ways to construct such a target. All that is needed is a tube that can be inserted into the focuser from below, of such diameter and surface as to fit snugly in place. The target is glued to the tube at its lower end. Our tube, with target attached, fits the 2" focuser on our XT10 Dob. Had we been using a 1.25" focuser, we could instead have used a plastic 35mm film container, such as is commonly used to make a collimating cap by cutting away the bottom and drilling a small hole at the center of the top [Hack #37].

Although you could be more elegant, we used a PVC plumbing fitting, one normally used to couple a threaded metal pipe to a smooth PVC pipe. We chose it to be loose enough in the focuser tube that its diameter could be shimmed out with self-adhesive foam, namely, by four Scotch Brand Mounting Squares, each 1/16" thick and 1" square, that were on hand. The foam has just enough give to allow the fixture to be easily slid in and out, to be held firmly in place, and to not scratch the inside of the focuser tube.

Our target, epoxied to the lower end of the PVC fixture, was made from a piece of index card stock, which was easy to trace the circle on, to cut out, and to punch through with a hand-held paper punch. The card stock proved to be a bit too thin: sufficient light from the illuminated back came through the front to make the shadow of the center spot a bit difficult to see. Rather than replacing the original target and constructing a new one, we simply painted over the front of the target with white latex paint. Latex dries flexible and would not flake off in use, depositing flakes on the optics as a hard-drying Krylon or spray paint might have done. Figure 3-21 shows our completed target, inserted in the bottom of the 2" focuser.

Figure 3-21. The completed Barlowed laser target in the focuser

Before first using the target to perform an actual collimation, we determined the direction on the target (e.g., toward the finder, toward the open end of the tube, etc.) toward which the shadow of the center spot moves when we tightened each of the primary collimation screws separately. With that information recorded on a handy reference card, we found that tweaking the primary into precise collimation takes only about a minute after inserting the laser-Barlow combination into the focuser tube from above. Figure 3-22 shows the appearance of the Barlowed laser on the target when the primary is slightly out of collimation. The donut-shaped shadow of the center spot on the primary mirror is shifted slightly away from the hole through which the laser exits.

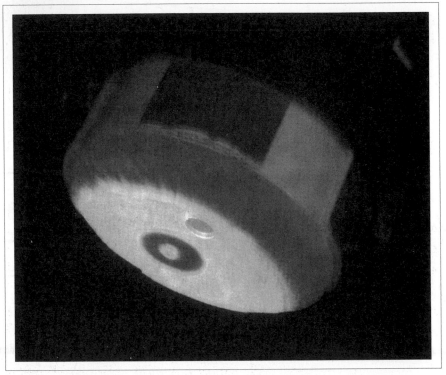

Figure 3-22. An offset center spot shadow indicates miscollimation

Figure 3-23 shows the appearance of the Barlowed laser on the target when the primary is collimated. The donut-shaped shadow of the center spot on the primary mirror is centered on the hole through which the laser exits.

You can even use the Barlowed laser method to collimate your primary mirror if you've forgotten to pack your target. Simply punch a small hole in a piece of heavy paper or thin cardboard stock to make a field-expedient target, as shown in Figure 3-24. Insert the Barlow into the focuser, the laser collimator into the Barlow, and turn on the laser. Slide the target around until the laser beam exits through the hole in the card, and examine the return pattern on the face of the card. When the donut-shaped shadow of the center spot is centered on the laser exit hole, your primary is collimated.

Using a Barlowed laser allows you to collimate your primary collimation very accurately, but not perfectly. For observing DSOs and other objects for which collimation is non-critical, this level of collimation may suffice. But, for planetary and other critical observing, don't forget to do a final tweak to the primary collimation using a defocused star **[Hack #40]**.

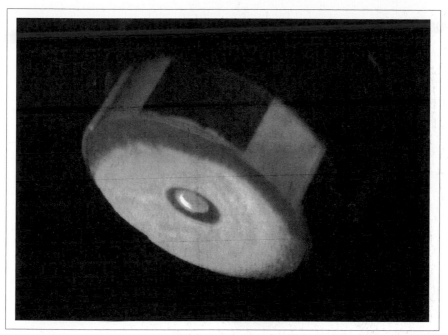

Figure 3-23. A center spot shadow concentric with the laser exit hole indicates the primary is collimated

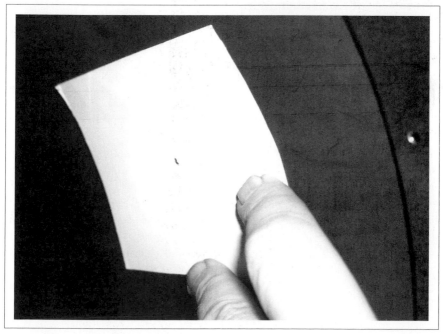

Figure 3-24. A field-expedient Barlowed laser target

Collimating the primary by using Carlin's Barlowed laser method does require that the batteries in the laser be reasonably fresh: the Barlowed return beam is much fainter (because of the large area it covers) than is the concentrated return beam of the un-Barlowed laser. But, as far as I can tell, changing the batteries in the laser a bit more often is the only price you pay to use what is clearly a much cleaner and faster way of collimating the primary than any other I have seen.

—*Gene A. Baraff*

HACK #40 Star-Collimate Your Scope
Align your scope perfectly using the properties of light.

Collimation is the process of aligning a telescope so that all of the mirrors and lenses share a common optical axis. In order to provide the best possible images, a scope must be collimated *perfectly*. Not just "collimated pretty well" or even "almost-perfectly collimated." An almost-perfectly collimated scope may show you 50% or less of the detail that is visible when collimation is dead-on.

Just as no pianist would play an untuned piano, no astronomer should observe with an uncollimated scope. And yet, very few astronomers collimate sufficiently often or sufficiently well to get the best performance possible from their scopes. Newbies never collimate. They're afraid they'll muck things up beyond repair. SCT and refractor owners seldom collimate. They're convinced their instruments don't require frequent collimation. Newtonian owners generally collimate fairly frequently, but most simply use a sight-tube/Cheshire and/or a laser collimator to get their scopes roughly collimated and then declare the job Good Enough. They're all wrong.

No physical collimation tool can ensure anything more than a rough collimation, and that's simply not good enough. The *only* way to collimate a scope properly is to defocus a star and observe the patterns that exist on both sides of proper focus. That's true for two reasons. First, physical collimation tools are accurate to only a few tenths of a millimeter, while star-collimation uses diffracted light patterns that are several orders of magnitude more precise. Second, the physical center of a lens or mirror does not always correspond to its optical center. In real-world telescopes, the difference may be only a fraction of a millimeter, but that is sufficient to prevent perfect collimation using only physical methods.

Don't confuse star-collimating with star-testing. The purpose of star-collimating is to get the scope's optics aligned precisely. The purpose of star-testing is to check the quality of the optics. Star-testing requires perfect seeing (atmospheric stability) if it is to provide meaningful results. Star-collimating can be done even under mediocre seeing conditions.

Fortunately, it's pretty easy to star-collimate a scope. To begin, get your scope roughly collimated using your sight-tube/Cheshire, laser collimator, or other physical collimation tools [Hack #37][Hack #38]. In a Newtonian reflector, it's important at this stage to get the secondary mirror collimated as closely as possible. Once the scope is roughly collimated, you're ready to start star-collimating. Take the following steps:

1. Point the scope at a 0th or 1st magnitude star. We generally use Polaris. It's bright enough for this step, and it has very little apparent motion. (It's critical to keep the star centered in the field of view during these tests.)

2. Use an eyepiece or eyepiece/Barlow combination that yields a magnification equal to the aperture in millimeters, plus or minus 20%. For example, for a 10" (250mm) scope, 250X magnification is ideal, but anything in the 200X to 300X range is acceptable.

3. Defocus the star until it becomes a huge white blob that fills the field of view. There will be a dark circle at or near the center of the field of view, which is the shadow of the secondary mirror. That shadow should be perfectly circular and perfectly centered in the field of view. If it is not, your rough collimation was rougher than you thought.

4. Use the primary mirror collimation screws to center the secondary shadow. Having a helper adjust the collimation screws while you observe the image makes this step much faster and easier. During this process, you'll have to refocus the star frequently, ensure it's centered in the field of view, and then defocus quickly to check the centering of the secondary shadow.

Once you complete these steps, the scope is roughly collimated and you're ready to fine-tune the collimation. (Most scopes, once collimated, hold collimation fairly well, so it shouldn't be necessary to repeat the above steps each time you collimate the scope.) To fine-tune the collimation, take the following steps:

1. Point the scope at a 2nd or 3rd magnitude star. (Once again, it's important to keep the star centered in the field of view during these tests.)

2. Use an eyepiece or eyepiece/Barlow combination that yields a magnification equal to two or three times the aperture in millimeters. For example, for a 10" (250mm) scope, we use 500X to 750X for tweaking the collimation.

3. Defocus the star *slightly* until it shows a complex diffraction pattern similar to that shown in **Figure 3-25**. The center image shows the point of exact focus and the images to either side show the patterns that result as you move the focuser increasingly far inside and outside focus. The exact pattern is not important. That will vary with differences in optical quality, magnification, and many other factors. What is important is that the diffraction rings appear circular and concentric. If the rings are oval and the center bright spot is off-center, tweak the primary mirror collimation screws until the pattern is as circular and concentric as you can get it. Check both sides of focus.

Figure 3-25. Stellar diffraction patterns from inside focus to focused (center image) to outside focus

We generated the patterns shown in **Figure 3-25** with the free software utility Aberrator (*http://aberrator.astronomy.net*). These patterns are only examples. They assume an optically perfect 250mm f/5 scope with a 25% central obstruction and perfect seeing conditions. The patterns you see when you collimate your scope will certainly be different.

Atmospheric turbulence may break the patterns up or cause jaggies or wavering. With even moderate turbulence, you'll still be able to judge when the patterns are close to being circular and concentric. If the atmosphere is too turbulent to allow that, critical collimation doesn't matter because you won't be able to see fine detail anyway.

If the patterns are concentric but different on opposite sides of focus, that means your mirror or objective lens is not optically perfect. Don't worry too much about that. Even the best-quality telescopes don't necessarily show identical patterns inside and outside focus.

Counterweight a Dobsonian Scope

HACK
#41   Balance your Dob perfectly without using springs, clutches, or other kludges.

Traditional equatorial and alt-azimuth mounts lock the scope into position and drive it using gears. For small changes in direction, you move the scope by driving the gears with motors or manual slow motion controls. For large changes in direction, some mounts allow you to disengage the gears, pivot the scope, and then re-engage the gears. But while you are observing, the scope is always clamped into position.

Conversely, a Dobsonian mount uses simple friction bearings that allow the *optical tube assembly (OTA)* to move freely. When you want to point the OTA in a Dobsonian mount to a different part of the sky, you simply push the OTA up or down, right or left. The bearing friction must be small enough to allow smooth motions, but at the same time the friction of the altitude bearings—the bearings that support vertical movement—must be large enough to keep the OTA from shifting position under its own weight.

The altitude bearings must also have sufficient friction to accommodate small changes in weight. For example, removing one eyepiece before inserting another unbalances the scope because the front of the scope is suddenly lighter. The altitude bearings must have sufficient friction to keep the scope from swinging upward. Conversely, if you replace a light 1.25" eyepiece with a heavy 2" eyepiece, the front of the scope may have an extra pound or more pressing it down. If the altitude bearings have insufficient friction, the front of the scope nosedives.

The best way to accommodate these opposing requirements—low friction for good motions versus higher friction for tube stability when the weight changes—is to balance the OTA precisely by moving the altitude bearings forward or backward on the OTA as necessary to achieve balance. For example, if your heaviest eyepiece weighs 18 ounces, you'd place the altitude bearings so as to exactly balance the scope with a 9 ounce eyepiece installed. That way, the front of the OTA would never be more than plus or minus 9 ounces from exact balance, with or without an eyepiece in place. The friction of your altitude bearings could be low enough to provide only 10 ounces or so of resistance—enough to keep the tube from drifting on its own, but light enough to provide very easy altitude motion.

The problem is that the Dob manufacturer can't balance the tube for you because they don't know what you're going to hang on the front or back end. For example, you may decide to add cooling fans to the primary mirror cell, which makes the OTA tail-heavy. Or you may decide to replace the light optical finder supplied with the scope with a heavy 50mm optical finder and a Telrad, which makes the OTA nose-heavy.

The best solution is to make the altitude hubs movable, and that is, in fact, the method used by high-quality tube Dobs. With movable altitude hubs, you can simply slide the OTA backward or forward to reach the balance point. But adjustable altitude hubs are a relatively costly feature to implement. The ubiquitous, inexpensive Chinese and Taiwanese Dobs—which are sold under numerous brand names including Orion, Celestron, Skywatcher, and others—use a cheaper and less desirable method. Instead of providing a mechanism to balance the tube, they use springs or clamps to increase the friction on the altitude bearings.

Figure 3-26 shows the spring used to adjust altitude bearing tension on our 2000-vintage Orion XT-10 Dob. (This scope was made by the Taiwanese company Guan Sheng; current Orion Dobs are made by the Chinese company Synta, and they use clamps rather than springs.) The two small white pads visible at left and right center of the image are the Teflon pads upon which the circular altitude hub rotates.

Figure 3-26. A spring pulling down on the altitude hub increases friction on the altitude bearing

But increasing altitude bearing friction doesn't eliminate balance problems; it simply conceals them. With the springs or clamps in use, the OTA doesn't drift under its own weight or when you change eyepieces, but this is achieved at the expense of smooth altitude motion. It's difficult or impossible to get decent altitude motion if you use these springs or clamps.

What are the alternatives, then? You could drill a series of holes in the OTA and use trial and error to locate the proper position for the hubs to achieve perfect balance. That may be a viable solution for you if your Dob is tail-heavy. Unfortunately, most econo-Dobs are nose-heavy. Their makers balance them with a light optical finder and the light eyepieces bundled with the scope. By the time you add a heavy 50mm RACI finder, a Telrad, and a heavy premium eyepiece, the OTA is nose-heavy, and often by a lot.

Figure 3-27 shows a typical Dob (ours) after upgrades. Left to right are a 50mm RACI finder, the Telrad, and a 14mm Pentax XL eyepiece that weighs about 13 ounces. All told, this scope has about 20 ounces more on the front end than the manufacturer balanced it for. That doesn't sound like much, but it's enough extra to make the scope nosedive unless it is counterweighted.

Figure 3-27. A front-end heavy Dob after typical upgrades

You might think the easy solution would be to relocate the altitude hubs nearer the front end of the OTA, which would indeed work. Unfortunately, doing that requires additional clearance at the rear of the OTA, and the bases supplied with econo-Dobs have almost zero extra clearance. When we measured our 10" Guan Sheng Dob, for example, we found that we could relocate the altitude hubs at most about 1/2" farther up the tube—not enough to do us much good. A change of more than 1/2" prevents the OTA from reaching vertical because the rear end of the OTA hits the baseboard of the Dob.

Time for Plan B. If the front of the OTA is too heavy and we can't shift the balance point, the only solution is to make the rear of the OTA heavier to bring things back into balance. The way to do that is to add counter-weights, ideally ones that are adjustable in weight and position to allow rebalancing as necessary.

> Dob owners use an incredible array of objects to counter-balance their scopes, from scuba diving weights to bean bags filled with lead shot to heavy magnets. We prefer magnets, which are cheap, compact, dense, and easy to move around as needed to rebalance the scope when you make changes to it. Magnets are a particularly good choice if you have one of the Chinese or Taiwanese Dobs that uses a steel tube. You just stick them to the OTA and they stay stuck. If you have a Sonotube (cardboard) OTA, you'll need to use tape, bungee cords, cable ties, glue, or other means to affix the weights.

Many Dob owners who add counterweights treat their OTAs like a see-saw, concerned only with balancing them on the longitudinal axis (over the length of the OTA). That's a mistake. You also need to consider the radial axis, which is to say where the various weights are located around the circumference of the tube.

A Dob can move from 0° altitude, when it is pointed at the horizon, to 90° altitude, when it is pointed at zenith. As you move the OTA through its range of vertical motion around the altitude bearing axis, the relative positions of the various weights and their moment arms change. For example, if you counterweight the OTA at the rear center of the tube so that the OTA is properly balanced when it is elevated 45°, you have too little counterweight when the OTA is below 45° and too much when it is above 45°.

Some cunning astronomers use a short, heavy chain attached to the rear of the OTA. When the scope is horizontal, the full weight of the chain counterweights the tube. As the OTA is elevated toward zenith, more and more of the chain comes to rest on the base of the Dob, reducing the amount of counterweight.

The problem with counterweighting at the rear center or rear top of the OTA is that the axis between the counterweight and the offsetting weight at the front of the OTA doesn't pass through the center of mass. Ideally, you want the center of mass to be located at the exact midpoint between the altitude bearings. The best way to approximate this ideal is to eyeball the weights on the front of the scope to determine the approximate position of their center of mass. In our case, looking from the front of the scope, that center is at about 1 o'clock, roughly 6" from the front of the tube.

Once you have determined the center of mass of the front-end components, imagine an axis passing from that center of mass through the balance point between the altitude hubs. Where that axis intersects the rear of the OTA is the point where the counterweight should be. Figure 3-28 shows our counterweight at about the 7 o'clock position (as viewed from the front of the scope).

Figure 3-28. A counterweight placed opposite the center of mass of the front-end components

We used a 19-ounce bucking magnet, wrapped in duct tape to prevent scratching. These magnets are the shape of a flat donut, and they have a very strong grip. We bought six of these magnets from Parts Express (*http://www.partsexpress.com*) for about $1.50 each, plus shipping, but you may be able to find them at a local electronics store. With the steel backing plate in the mirror cell, one 19-ounce magnet suffices. With the backing plate removed, which is how we normally use the scope, a second magnet is needed.

> A magnet works well if just one is heavy enough. If you need two or more for extra weight, magnets can become problematic because they attract or repel each other strongly, depending on their orientation. Stacking two magnets with the north pole of one against the south pole of the other is an easy solution, if the stack fits the available space. (We couldn't use a double stack at the 7 o'clock position, because the stack wouldn't clear the Dob base when we swung the OTA vertical.) In that situation, simply move the magnets around until you find a configuration where the scope balances properly and the magnets do not interfere with each other or with moving the OTA to vertical.

HACK #42 Improve Dobsonian Motions with Milk Jug Washers

Eliminate the jerky azimuth motions that start to plague your scope after a while.

The azimuth (right and left) motions of inexpensive Dobs may be adequate when they are new, but often the azimuth motions become rough and heavy as the scope breaks in. This occurs because inexpensive Dobs use hard plastic or mechanical Teflon pads riding on a smooth laminate surface, and the friction and stiction (static friction) of this combination are difficult to control.

The real fix is to replace the bearings with virgin Teflon pads riding on bumpy Ebony Star laminate, but that requires spending money and devoting several hours' work to the project. But there is a fast, free solution, albeit a temporary one. You can supplement the existing azimuth bearings by adding a center bearing made up of several thicknesses of plastic washers cut from flat-sided milk jugs. (Although milk-jug washers are traditional, some astronomers use old AOL CDs or even LP records; they're called milk-jug washers regardless.)

The improvement in azimuth motions can be significant and, depending on the material you use for the washers, may last for dozens of observing sessions. When the washers begin to lose effect, it takes only a few minutes to remove the old washers and replace them with fresh ones.

To make the washers, cut full circles from the flat portion of the milk jug, or simply use recycled CDs. (In fact, a CD also makes an excellent template for cutting milk jug washers.) Don't worry too much about the size. We used washers about the diameter of a CD, but even 2" washers work fine. (The sole purpose of the washers is to bear some of the weight that would otherwise be borne by the azimuth pads; central washers contribute almost no friction.) Punch a hole in the center of the washer that's large enough for the Dob center bolt to pass through without binding. If you use CDs and the center hole is too small, use a file or reamer to enlarge it until the Dob center bolt slides freely within the hole.

You'll need to experiment to determine how many washers to use. The number you need varies with the thickness of the material and your particular scope. (Our scope needed five washers cut from milk jugs or two CDs.) If the washer assembly is too thin, it makes no contact and accomplishes nothing. If it is too thick, it takes all the weight off your existing azimuth bearing pads and can actually make the motions worse. The goal is to let the center bearing washer take just part of the weight.

To install the washers, remove the nut from the center bolt and lift the rocker box free of the baseboard. Place a few of the washers over the center bolt, as shown in **Figure 3-29**, replace the rocker box, and test the azimuth motion. (You needn't replace the nut on the center bolt each time; it's there simply to keep the rocker box connected to the baseboard, but it applies no tension.)

When you reach the proper number of washers, the improvement in azimuth motion should be obvious immediately. You should be able to start the scope moving with the gentlest of finger pressure, and keep it moving smoothly with almost no discernible pressure. When you stop pressing, the scope should stop moving instantly, without any overshoot.

It's best to try the washers without any lubrication first. If they work to your satisfaction, great. If you'd like the motions to be lighter still, try lubricating the washers slightly. We've tried everything from machine oil to Vaseline to WD-40 to paste wax to spray-on furniture polish to graphite lock lubricant, with both milk jug washers and CDs.

Most lubricants seem to reduce friction well enough, but at the expense of increasing stiction. Stiction is static friction, which gives the washers a kind of undesirable inertia. Although they move freely once they are moving, it takes significant pressure to get them moving in the first place. Stiction makes it very difficult to track objects, particularly when you are using high magnifications. We got the best results with graphite lock lubricant, but, as always, your mileage may vary.

Figure 3-29. Installing milk jug washers (actually, CDs in this case)

If milk jug washers work for you, keep a spare set or two in your observing kit. They stop working as they become worn, and once they start degrading they go downhill fast. Still, it takes literally a couple of minutes to replace them, and they may be all you need.

HACK #43 Upgrade Your Dobsonian Bearings

Use Ebony Star and Teflon for butter-smooth motions.

Dobsonian telescope mounts move in altitude and azimuth on friction bearings. There are two double *altitude bearings*, one on each side of the telescope tube or mirror box. The scope pivots vertically from horizon to zenith on these altitude bearings, one of which is shown in **Figure 3-30**.

The scope rotates horizontally on one triple *azimuth bearing*, shown in Figure 3-31. On the left is the bottom of the circular baseboard, which is covered in Ebony Star laminate. The triangular groundboard, the top surface of which is shown on the right, has Teflon bearing pads applied directly over its feet. The two boards are loosely joined by the azimuth bolt, visible in the center of the groundboard. When the two components are assembled, the groundboard sits flat on the ground, with the baseboard above it.

When the Dobsonian scope is rotated in azimuth, the laminate surface rides on the Teflon pads, providing smooth movements.

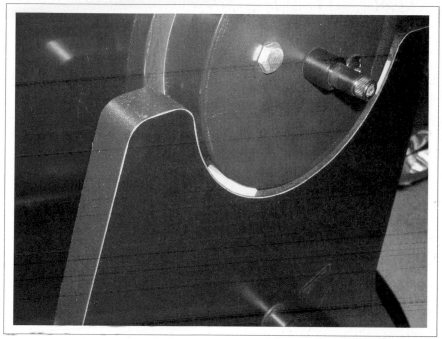

Figure 3-30. An altitude bearing

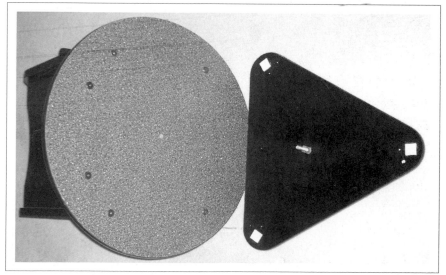

Figure 3-31. The azimuth bearing

This arrangement seems simple enough in theory, but in practice there are three requirements for ideal Dob bearings:

- Ideal bearings have perfect friction, not too much and not too little. The friction must be low enough to allow you to start the scope moving with the gentlest of fingertip pressure, but high enough to stop the scope's motion instantly when you stop applying pressure. Also, altitude bearing friction must be high enough to allow you to swap eyepieces in and out without the scope moving as the weight on the front of the tube changes.

- Ideal bearings exhibit zero *stiction* (STatic frICTION). Stiction is, in effect, variable friction. In a scope that suffers from stiction, you must apply significant pressure to start the scope moving. But as the scope begins to move, it "breaks loose" and suddenly begins moving much more freely. The motions of such a scope feel jerky. It is very difficult to track objects, particularly at high magnification, in a scope that suffers from stiction.

- Ideal bearings require exactly the same effort to move the scope in altitude as in azimuth. There should be no feeling of having to move the scope discretely on two axes, but rather a sense that the scope pivots freely in any direction.

Numerous combinations of materials have been used for Dob bearings—in fact, some astronomers experiment constantly with different bearing materials and surface treatments—but the most commonly used and recommended combination is virgin Teflon pads bearing on Ebony Star laminate. The combination of virgin Teflon and Ebony Star has an unusual property: its dynamic coefficient of friction and static coefficient of friction are nearly the same and can be made almost identical by applying paste wax to the Ebony Star surface.

Unfortunately, virgin Teflon and Ebony Star laminate are more costly than some alternatives, so the makers of inexpensive Dobs make do with cheaper materials, such as nylon or mechanical Teflon pads riding on some sort of smooth laminate surface. Alas, those combinations have inferior friction/stiction characteristics, so inexpensive Dobs nearly always have inferior motions. Their manufacturers use several tricks to cover up bearing (and balance) problems, including springs, clamps, and other workarounds. Ultimately, though, none of those workarounds actually works very well because cheap bearings simply don't have the proper friction/stiction characteristics.

The real solution is to balance your Dob properly [Hack #41] and to replace the bearings with virgin Teflon riding on Ebony Star laminate. Fortunately, it's easy to do that, and it costs only $20 to $40 for a typical small to mid-size Dob. Here are the materials and tools you'll need:

- A sheet of Ebony Star laminate large enough to provide a circle the size of your baseboard and two strips wide enough to cover the bearing surfaces on your altitude hubs and long enough to cover most or all of their circumference. For a typical small or mid-size Dob, a 2-foot square sheet suffices. That will provide a 20" or 22" circle for the baseboard and the two 3/4" or 1" strips needed for the altitude hubs.

- Seven or eight small virgin Teflon pads, 3/16" thick. You'll need three pads for the azimuth hubs, four for the altitude hubs, and (optionally) a center pad [Hack #42]. For small to mid-size Dobs, the azimuth pads should be about 1" square and the altitude pads 3/4"×1". That means you can cut all seven pads from one 1"×6" strip of virgin Teflon. If you use a center pad, make it 1.5" square.

> For larger Dobs, increase the size of the pads and use a center pad. For Dobs from 75 to 125 pounds, use 1.5" square azimuth pads, 1"×1.5" altitude pads, and a 2.5" square center pad. For Dobs from 125 to 250 pounds, use 2" to 2.5" square azimuth pads, 1.5"×2" altitude pads, and a 3.5" to 4" square center pad. The goal is to load the azimuth and altitude pads at roughly 15 pounds per square inch and to have a center pad with roughly the same area as the combined area of the three azimuth pads.

- Laminate adhesive to secure the Ebony Star laminate to the baseboard and altitude hubs.

- Large hose clamps or cable ties to hold the Ebony Star laminate in place on the altitude hubs until the adhesive sets. (Ebony Star laminate is reasonably flexible, but when you bend it at the tight radius of the altitude hubs it wants to spring loose.)

- Small finishing brads to secure the Teflon pads to the Dob.

- A tack hammer and nail set to drive the brads and countersink their heads well below the surface of the Teflon.

- Sandpaper to smooth and bevel the edges of the Teflon pads.

Ebony Star (*http://www.wilsonart.com*) can be purchased at most hardware stores and home supply centers, although it is often a special-order item. Don't be surprised at the price. When he was building his 17.5" Dob, our observing buddy Steve Childers ordered a 4×8 foot sheet of Ebony Star. It cost $80. Fortunately, many home supply centers will sell you a 2×2 foot piece for $20 or so.

A local cabinetry installer may be willing to give you scrap pieces large enough for your needs. Sink cutouts are often large enough, for example. A cabinetry installer also has laminate cutting tools and may be willing to trim your Ebony Star to size for a small charge.

Actually, it's not critical that you use Ebony Star laminate specifically. What's important is not the brand name but the surface of the laminate. Standard laminates such as Formica have a smooth surface that produces stiction. Ebony Star and similar laminates have a "nubby" surface that nearly eliminates stiction. Although we've never used anything but Ebony Star laminate, by all reports any laminate with a similar surface texture works about as well. In fact, many astronomers substitute the inexpensive "glass board" laminate or fiber-reinforced plastic (FRP) laminate that is commonly used in public bathrooms.

The type of Teflon you use for the pads is critical. There are two types available. *Mechanical Teflon* (also called *recycled Teflon* or *remanufactured Teflon*) is relatively inexpensive but has unsuitable friction/stiction characteristics. You want the second type, called *virgin Teflon*, which is considerably more expensive than mechanical Teflon. There are two ways to get it:

- If you're willing to pay for convenience, buy a "Teflon kit" with the pads already cut to size and pre-drilled for mounting screws. Astro-Systems (*http://www.astrosystems.biz*), for example, sells a small Dob kit for $25 shipped that includes three 1" square azimuth pads, four 3/4"×1" altitude pads, and stainless steel mounting screws. They also offer a large Dob kit and a custom kit with larger pads. All pads are 3/16" thick.

- If you're cheap, like us, buy a raw chunk of virgin Teflon and cut your own pads. The best source we've found for bulk Teflon is McMaster-Carr (*http://www.mcmaster.com*). We ordered a 6" square piece of 3/16" virgin Teflon—the equivalent of six AstroSystems small Dob kits—for $16.08 shipped.

Teflon is hard to cut. The best method we found was to use a fine point marker to mark the measurements, score the lines with a razor knife, and then use a fine hacksaw blade to make the actual cuts. After you cut them, use a sharp knife or sandpaper to smooth and bevel the edges. You may also want to use sandpaper to roughen the surface of the Teflon pads slightly.

Installing the upgraded bearings is straightforward. The hardest part is getting the Ebony Star trimmed to size and the center hole drilled in the azimuth portion. You can do the trimming with a razor knife, but a router and laminate bits make the job much easier and neater. If you've never cut laminate before and don't have a friend with the necessary tools and skills, we recommend you pay a kitchen installer a few bucks to make the necessary cuts and drill the center hole. Once the Ebony Star is prepared, proceed as follows:

1. Disassemble your Dob, removing the baseboard and rocker box assembly from the ground board. Clean the bottom surface of the baseboard thoroughly, using Formula 409, Fantastic, or a similar household cleaner. Make sure you've removed all oil, lubricant, or other contaminants that might prevent the laminate adhesive from adhering.

2. Remove the old nylon or plastic azimuth pads from the ground board and the altitude pads from the rocker box. These pads are usually secured with small countersunk brads or staples. Use a screwdriver or pry bar to pop them loose.

3. Apply the laminate adhesive to the baseboard and/or the rear surface of the Ebony Star laminate, following the directions on the can.

4. Insert the center bolt temporarily in the baseboard, and use it as a guide to press the Ebony Star laminate into position, centered on the baseboard and with the center hole in the laminate matching the center hole in the baseboard. With everything aligned, press the Ebony Star firmly into contact with the baseboard. Be careful when doing this. Once the two surfaces make contact, it may be difficult or impossible to correct any misalignment.

5. Remove the center bolt and place the rocker box assembly on a flat surface. The weight of the rocker box is sufficient to clamp the Ebony Star laminate into close contact with the baseboard. Leave the assembly undisturbed for at least the time specified for the adhesive to set.

6. Following the directions on the can, apply laminate adhesive to the back of each Ebony Star laminate strip and/or the surface of the altitude hubs. Press the Ebony Star laminate strips into position around the full circumference of the altitude hubs, making sure that the seam is located where it will never contact the altitude bearing pads.

 The strips needn't be exactly the circumference of the altitude hubs. It's OK to trim them a bit shorter than the full circumference because the altitude pads don't bear on all parts of the laminate strip as the Dob is elevated from the horizon to zenith. To calculate the length of the strips, multiply the diameter of your altitude hubs by π (about 3.14). For example, the 5" altitude hubs on our Guan Sheng 10" Dob have a circumference of about 15.7".

7. If your Dob has typical 4" or 5" diameter altitude hubs, you'll find that the Ebony Star doesn't want to stay stuck when bent to such a tight radius. You'll need to clamp it into place until the adhesive sets. The best way to do that is to use hose clamps large enough to fit the diameter of your altitude hubs. If you don't have such clamps and don't want to buy them, you can instead use nylon tie-wraps. Pull them as tight as possible, putting the junction opposite the ends of the Ebony Star laminate strips.

8. Attach the three azimuth pads to the upper surface of the ground board and the four altitude pads to the rocker box. Use two finishing brads to secure each pad, and countersink the heads of the brads well below the surface of the Teflon.

 The Ebony Star laminate increases the diameter of the altitude hubs slightly. Depending on the thickness of the original pads and how they were installed, you may need to file down the mounting locations a bit to allow the altitude hubs to sit fully in the cutouts in the Dob base.

9. Apply a coating of silicone paste wax to all of the Ebony Star laminate surfaces.

Although the amount of friction provided by the pads is determined solely by their sizes and the force bearing on them, the placement of both the altitude pads and azimuth pads determines how freely the bearing moves. Increasing the separation between the pairs of altitude bearing pads stiffens altitude motions. Increasing the radial distance between the center bolt and the azimuth pads stiffens azimuth motions.

For altitude bearings, the amount of normal force on the four altitude pads determines the amount of friction provided by those pads. As you increase or decrease the distance between the altitude pads, you also change their angle with respect to the vertical, which in turn changes the force on the bearing pads. Consider the normal force the pad exerts on the hub. That normal force can be resolved into two components, a vertical component, which acts to support the scope, and a horizontal component, which is equaled and opposed by the other pad on the same hub.

- The vertical component is independent of angle because each of the four altitude pads must support one quarter of the (constant) weight of the tube.

- The horizontal component depends on the angle. Because we are talking about a force normal to the surface, the horizontal force must change if the vertical force remains constant but the angle of the normal with respect to the vertical changes. If the pads are at the bottom of the hub, there is no horizontal component: the normal force on each pad is one-quarter the weight of the tube. As you increase the separation between the pads, the vertical component remains constant, but the horizontal component increases, increasing the normal force on the pads and therefore the amount of friction.

For azimuth pads, the radial distance of the pads from the center bolt determines the amount of mechanical advantage (leverage) that is applied when you move the tube. The amount of force needed is determined by the ratio between the separation between the bearings and the length of the lever. Unfortunately, although the position of the azimuth pads remains fixed, the length of the lever (and therefore the mechanical advantage it provides) varies with the elevation of the tube. That is, when the scope is pointed at the horizon, the effective lever length is quite long, and the scope is very easy to move in azimuth. Conversely, when the scope is pointed at zenith (straight up), the length of the lever is zero, and you have no mechanical advantage. (The zenith is called Dobson's Hole for just this reason. When the tube is pointed straight up or nearly so, it becomes almost impossible to move in azimuth.)

There are a couple of steps you can take to minimize the problem of variable azimuth motions:

- Alter the distance of the azimuth pads from the center bolt. Traditionally, the azimuth pads are placed near the outer edge of the ground board, directly over its feet, with the goal of transferring the scope's weight directly to the feet, avoiding flexure of the ground board. But the farther the azimuth pads are from the center bolt, the stiffer the scope's azimuth motions will be. If your azimuth motions are too stiff, you can improve them by moving the azimuth pads closer to the center bolt.

- Use a center pad to shift some of the weight of the scope off the azimuth pads. If you use a center pad with an area about equal to the combined area of the three standard azimuth pads, you effectively cut the friction of those pads in half, making the scope much easier to move. The total friction remains the same, of course, but half of it is produced by the center pad. Because the center pad is located right at the center bolt, even a small lever arm gives a huge mechanical advantage, for all intents and purposes, eliminating the friction of the center pad.

Most Dob users do nearly all of their observing between 20° and 70° elevation. Below 20° the object is down in the muck and the views are generally poor. (Of course, you may have no choice, because a particular object may never rise very high at your latitude.) Above 70° Dobson's Hole becomes a problem, but it's easy enough just to wait until the object is a bit lower in the sky. Accordingly, it's usually best to optimize azimuth motions for 45° elevation, the midpoint between 20° and 70°.

Ideally, you want the effort required to move the scope in azimuth when it is at 45° elevation to exactly match the effort required to move it in altitude. If you strike that balance, there will be no sense of having to move the scope up and down versus right and left. The scope simply moves smoothly in any direction, and controlling it becomes entirely automatic on your part. Even when the scope is pointed as low as 20° or as high as 70°, the effort required to move it in azimuth is close enough to ideal that you'll likely be unaware of any differences in motions. Getting your scope to this point may require some trial and error, but the results are worth it.

For another take on this issue, see Gene Baraff's description of his experiments with Ebony Star, virgin Teflon, and PFA (an alterative to Teflon) at *http://groups.yahoo.com/group/ telescopes/message/103010.*

CHAPTER FOUR

Accessory Hacks

Hacks 44–65

We sometimes think that the popularity of a hobby depends on how many accessories you can buy for it. Hunting, fishing, golf, woodworking, photography, art, model trains, birdwatching, camping—just about every popular hobby offers lots of opportunities for accessorizing. On that basis, it's no surprise that amateur astronomy is very popular indeed.

In perhaps no other hobby is the divide between how many accessories you *need* and how many accessories you *want* so pronounced. What you *need* is a telescope, two or three eyepieces, a red LED flashlight, some charts, and a dark place to observe. What you *want*, well, the sky's the limit. Well-to-do amateur astronomers sometimes spend thousands of dollars—even tens of thousands—on accessories. (We confess that we're guilty of some excess; our most expensive accessory is an Isuzu Trooper SUV that we've dedicated to storing and transporting our observing equipment.)

Fortunately, you don't need to spend a lot of money to be well equipped for observing. The trick is to know how to choose accessories that provide the maximum bang for the buck. That doesn't mean buying a lot of cheap accessories, either. It means choosing a few good-quality accessories that pay off in usefulness all out of proportion to their cost. It also means making do with what you have, doing inexpensive upgrades and modifications that greatly improve utility at minimal or no cost, repurposing inexpensive products to substitute for expensive, specialized astronomy products, and learning to take advantage of the full features of each accessory you add to your kit.

In this chapter, we'll look at how to choose, use, build, and maintain essential astronomy accessories (and a few non-essential ones, but don't tell your spouse that...).

Dark Adapt Your Notebook Computer

HACK #44

Use your notebook while observing without damaging your night vision.

A notebook computer is an invaluable observing aid, but notebook displays produce enough light to destroy your night vision, not to mention that of everyone else in the vicinity **[Hack #5]**. Fortunately, it's easy to modify your notebook to be suitable for observing sessions without making any permanent changes that would make it unsuitable for daytime use. Here are the changes we recommend.

Dim the display

All notebook computers make some provision for adjusting the brightness of the display. The first step in dark adapting your notebook is to use these controls to set the notebook to the lowest possible screen brightness. Unfortunately, notebook computers are not designed with astronomers in mind, so the lowest available screen brightness is usually still much too bright.

> Some notebooks, including the Compaq Armada E500 we use, allow you to change screen brightness by moving a physical slider or by using some combination of function keys. These models allow you to dim the screen whether you're running Windows or Linux. Other notebooks require a Windows-only software utility to change screen brightness, so you may be out of luck if you're running Linux. If your notebook is in the latter category and you're running Linux, search the Web using the string *<maker-name> <model-name> linux screen brightness* or a similar search phrase. Linux-based screen brightness utilities are available for some popular notebook models.

Use night-vision mode

All serious planetarium programs offer some form of night-vision mode. Figure 4-1 shows the night-vision mode of KStars. It is typical of the night-vision mode of most planetarium programs in that it uses a dim red color scheme for the main display, but leaves bright, colorful icons, title bars, and task bars visible. Just one of those items is enough to damage night vision noticeably, and all of them suffice to reduce your night vision severely. Night-vision mode is helpful, but used alone it is not nearly enough, even with your notebook display fully dimmed.

Create an appropriate desktop theme

Standard desktop themes are bright and colorful, which is exactly what you don't want in a notebook used for astronomy. Windows and Linux both include numerous standard themes, but none is optimized for astronomy. We suggest creating your own theme, limiting your palette to black and dark reds.

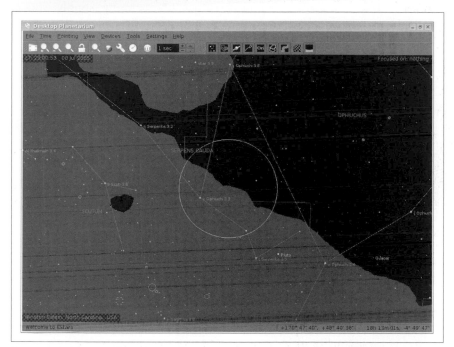

Figure 4-1. Kstars night-vision mode

> You can also use a free gamma-control utility such as Dark-Adapted (*http://www.adpartnership.net*) or NightMaster (*http://www.ilangainc.com*) to dim and adjust the colors of your notebook screen without changing your theme.

Screen your display with red film

Covering your notebook screen with red plastic film eliminates nearly all of the actinic light that damages night vision and blocks the bright white light that leaks around the edges of most notebook displays. But not just any red plastic will do. You need something that is transparent, a deep ruby-red color, optically clear, stiff enough to mount and remove easily, thin enough to allow you to close the display, and robust enough to stand up to heavy use and dew.

We've tested various materials and found two that are suitable. *Rubylith film* is used by commercial printing companies and can be purchased in sheets of various size from most printing supply companies. (Avoid the similar Amberlith, which is an orangish color rather than the deep red of

Rubylith.) *Theatrical acetate gels* are available in hundreds of colors, including deep red, and can be purchased in various sizes from theatrical supply companies and professional photography supply stores. Make sure to buy the transparent, clear version; theatrical gels are also available in frosted, translucent versions.

Figure 4-2 shows our Compaq Armada notebook computer set up for an observing session.

Figure 4-2. Our Compaq Armada notebook, equipped for nighttime use

Rather than modify your primary notebook computer, we suggest you buy a used notebook and dedicate it to astronomy. You can buy a suitable used notebook on eBay for $500 or less, sometimes much less. Using a "disposable" notebook avoids exposing your expensive primary notebook to the risks of a typical observing session. In the dark, we've more than once come close to knocking our notebook off the chart table, and as often as not it's literally dripping with dew when we pack it up at the end of a session.

 If you do buy a used notebook, plan to spend an extra $50 to $100 or more to replace the battery. We've never yet seen a used notebook offered with a good battery.

Most planetarium programs have minimal hardware requirements and run acceptably on even slow older notebooks. For example, we used to run Cartes du Ciel under Windows 2000 on our Compaq Armada E500, which has only a Pentium III/700 processor and 256 MB of memory. We now have Xandros Linux installed on the Armada, which now runs KStars and XEphem [Hack #64] with no problems at all.

HACK #45   Dark Adapt Your Vehicle

Keep the observing site dark by making your vehicle less of a beacon.

Nowadays, even many dark-sky observing sites are plagued with light pollution from streetlights and other local sources. You can't do much about most of those local light sources other than use screens or move your scope to a shaded spot. But there is one common source of local light pollution you can do something about: your vehicle.

Unmodified vehicles are astronomy-hostile. If you open a door, a bright light comes on. If you shift into reverse, a bright light comes on. If you step on the brake, a bright light comes on. Any of these is sufficient to destroy the dark adaptation of anyone nearby. (Yes, even the brake lights, which appear red but in fact have enough white light to destroy night vision.) Fortunately, it's cheap and easy to modify your vehicle to avoid these problems.

Some astronomers take radical steps, such as replacing all of the interior incandescent lamps with red LED arrays and installing switches and cabling to control their brake and backup lights manually. We've heard of one guy who installed a single switch on his dash that toggles relays to switch his entire vehicle between "normal mode" and "astronomy mode"—all interior lamps, including the dashboard lights, switch to red LED and all exterior lights are disabled. That's a bit much, even for us.

Fortunately, you don't have to go to those lengths to make your vehicle astronomy-friendly. Here's what we recommend:

- Learn the location of your vehicle's fuse box and which fuses control which lights. When you arrive at a dark-sky observing site, pull the fuses for your backup lights and brake lights. That way, you can depart without ruining other observers' dark adaptation. If your vehicle turns on its headlights automatically when you start it, pull that fuse as well. (When you head home, don't forget to reinstall the fuses once you're a safe distance from the observing site.)

- Either turn off your interior lights—if necessary, pull their fuses—or install red filters on them. We prefer to install red filters because we often work in or near our vehicle and it's convenient to have safe illumination inside.

There are several popular ways to filter the overhead and courtesy lights:

- Replace the bulbs with red LED or red-filtered incandescent bulbs. Crutchfield (*http://www.crutchfield.com*) and other car audio dealers sell these replacement bulbs in sizes to fit various vehicles. The LED models may be quite expensive, though—$20 or more per bulb—and you'll need several to do the whole vehicle.

- Paint the bezels with red fingernail polish. Use the deepest red polish you can find, and apply several coats. The drawback to this method is that it's irreversible, at least without replacing all of the bezels. (And you won't believe how much those little pieces of plastic can cost. Our local Isuzu dealer quoted $47 for a replacement bezel for the overhead light in our Trooper.)

- Paint the bulbs with red fingernail polish, again using several coats of the deepest red polish you can find. The drawback to this method is that the bulbs become quite hot, and the red coating may blister and peel.

- Use Rubylith film or ruby-red theatrical gels **[Hack #44]**. This method may not work, depending on the physical configuration of your courtesy lights. If the bulb is recessed and the bezel mounts flat against a surface, clamp the ruby-red film between the bezel and the surface, sealing any gaps through which white light might escape. If the bulb is exposed, as is common for overhead lights, it's difficult to get the red film to conform to the shape of the bezel, and small gaps are likely.

- If the bulb is exposed or the bezel otherwise makes it difficult to seal all gaps with ruby-red film, use brake-light tape. You can buy a roll large enough to do your entire vehicle at any auto supply store for a couple of bucks. Simply apply the brake-light tape to the entire interior surface of the bezel, pressing the sticky side into place to conform to the shape of the bezel. Use more than one layer. Brake-light tape appears red, but actually transmits quite a bit of non-red light, which impairs dark adaptation. We've found that two or three layers is usually a good compromise between the amount of light transmitted and its color.

Figure 4-3 shows how we filtered the courtesy lights in our Isuzu Trooper. We applied one layer of brake-light tape first, pressing it down around the edges to make sure any gaps were covered. We then added a second layer, this one a small piece of ruby-red theatrical gel, to stop the non-red light passed by the brake-light tape.

Figure 4-3. Installing red filtering on an interior light

In Figure 4-3, it appears that the ruby-red film is skewed to the right, not covering some of the illuminated area at the left. In fact, once the bezel is screwed back into place, the second layer of film entirely covers the light emitting area. What's interesting is the bright vertical strip at the left, near the screw, where the light passes through only the brake-light tape. That additional brightness is due solely to the orange, yellow, and even green light being passed by the brake-light tape. The ruby-red film has a much sharper transmission cut-off, and stops that non-red light.

HACK #46 Use a Barlow

Double your eyepiece collection on the cheap.

A Barlow lens (named for the 19th-century British mathematician and optician Peter Barlow) is one of the most useful accessories you can have in your eyepiece case. A Barlow fits between the telescope's focuser and the eyepiece, where it increases the magnification provided by the eyepiece [Hack #47]. A Barlow lens has negative focal length, which means it reduces the convergence of the light cone arriving from the primary mirror or objective lens, effectively increasing the focal length of the scope. Because the magnification provided by an eyepiece is directly proportional to the focal length of the scope it is used in, using a Barlow has the effect of increasing magnification.

Although it's possible to describe a Barlow by its negative focal length—just as an eyepiece is described by its positive focal length—it's more convenient simply to describe it by its strength or amplification factor. Barlow amplification factors range from 1.5X to 5X or more. A 2X Barlow, for example, effectively doubles the magnification of any eyepiece it is used with. (So-called "zoom Barlows" with variable amplification factors exist, but most of these are of low quality.) Figure 4-4 shows a selection of high-quality Barlows. From left to right are a Tele Vue 3X, an Orion Ultrascopic 2X, and a Tele Vue Powermate 2.5X.

Figure 4-4. Tele Vue 3X, Orion Ultrascopic 2X, and Tele Vue Powermate 2.5X Barlows

Barlows are available in 1.25" models, which accept 1.25" eyepieces and fit 1.25" focusers, and in 2" models, which accept 1.25" and 2" eyepieces and fit 2" focusers. 1.25" models are far more popular than 2" models. This is true because most astronomers use Barlows with mid-power and high-power eyepieces—which are nearly all 1.25" models—to reach the high magnifications needed for Lunar and planetary observing, as well as observing small DSOs.

> It is *possible* to use a Barlow with 2" eyepieces—for example, a 27mm Tele Vue Panoptic used with a 2" 4X Powermate yields the equivalent of a 6.75mm Panoptic—but few astronomers choose to do so. For various reasons, it is generally preferable to use a 1.25" Barlow with a shorter 1.25" eyepiece, not least because 1.25" eyepieces and Barlows can also be used with secondary scopes that have only 1.25" focusers.

Although image amplification is the most obvious characteristic of a Barlow, it is not the only effect, nor necessarily even the most important one. Here are other good reasons to use a Barlow:

Double your eyepiece selection

The most compelling reason to buy a Barlow is that it effectively doubles the number of eyepieces you have available. For example, if you have 32mm, 20mm, and 12mm eyepieces, using a 2X Barlow adds the equivalent of 16mm, 10mm, and 6mm eyepieces to your collection. Because a good Barlow costs no more than a mid-range eyepiece, this is an extremely efficient way of adding to your eyepiece collection.

Reach high magnifications with short focal length scopes

Short focal length scopes, such as short-tube refractors and many Dobs, require very short focal length eyepieces to reach high magnifications. For example, you need a 2.5mm eyepiece to reach 160X with an 80mm f/5 **[Hack #9]** short-tube refractor or a 3mm eyepiece to reach 400X with an 8" f/6 Dob. Inexpensive eyepieces in these focal lengths are nearly unusable because of their tiny eye lenses and nonexistent eye relief. Using a Barlow with a longer focal length eyepiece allows you to reach high magnifications more conveniently.

Increase the focal ratio of the scope

By increasing the focal length of a telescope, a Barlow also increases its focal ratio. For example, using a 2X Barlow with a 250mm f/5 telescope (focal length 1,250mm) effectively converts that scope to a 250mm f/10 scope (focal length 2,500mm). If you are imaging, this increase in focal ratio increases exposure times by a factor of four, albeit at a larger image scale.

More important to visual observers is the effect of focal ratio on eyepieces. The obtuse converging light cone from a fast scope (such as an f/5 model) is very difficult for eyepieces to handle without showing obvious aberrations, particularly near the edge of the field with wide-field eyepieces. The acute converging light cone from a slower scope is much easier on eyepieces. Only expensive premium eyepieces handle the light cone from f/6 or faster scopes well, but even inexpensive eyepieces handle an f/8 or slower light cone well.

So, for example, to view at 125X with the 250mm f/5 scope, you need a 10mm eyepiece or its Barlowed equivalent. If you want a 10mm wide-field eyepiece that you can use natively with good edge performance in an f/5 scope, you're in premium eyepiece territory—a $240 Tele Vue Radian or a $310 Pentax XW. If, instead, you use a 2X Barlow, you can mate it to a less expensive 20mm wide-field eyepiece and still get excellent image quality.

Preserve eye relief

Other than premium eyepieces, many of which feature fixed 20mm eye relief regardless of their focal lengths **[Hack #49]**, most eyepiece designs have eye relief that is some fraction of the focal length. Plössls, for example, typically have eye relief of at most 60% to 70% of their focal lengths, and even that may be reduced by the physical design of the eyepiece. A 32mm Plössl might have eye relief of 20mm, a 20mm Plössl 12mm, and a 11mm Plössl only 6mm.

Most observers find an eyepiece with eye relief shorter than 12mm uncomfortable to use. (Those who wear eyeglasses require 20mm or so.) But a Barlowed eyepiece retains its original eye relief despite its effectively shorter focal length and higher power. So, for example, rather than use an 11mm Plössl natively with its short 6mm eye relief, you can use a 3X Barlow to effectively convert a 32mm Plössl to an 11mm Plössl while maintaining the 20mm eye relief of the 32mm Plössl.

Using a Barlow actually *extends* the eye relief of most eyepiece designs, including Plössls. The amount of extension varies with eyepiece design and the type and power of the Barlow. Unfortunately, the amount of extension varies proportionately to the focal length, which is the exact opposite of what we'd prefer.

Short focal length eyepieces, which need more eye relief, show little increase. Long focal length eyepieces, which already have lots of eye relief, gain lots of eye relief when used with a Barlow. In fact, Barlowing a long focal length eyepiece can make the eye relief uncomfortably long. It becomes difficult to find and hold the exit pupil of the eyepiece, which causes blackouts.

Improve filter performance

Interference filters, such as narrowband and line filters [Hack #59] work best at longer focal ratios, where the acute light cone from the primary mirror or objective strikes the filter nearly perpendicularly. The obtuse light cone from a fast scope, such as an f/5 model, strikes the interference layers of the filter at a more oblique angle, which effectively increases the thickness of the layers, degrading filter performance and shifting its transmission curve slightly. Line filters in particular work better with slower focal ratios, and the easy way to accommodate their needs in a fast scope is to use a Barlow between the scope and the filter.

Broadly speaking, there are two types of Barlows:

- Standard Barlows, such as the Orion Ultrascopic 2X and Tele Vue 3X models shown in Figure 4-4, are 5" to 6" long, and are used primarily in Newtonian reflectors, including Dobs. You can use a standard Barlow in a refractor, SCT, or other scope that uses a diagonal, either by inserting it between the telescope and diagonal or by (carefully) inserting it between the diagonal and the eyepiece. (The danger is that the long Barlow may protrude too far into the diagonal, damaging the mirror.)

- Short Barlows are about half the length of standard Barlows and may be used in any type of scope, including Newtonian reflectors with very low-profile focusers, in which a standard Barlow may protrude into the light path.

> Tele Vue makes a series of Barlow-like devices called Powermates. In effect, a Powermate is a standard 2-element Barlow with a second doublet lens added to minimize vignetting (darkening the edge of the image) and excessive extension of eye relief when used with long focal length eyepieces. Powermates are available in 1.25" 2.5X and 5X models ($190) and 2" 2X and 4X models ($295). We don't doubt that Powermates are excellent products, but we've never been able to tell any difference in image quality between a Powermate and a high-quality standard Barlow.

Standard and short Barlows of comparable quality sell for similar prices, but there are significant optical differences. The shorter tube of a short Barlow means it must use a stronger negative lens to achieve the same level of amplification as a longer Barlow. That has three disadvantages:

- Inferior image quality. Although the best short Barlows are very good indeed, the laws of physics dictate that they must be inferior optically to a full-length Barlow that uses lenses of similar quality. This inferiority most commonly manifests as lateral color (fringing), particularly near the edge of the field.

- Vignetting. Because a short Barlow must bend light much more sharply than a full-length Barlow, short Barlows are subject to vignetting, particularly when used with longer focal length eyepieces.

- Excessive eye relief. Although any Barlow extends the eye relief of most eyepiece designs, the stronger negative lens of a short Barlow exaggerates this effect. For example, we have no problem using our Orion Ultrascopic 30mm eyepiece with either our Orion Ultrascopic 2X Barlow or our Tele Vue 3X Barlow. But the Ultrascopic 30mm used with a short Barlow has its eye relief extended so far that we have trouble holding the exit pupil.

As you might have guessed, we're not fans of short Barlows. In fact, we don't own one. We use only full-length Barlows in our scopes, including our refractor. We freely confess, though, that many very experienced observers use and recommend short Barlows such as the Orion Shorty Plus Barlow and the Celestron Ultima Barlow (which are identical except for the brand name).

There are many cheap Barlows available, but we suggest you avoid them. A Barlow is a lifetime investment, and the difference in price between a mediocre model and an excellent one is not great. For a full-length Barlow, we recommend the $85 Orion Ultrascopic 2X model—which is often on sale for $75—and the $105 Tele Vue 3X model. (Tele Vue also makes a superb 2X Barlow, but it sells for $20 or $30 more than the Orion Ultrascopic, and we can discern no difference in image quality between them.) If you must have a short Barlow, get the $80 Celestron Ultima or the identical $70 Orion Shorty Plus, which is often on sale for $60.

> Ignore the marketing hype. It doesn't matter if a Barlow has two or three elements or is described as "apochromatic" (which is marketing-speak for a 3-element Barlow). What matters is the figure and polish level of the lenses and their coatings and the mechanical quality of the Barlow. There are superb 2-element Barlows, including both Tele Vue models, and very poor 3-element Barlows.

Choose the amplification factor of your Barlow with your current eyepiece collection in mind, as well as any plans you have for expanding it. Avoid duplication between Barlowed focal lengths and native focal lengths. For example, if you have 32mm and 16mm eyepieces, using a 2X Barlow with your 32mm eyepiece effectively duplicates your 16mm eyepiece. Using a 3X

Barlow instead provides the equivalent of 10.7mm and 5.3mm eyepieces, both of which are useful extensions to your arsenal. Conversely, if you have the 25mm and 10mm eyepieces commonly bundled with inexpensive scopes, a 2X Barlow adds the equivalent of 12.5mm and 5mm eyepieces, again a useful expansion of your selection.

Don't hesitate to buy more than one Barlow. Just as having one Barlow can effectively double your eyepiece collection, having two Barlows can effectively triple it, if you choose your native eyepiece focal lengths carefully to avoid overlap. We carry 2X and 3X Barlows in our eyepiece case, so we're always prepared for any eventuality.

If you have two Barlows, you can stack them to achieve "stupid high powers" on those extraordinary nights when the atmosphere is stable enough to support them. It doesn't happen often, but when it does, stacking two Barlows has allowed us to view Jupiter at 800X and Luna at 1,200X.

HACK #47 Determine Actual Barlow Magnification

Find out why 3X doesn't always mean 3X.

A Barlow [Hack #46], named for the English optician Peter Barlow (1776–1862), is a lens of negative focal length that astronomers use to increase the magnification provided by a telescope and eyepiece. In effect, a Barlow serves as an image amplifier. The Barlow is inserted into the focuser, and the eyepiece is inserted into the Barlow.

Barlows are made in 1.25" models, which accept any eyepiece with a 1.25" barrel, and in 2" models, which which accept both 2" eyepieces and, with an adapter, 1.25" eyepieces as well. Because one Barlow can be used with any number of eyepieces, Barlows are very popular among amateur astronomers as an inexpensive means to multiply their range of available eyepiece focal lengths. Figure 4-5 shows a typical 1.25" Barlow; this one is an Orion Ultrascopic 2X model, with an Orion Ultrascopic 30mm eyepiece.

A Barlow functions by effectively increasing the focal length of the telescope, and thereby its magnification with any given eyepiece. For example, if you use the Ultrascopic 30mm eyepiece in a telescope of 1,200mm focal length, the combination provides 40X magnification (1,200/30=40). If you insert the Tele Vue 3X Barlow between the eyepiece and focuser, it has the effect of increasing the focal length of the scope from 1,200mm to 3,600mm, and thereby the magnification from 40X to 120X.

Figure 4-5. An Orion Ultrascopic 2X Barlow with an Orion Ultrascopic 30mm eyepiece

Alternatively, you can think of a Barlow as reducing the effective focal length of any eyepiece it is used with. For example, the 30mm Ultrascopic eyepiece used with the Tele Vue 3X Barlow has an effective focal length of 10mm. A 10mm eyepiece used with a 1,200mm focal length scope provides 1,200/10=120X magnification.

So far, so good. The problem is that Barlows don't actually have a fixed magnification, despite the fact that they're clearly labeled as such. The actual magnification provided by a particular Barlow depends on many factors, including the depth to which the Barlow is seated in the focuser, the depth to which the eyepiece is seated in the Barlow, the focal plane position of the eyepiece, the optical design of the eyepiece, and so on.

A nominal "3X Barlow" may actually provide anything from 2.5X to 5X magnification or more, depending on how it is used. A 3X Barlow may in fact provide 3X amplification, for example, when used between the diagonal of a refractor and the eyepiece, as shown in Figure 4-6. But if you insert that same Barlow between the focuser of the refractor and the diagonal, as shown in Figure 4-7, it may provide 4.5X or 5X amplification because of the additional separation between the Barlow and eyepiece.

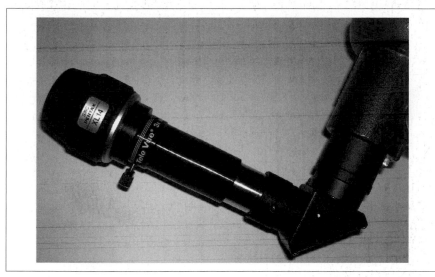

Figure 4-6. A Barlow inserted between the diagonal and eyepiece provides near-nominal amplification

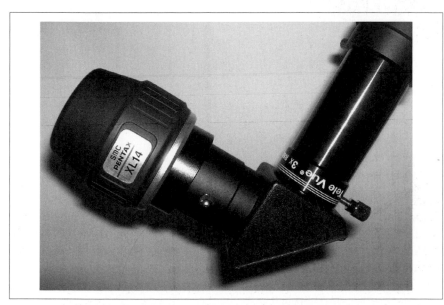

Figure 4-7. A Barlow inserted between the focuser and diagonal provides greater amplification

It's important to know the actual amplification factor of a Barlow because it affects the true magnification and field of view of any eyepiece it is used with. The only way to determine the actual amplification factor of a Barlow with a particular scope and eyepiece is to measure it.

To do so, you need a meter stick or other accurate measuring device and a large, flat open area. (You can use a yard stick, but a meter stick makes calculations much easier.) Set up your scope with the eyepiece inserted but without the Barlow. Orient the scope horizontally. By trial and error, place the meter stick at a distance where, when it is sharply focused, the eyepiece field of view includes most, but not all, of the meter stick.

The necessary distance varies according to the focal length of the scope and the eyepiece characteristics. For our 1,255mm Dob and the 30mm Ultrascopic, it was just under 50 meters. With a longer focal length scope and/or a higher power eyepiece, the distance may be several hundred meters, which makes it helpful to have an assistant and a pair of walkie-talkies or cell phones.

You can determine a ballpark starting distance by calculating the field of view of your scope and eyepiece **[Hack #57]**. For example, the 30mm Ultrascopic in our 1,255mm Dob has a TFoV of about 1.25°. A 1° field at 100 meters corresponds to about 1.75 meters, so a 1.25° field would be about 2.18 meters, or 1.09 meters at 50 meters. Accordingly, Robert had Barbara walk out about 50 meters to start and then adjusted the distance from there.

Record the total number of millimeters visible from one edge of the field to the other. Then insert the Barlow, refocus on the meter stick, and record the number of millimeters visible. Divide the first number by the second to determine the actual amplification factor of the Barlow when used with that scope and eyepiece. When we tested, 95.8 cm (958mm) was visible on the meter stick with the eyepiece alone. With the Barlow inserted, only 31.2 cm (312mm) was visible. The amplification factor of that Barlow, used with that scope and eyepiece at that distance is 958/312 or 3.07X.

Repeat the process for each scope/eyepiece combination you plan to use with your Barlow.

If you have an SCT or other telescope that focuses by moving the primary mirror, the actual amplification factor of a Barlow is different when the scope is focused on a relatively nearby object than when it is focused at infinity. For best accuracy, you should test no closer than 100 times the focal length of the scope.

Alternatively, you can calculate the actual amplification factor of your Barlow by drift testing an eyepiece [Hack #57] with and without the Barlow in place. Divide the time required for the star to drift across the field without the Barlow in place by the time required with the Barlow in place to give the actual amplification factor of the Barlow.

HACK #48 See More of the Sky

Find your way around the sky with a low-power, wide-field "finder" eyepiece.

To a beginner, looking at the night sky through a telescope feels like reading a newspaper through a straw. Telescopes, even "wide-field" models, have relatively narrow fields of view. (To get an idea of how much sky an "average" telescope shows, hold up a quarter at arm's length. Not much, is it?) This narrow window on the sky can make it very difficult to locate objects.

To minimize the problem, you need a "finder" eyepiece to provide the widest possible true field of view in your telescope. How wide that field can be is determined by the focal length of your scope and the focuser size. For example, a typical 8" SCT with a focal length of 2,032mm and a 1.25" focuser has a maximum possible field of just under 0.9°. Conversely, a short-tube refractor with a 400mm focal length and a 2" focuser has a maximum possible field of about 7° (which means the scope can serve as its own finder).

To determine approximately the widest possible field of view your telescope can provide, make the following calculation:

- If your telescope has a 1.25" focuser, divide 1,750 by the focal length of the scope in millimeters. For example, if your scope has a focal length of 1,200mm, the widest possible true field of view in degrees is about 1,750/1,200=1.46°.

- If your telescope has a 2" focuser, divide 2,800 by the focal length of the scope in millimeters. For example, if your scope has a focal length of 1,200mm, the widest possible true field of view in degrees is about 2,800/1,200=2.33°.

Although the difference between 1.46° and 2.33° doesn't seem great, the amount of sky visible is proportional to the square of the field of view. That means the 2.33° eyepiece shows more than 2.5 times as much sky as the 1.46° eyepiece, a very significant difference. For example, Figure 4-8 shows a simulation of the star field around the Crab Nebula (M1) with a 1,200mm scope using a 2" 40mm eyepiece with a 70° apparent field of view (outer circle), a 32mm 52° Plössl (middle circle), and a 1.25" 25mm 50° Plössl (inner circle).

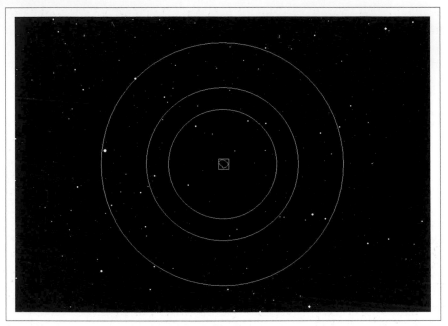

Figure 4-8. The fields of 2" (outer circle) and 1.25" (inner circles) finder eyepieces

Obviously, it's much easier to locate the object with the 2" wide-field eye-piece, but even the 1.25" 32mm eyepiece shows significantly more sky than the 25mm Plössl eyepiece typically bundled with inexpensive scopes. When you're first getting started, you can get by with the bundled 25mm or 26mm Plössl eyepiece, if your telescope came with one. If you observe mostly Luna and the planets, your first priority will probably be to buy additional high-power eyepieces. But, if you have any interest in observing DSOs **[Hack #22]**, you should probably make it a high priority to buy a better finder eyepiece.

> When you are evaluating finder eyepieces, a quick way to determine their suitability as a finder eyepiece is to multiply the focal length of the eyepiece in millimeters by the apparent field of view in degrees and compare that result to the numbers given previously. For example, a 1.25" 35mm Celestron Ultima Plössl with a 49° apparent field of view yields 35×49=1,715, very close to the 1,750 figure for the maximum possible true field of view in a 1.25" focuser. Similarly, a 2" 40mm Pentax XW eyepiece with a 70° apparent field of view yields 40×70=2,800, which is right at the limit for the widest possible true field of view in a 2" focuser, as is a 41mm 68° Tele Vue Panoptic at 41×68=2,788.

If you are limited by a 1.25" focuser, choose a finder eyepiece according to your budget, as follows:

- For a premium finder eyepiece, the $295 24mm Tele Vue Panoptic is hard to beat. Its true field is nearly the widest possible in a 1.25" focuser, and its relatively high power darkens the sky background relative to a longer focal length eyepiece.

- For a finder eyepiece in the $100 range, choose the 30mm or 35mm Orion Ultrascopic (also sold under the Celestron Ultima brand name) or the Tele Vue 32mm Plössl. All of these eyepieces are superb optically and have excellent fit and finish. They're less expensive than the Panoptic because their apparent fields are in the 49° to 52° range versus the 68° of the Panoptic.

- For a finder eyepiece in the $50 range, choose a 32mm Chinese Plössl, such as those sold by Orion under their Sirius brand name. These eyepieces are a step behind the $100 eyepieces. Their coatings and lens polish aren't as good, which means they show lower contrast and some scatter. Mechanically, they're obviously not up to the Japanese standard of the $100 eyepieces, but they're not terrible either.

 Avoid 40mm 1.25" eyepieces. The 1.25" barrel diameter limits their apparent field of view to about 42°, which is like looking through a drinking straw. A 32mm Plössl provides almost exactly the same true field of view at much higher magnification, which darkens the background sky, making it easier to locate objects. A 40mm 1.25" eyepiece is entirely worthless. We don't understand why anyone makes them, let alone why anyone would buy one.

If you have a 2" focuser, choose a finder eyepiece according to your budget, as follows:

- For a premium finder eyepiece, the $500 Pentax 40mm XW and the $495 Tele Vue 41mm Panoptic are superb choices, as they should be at that price.

- For a finder eyepiece in the $225 to $300 range, buy a Celestron 40mm Axiom or a University Optics 40mm MK-70 (*http://www.universityoptics. com*). At $226, the MK-70 is about $50 cheaper than the Axiom, has less eye relief, and is less suitable for scopes with focal ratios of f/8 **[Hack #9]** or faster. If you are comfortable buying a used eyepiece, a 40mm Pentax XL is better optically than either of these eyepieces, and it sells for about $225 on Astromart (*http://www.astromart.com*). In 2004, Pentax replaced their superb XL series eyepieces with the even more superb XW series, but XL eyepieces are still widely available on the used market.

- In the $150 range, the best bet is the Orion 40mm Optiluxe, which is Japanese-made and has excellent fit and finish. Optically, it's a mixed bag. With an f/8 or slower scope, such as an SCT, the Optiluxe provides excellent image quality. At the f/4.5 to f/6 focal ratios common on Dobs, the Optiluxe is extremely soft at the edges of the field. With a fast scope, you probably won't want to use the Optiluxe to actually observe large objects, but some edge softness is acceptable if you are using the eyepiece primarily for locating objects rather than viewing them.

- In the $75 range, several companies resell 3-element Chinese 2" eyepieces with varying focal lengths and apparent fields of view. Orion, for example, resells these eyepieces as their DeepView series. The 42mm DeepView has a 52° apparent field. Although this is not even close to the maximum possible field in a 2" eyepiece, it is still much larger than the true field available in any 1.25" eyepiece. The downside is edge performance in fast scopes, which is simply hideous. Stars near the edge of the field appear as elongated blobs rather than as sharp points.

Figure 4-9 shows two top-notch finder eyepieces, a 2" 40mm Pentax XL (left) and a 1.25" Orion 30mm Ultrascopic.

There are several myths and misconceptions about finder eyepieces, so it's worth taking a moment to set the record straight. Here are some things to be aware of:

- Many astronomers refuse to use very wide finder eyepieces because they believe that such eyepieces "waste light" as a result of their very large exit pupils. For example, a 40mm eyepiece used in an f/5 scope provides an 8mm exit pupil, much larger than the maximum entrance pupil of your eye [Hack #7]. If your eye can dilate only to 5mm, that 8mm exit pupil reduces the effective aperture of your scope by the same factor. Although that is true, it matters little because you are seeing the brightest possible image at that image scale regardless of the exit pupil size. If you substitute an eyepiece that provides the same true field of view with a shorter focal length and a wider apparent field, objects appear larger but dimmer as a result of the higher magnification.

- The myth persists that finder eyepieces produce low contrast. As with many myths, there's a kernel of truth here. Finder eyepieces are of long focal length, which yields low magnification. Like any extended object, the background sky is dimmed by higher magnification. If you observe from a light-polluted site, using low magnification (and an accordingly large exit pupil) means the background sky appears bright and washed out. This has nothing to do with the inherent contrast of the eyepiece, but only with the observing conditions and low magnification. From a

Figure 4-9. Two excellent "finder" eyepieces: 2" Pentax 40mm XL (left) and 1.25" Orion 30mm Ultrascopic

dark site, that same eyepiece provides a pitch-black sky background and all of the contrast it's inherently capable of providing.

If you've ever wondered why anyone would pay $295 for a 24mm Panoptic that provides almost the same true field as a $100 32mm Plössl, this is why. The higher magnification of the 24mm eyepiece provides a darker sky background, which is prettier and makes it easier to locate and see objects. Similarly, some well-heeled observers pay $620 for a Tele Vue 31mm "TermiNagler." The true field of view of the 31mm Nagler is actually slightly narrower than that of the Pentax 40mm XW or the 41mm Panoptic, but the 31mm Nagler provides much higher magnification and a more pleasing view.

- Many astronomers believe that wide-field eyepieces produce coma, the aberration that makes stars near the edge of the field appear as short streaks instead of pinpoints. In fact, this aberration in the form of off-axis coma is inherent in any parabolic mirror, which means that every Newtonian primary mirror suffers from coma in proportion to its focal ratio. Even an imaginary "perfect" eyepiece would show coma in a Newtonian reflector near the edge of its true field. Lenses, including eyepieces, can also suffer from coma as well as other edge aberrations, and it is these eyepiece aberrations that most astronomers are referring to when they speak of coma. In fact, a well-corrected wide-field eyepiece like the Pentax 40mm XW or the Tele Vue 41mm Panoptic produces almost no coma of its own, and will show off-axis coma from the primary mirror only at focal ratios below f/5. Very fast scopes, such as f/4.2 and f/4.5 Dobsonians require a coma corrector such as the Tele Vue Paracorr when used with wide-field finder eyepieces.

HACK #49 Optimize Your Eyepiece Collection

Choose eyepieces that match your budget and observing habits.

Astronomers tend to collect eyepieces like dogs collect fleas. We know astronomers who have owned 50 and even 100 eyepieces over their observing careers. Some constantly buy and sell eyepieces in an eternal search for the perfect set. That gets expensive fast, though, and there is a better way to do it. Read on to discover what you need to know to assemble a perfect eyepiece set for your own scope and your own needs at a price you can afford.

Eyepiece Characteristics

Here are the important characteristics of eyepieces:

Barrel size

Eyepieces are made with barrel sizes designed to fit standard telescope focusers. For some reason, eyepiece barrel sizes are always denominated in inches, even outside the U.S. The most common barrel size is 1.25". Low-power, wide-field models use 2" barrels. The larger diameter of 2" eyepieces allows them to show more of the sky, but 2" eyepieces are more expensive to produce than 1.25" models. Accordingly, with few exceptions, 2" models are available only in longer focal lengths.

A telescope with a 1.25" focuser accepts only 1.25" eyepieces. A telescope with a 2" focuser accepts 2" eyepieces, and, with an adapter, 1.25" models. Figure 4-10 shows a 1.25" Tele Vue Radian eyepiece on the left and a 2" Tele Vue Panoptic eyepiece on the right.

Figure 4-10. Tele Vue 1.25" Radian (left) and 2" Panoptic eyepieces

Eyepieces with 0.965" barrels were formerly common, and remain popular in Japan. In the U.S. nowadays, only the very cheapest of telescopes have 0.965" focusers. We recommend avoiding 0.965" eyepieces because their small barrel diameter provides a very constricted view at anything less than medium magnification.

Focal length

The focal length of an eyepiece is stated in millimeters and determines the magnification that eyepiece provides with any particular telescope. To determine the magnification, divide the focal length of the telescope by the focal length of the eyepiece. For example, a 25mm eyepiece used in a scope of 1,000mm focal length yields 1,000/25=40X.

Mainstream eyepieces are available in focal lengths from 2.5mm to 56mm. The optimum range of eyepiece focal lengths for a particular scope depends on the focal ratio of that scope and the size of your dark-adapted entrance pupil **[Hack #7]**. The exit pupil provided by an eyepiece is calculated by dividing the eyepiece focal length by the focal ratio of the scope. For example, a 10mm eyepiece used in an f/5 **[Hack #9]** Dob provides a 2mm exit pupil, while that same eyepiece used in an f/10 SCT provides a 1mm exit pupil.

The ideal range of eyepiece focal lengths for a particular scope is one that provides exit pupils no smaller than about 0.7mm—anything less than that greatly reduces visual acuity—and no larger than the size of your fully dark-adapted entrance pupil. For example, if you have an f/5 scope and your maximum entrance pupil is 6mm, the ideal range of eyepiece focal lengths for you is 3.5mm (3.5mm/5=0.7mm) to 30mm (30mm/5=6mm). An exit pupil in the 2mm to 3mm range provides maximum visual acuity, which allows you to see the most detail.

When you select focal lengths for your eyepieces, keep in mind the effect of your Barlow lens **[Hack #46]**. For example, if you already own a 20mm eyepiece and a 2X Barlow, it makes little sense to buy a 10mm eyepiece because a 10mm eyepiece duplicates the effect of using the 20mm eyepiece with the Barlow.

Apparent field of view

The *apparent field of view* (AFoV) of an eyepiece is the angular size of the circular image it presents. At any given magnification, an eyepiece with a wide AFoV shows more of the sky than an eyepiece with a narrower AFoV. For example, Figure 4-11 shows simulated views of M42, the Great Orion Nebula, in two eyepieces of the same focal length but with different apparent fields. Because the focal length of the two eyepieces is the same, so is the magnification, and the nebula appears to be the same size. But the eyepiece on the left has an 82° AFoV, and so it shows much more of the surrounding sky than the image from the 50° eyepiece, shown on the right.

Eyepieces that provide a wide AFoV are popular for two reasons. First, many people find the larger field of view pleasing esthetically. Second, if you are using an undriven scope, such as a Dobsonian, the larger AFoV allows the object to remain in view longer before you have to "bump" the scope to recenter it. The AFoV of mainstream eyepieces ranges from 40° to more than 80°. All other things being equal, eyepieces with a wide AFoV cost more than those with a narrow AFoV.

Eye relief

Eye relief is the distance an eyepiece projects its exit pupil from the outer surface of the eyepiece eye lens. When you view with an eyepiece that has short eye relief, you have to press your eye right up against the

Figure 4-11. The Great Orion Nebula (M42) with an 82° eyepiece (left) and a 50° eyepiece

eyepiece. Using an eyepiece with long eye relief allows you to maintain some separation between the eyepiece and your eye. If you must wear eyeglasses while observing, or if you just find longer eye relief more comfortable, look for an eyepiece with 20mm or so of eye relief. If you observe without glasses, 12mm or so of eye relief is adequate.

 Too much eye relief can be as bad as too little. Some long focal length eyepieces have excessive eye relief, sometimes 40mm or more. When eye relief becomes longer than 25mm or so, it becomes very hard to get and hold the exit pupil, which causes blackouts as your eye moves around, seeking the exit pupil. Eye cups and other aids can minimize this problem, but in general we recommend avoiding any eyepiece with eye relief greater than 25mm.

Mechanical and optical quality

The mechanical fit and finish of eyepieces varies dramatically. In general, all Japanese-made eyepieces and some Taiwanese models have excellent mechanical quality. Name-brand eyepieces, regardless of where they are made, also have excellent fit and finish. Low-cost eyepieces, referred to generically as "Chinese eyepieces," are of noticeably lower quality. Labels are painted on rather than engraved, tolerances are looser, lens edges are not blackened, internal baffling is poor or absent, and so on.

Optical quality is even more important than mechanical quality. The major reason for the price difference between cheap generic eyepieces and expensive name-brand models is the level of attention given to details such as lens polish and coatings. More expensive eyepieces generally have much better lens polish and coatings, which translates to sharper images, higher contrast, and less ghosting and flaring.

Edge performance in fast scopes

In fast focal ratio scopes, the light cone converges over a shorter distance than in slower scopes. This fast-converging cone of light means that individual light rays arrive from significantly different angles, which is difficult for an eyepiece to handle without showing visible aberrations. Those aberrations are minimal at the center of the field, but become increasingly apparent as you approach the edge of the field, where the incoming light rays must be bent more sharply by the eyepiece. Edge performance is particularly problematic for eyepieces that provide a wide apparent field of view.

In an f/10 or slower scope, for example, nearly any well-made, wide-field eyepiece provides a sharp, pleasing image across most or all of its field. In an f/5 or faster scope, only top-notch eyepieces of modern design—such as Pentaxes, Naglers, Radians, and Panoptics—are capable of providing a good image across a wide apparent field of view.

If you have a slow focal ratio scope, even older wide-field designs such as Erfles and Königs work reasonably well (although the modern premium eyepieces are still better). If you have a fast focal ratio scope, you basically have three choices: (1) pay the price for premium wide-field eyepieces; (2) limit yourself to Plössls and similar older designs, which have narrower apparent fields but are sharp to the edge even in fast scopes; or (3) buy inexpensive wide-field eyepieces and resign yourself to very poor edge performance.

Physical size and weight

Even within one barrel size, eyepieces differ dramatically in physical size and weight. Physical size is usually not an issue, but it may become important if you need to use an eyepiece on a small telescope with little physical clearance or if you plan to install two identical eyepieces in a binoviewer. For example, the body diameter of Pentax XL and XW eyepieces is so large that even if you adjust the binoviewer so that the eyepieces touch, most people's eyes are too close together to see through both eyepieces simultaneously.

Weight may be an issue with a Dobsonian scope, where changing from a very light eyepiece to a very heavy one may cause balance problems.

Mainstream eyepieces range in weight from a few ounces for a typical 1.25" eyepiece to 35 ounces for the 31mm Tele Vue Nagler. Various companies manufacture eyepiece weight rings that slide onto the barrel and lock with a set screw. These rings allow you to increase the weights of your lighter eyepieces to avoid this balance problem.

Parfocality

Different eyepiece designs come to focus at different points, which is to say that you may have to rack the focuser in or out a significant distance when changing eyepieces. What you want are eyepieces that you can simply swap in and out without having to refocus. Such eyepieces are referred to as being *parfocal* with each other.

Most eyepiece series are parfocal within the series. For example, all Tele Vue Radians are parfocal with each other, as well as with most Tele Vue Plössls, most Tele Vue Naglers, and some Tele Vue Panoptics. There are often exceptions at very short or very long focal lengths. By happenstance, an eyepiece from one manufacturer may be parfocal with an eyepiece from a different manufacturer. For example, our 30mm Orion Ultrascopic is parfocal with our Pentax eyepieces.

Just as you can adjust the weights of your eyepieces by adding weights, you can adjust them to be parfocal by adding parfocalizing rings, which are available from various manufacturers. To parfocalize your eyepieces, you first determine which of them requires the most in-focus. With that eyepiece focused sharply on a star, you lock your focuser. You then add a parfocalizing ring to each of your other eyepieces and physically slide them in and out in the focuser until they reach focus. At that point, you slide the parfocalizing ring down to the top of the focuser and lock the set screw. Other than the cost of the parfocalizing rings, the only drawback to this workaround is that some of your eyepieces may end up not having much of their barrels held by the focuser and so may be at risk of falling out.

Filter threading

Most eyepieces are threaded for filters on the field lens end. Some are also threaded for filters on the eye lens end. Filter threads are generally standardized for 1.25" and 2" eyepieces, but there are exceptions. Meade eyepieces and filters, for example, use a thread that is not quite standard. Meade filters appear to fit standard eyepieces, and vice versa, but they don't, really. It's quite easy to strip or crossthread a Meade filter in a standard eyepiece or a standard filter in a Meade eyepiece, which for us is reason enough to avoid Meade eyepieces and filters. There are other minor exceptions to thread standardization, but all of those are "system" designs, such as the Questar scopes and Brandon eyepieces.

Real-World Eyepieces

Obviously, we'd like the best of all worlds—a wide apparent field of view, plenty of eye relief, an image that's sharp all the way to the edge of the field even in a fast scope, and top-notch mechanical and optical quality. Oh, yeah, and it shouldn't cost much, either. Unfortunately, no such eyepiece exists.

There is a class of eyepieces that comes close, though, on everything except price. These eyepieces—Tele Vue Naglers, Panoptics, and Radians, and the Pentax XWs, along with the recently discontinued Pentax XLs, which are still widely available to purchase used—are what astronomers call premium eyepieces. They are world-class in every respect, including price. Many would also include the Celestron Axiom, Meade Super Wide, Meade Ultra Wide, and Vixen Lanthanum Super Wide eyepieces in this category, but we consider them a small step down from the other premium models. Table 4-1 lists the important characteristics of premium eyepieces.

Table 4-1. Important characteristics of premium eyepieces

Model	Focal lengths (mm)	AFoV	Eye relief	Price
Celestron Axiom	15, 19, 23, 34, 40, 50	70°[a]	7–38mm	$140–$300
Meade Super Wide Angle	16, 20, 24, 28, 34, 40	68°	[b]	$179–$399
Meade Ultra Wide Angle	4.7, 6.7, 8.8, 14, 18, 24, 30	82°	[b]	$199–$449
Pentax XL	5.2, 7, 10.5, 14, 21, 28, 40	65°[c]	20mm	[d]
Pentax XW	3.5, 5, 7, 10, 14, 20, 30, 40	70°	20mm	$310–$500
Tele Vue Nagler	2.5, 3.5, 4.8, 5, 7, 9, 11, 12, 13, 16, 17, 20, 22, 26, 31	82°	6–19mm	$155–$620
Tele Vue Panoptic	15, 19, 22, 24, 27, 35, 41	68°	5–28mm	$210–$495
Tele Vue Radian	3, 4, 5, 6, 8, 10, 12, 14, 18	60°	20mm	$240
Vixen Lanthanum Super Wide	3.5, 5, 8, 13, 17, 22, 42	65°[e]	20mm	$240–$360

[a] 50mm Axiom has 52° AFoV.

[b] The original Meade Series 4000 Super Wide Angle and Series 4000 Ultra Wide Angle eyepieces have been replaced by the Series 5000 models. Meade specifies only "long eye relief" for these new models. The 5000-series eyepieces had not yet shipped when we wrote this, so we have been unable to determine eye relief figures for them.

[c] All Pentax XL eyepieces have a 65° AFoV, except the 28mm, which has only a 55° AFoV.

[d] Pentax XL eyepieces were discontinued in 2004, but remain widely available used. The 2" 40mm Pentax XL typically sells for $225 to $250 used, if in good condition. All other focal lengths are 1.25" models, and typically sell for $200 to $225 used.

[e] The 42mm Vixen Lanthanum Super Wide has a 72° AFoV.

Although astronomers argue about which of the premium eyepieces are "best," the truth is that all of them provide superb images. Whatever their preferences, most would agree that the Pentax and Tele Vue lines are the best of the best, and those are the eyepiece lines we recommend unreservedly.

When choosing among them, we suggest you not worry too much about image quality. There are very minor differences, yes, but the similarities are much greater. Where these premium eyepieces differ is more a matter of emphasis. Naglers, for example, offer an extremely wide 82° apparent field, but have only 12mm of eye relief in most of the current models. Pentax XWs, on the other hand, offer a narrower (but still very wide) 70° apparent field, but have a generous 20mm of eye relief across the line. Radians have the same 20mm of eye relief as the Pentaxes, but only a 60° apparent field. On the other hand, Radians cost less than Pentax XWs. And so on.

> There are variations within the lines, as well, which is why many astronomers have a mix of eyepieces. For example, some prefer Pentax eyepieces for high magnification, but use Panoptics or Naglers for medium and low magnification. It's not unusual for one astronomer to have a mix of Pentax, Nagler, Radian, and Panoptic eyepieces. It's more common, though, to have only one or two lines. For example, many astronomers use all Naglers, all Pentaxes, or Radians for medium and high magnifications and Panoptics for medium and low magnifications.

A step below the premium eyepiece lines are what we call "semi-premium" eyepieces. Like premium eyepieces, these semi-premium eyepieces have world-class fit and finish, and their optics have excellent polish and coatings. But they lack one or more of the important (and expensive) characteristics of premium eyepieces, such as very wide apparent fields, long eye relief, or suitability for use with fast focal ratio scopes. Table 4-2 lists the important characteristics of semi-premium eyepieces.

Table 4-2. Important characteristics of semi-premium eyepieces

Model	Focal lengths (mm)	AFoV	Eye relief	Price
Antares Speers-WALER	7.5, 10, 14, 14L, 18, 18L	82°	12mm	$120
Antares Elite Plössl	5, 7.5, 10, 15, 20, 25	52°	3–15mm	$75–$85
Celestron Ultima	5, 7.5, 10, 12.5, 18, 30, 35, 42	49°–51°[a]	3–24mm	$80–$110
Orion Ultrascopic	3.8, 5, 7.5, 10, 15, 20, 25, 30, 35	52°[b]	2–24mm	$90–$130
Tele Vue Plössl	8, 11, 15, 20, 25, 32, 40	50°[c]	5–23mm	$82–$110

Table 4-2. *Important characteristics of semi-premium eyepieces (continued)*

Model	Focal lengths (mm)	AFoV	Eye relief	Price
University Optics König	12, 16, 24, 25, 32, 40	60°–82°	8–26mm	$80–$296
University Optics Orthoscopic	4, 5, 6, 7, 9, 12.5, 18, 25	40°–47°	3–20mm	$60–$120
Vixen Lanthanum	2.5, 4, 5, 6, 7, 9, 10, 12, 15, 18, 20, 25, 30, 40, 50	45°–60°[d]	20mm	$130–$200

[a] The 35mm Ultima has a 49° AFoV. The oddball 42mm Ultima has an AFoV of only 36° and eye relief of 32mm.

[b] The 35mm Ultrascopic has a 49° AFoV.

[c] The 40mm Tele Vue Plössl has only a 43° AFoV and, like all 1.25" 40mm Plössls, should be avoided. Tele Vue also offers a $228 2" 55mm Plössl, with a 50° AFoV and 20mm of eye relief. That eyepiece is suitable only for f/8 or slower scopes with 2" focusers, such as SCTs.

[d] The 2.5, 4, 5, 6, 7, and 50mm Vixen Lanthanums have a 45° AFoV. The 42mm Vixen Lanthanum has a 42° AFoV. The 30mm Vixen Lanthanum has a 60° AFoV.

The Antares Speers-WALER (wide-angle, long eye relief) eyepieces are a mixed bag. Speers-WALER eyepieces have a reputation for spotty manufacturing quality, and the fit and finish level is a bit lower than the other eyepieces in our semi-premium group. The 7.5mm and 10mm are reasonably good optically even in fast scopes, although not quite up to the optical standards of premium eyepieces. The 14mm is somewhat inferior optically to the 7.5mm and 10mm, particularly in fast scopes, and the 18mm more so. The 14mm and 18mm aren't bad eyepieces by any means, but they don't have the sharpness, contrast, and edge performance we expect from a semi-premium eyepiece. On the other hand, no other eyepieces in this price range come close to having the apparent field of view and eye relief of a Speers-WALER in a design that's usable down to f/5 or faster.

The 14mm and 18mm also require a great deal of in-focus, so much so that they will not come to focus in most Dobs (although they work fine in SCTs). For this reason, Antares introduced "L" models of the 14mm and 18mm SW eyepieces, which are designed for low-profile focusers and differ optically from the standard models. Unfortunately, the "L" models are optically inferior to the standard 14mm and 18mm models. If you're on a budget and want a 7.5mm or 10mm wide-field eyepiece, and if 12mm of eye relief is sufficient for you, the Antares Speers-WALER eyepieces are a good choice. We'd consider the 14mm and possibly the 18mm acceptable choices if we owned an SCT or a Dob with sufficient in-focus to handle them. We would avoid the "L" models entirely.

We consider the Antares Elite Plössl, Celestron Ultima, Orion Ultrascopic, and Tele Vue Plössl eyepieces essentially interchangeable. All have excellent fit and finish, top-quality optics and coatings, a similar range of focal lengths available, and similar apparent fields of view. The Antares Elite Plössl, Celestron Ultima, and Orion Ultrascopic eyepieces are in fact identical except for their labels and minor differences in focal length. They are Japanese-made, 5-element Masayuma designs, which add a fifth correcting element to the standard 4-element Plössl design. The Tele Vue Plössl is a standard Plössl design, superbly executed. All of these eyepieces offer sharp, flat fields and excellent edge performance in scopes as fast as f/4. They are less costly than premium eyepieces because they have only 50° apparent fields and very short eye relief in shorter focal lengths.

The University Optics Königs have the excellent fit and finish, polish, and coatings typical of most Japanese optics. The inexpensive models—the 12, 16, and 24mm 1.25" models and the 32mm 2" model—typically have eye relief of about 2/3 to 3/4 their focal lengths, and 60° to 68° apparent fields of view. Their downfall is their short eye relief in shorter focal lengths and their hideous edge performance in fast scopes. If you have an f/8 or slower scope, such as an SCT, these eyepieces provide excellent value for money, but don't even think about using them in a fast Dob. The more expensive 2" models—25mm, 32mm, and 40mm—are much better corrected for fast focal ratios, although their edges are still soft in fast scopes.

> If you wonder why we class a $296 eyepiece as only "semi-premium," it's because that eyepiece, the 32mm MK-80, competes head-on with the $620 31mm Tele Vue Nagler. Like the 31mm Nagler, the 32mm MK-80 provides an 82° apparent field and is of excellent mechanical and optical quality, but a $296 eyepiece simply can't match Al Nagler's "TermiNagler" in providing crisp images edge to edge in fast scopes. There ain't no such thing as a free lunch.

University Optics (UO) has been selling their Abbe Orthoscopic eyepieces for 45 years now, and they're as good as ever. Based on a 19th-century design, but using modern production methods and coatings, the UO Orthos provide absolutely top-notch optical performance and are the favorite of many observers for planetary work. The downside to Orthos is their relatively narrow fields of view, typically 40° to 45°, and their limited eye relief. The popularity of Orthos began to wane with the advent of Dobsonian reflectors and other undriven scopes, in which using a high-power eyepiece with a narrow apparent field makes it difficult to keep the object in the field of view. Still, for driven scopes—or Dobs with very smooth motions—an Ortho is an excellent and economical choice for high-power observing. In the 4mm to 12.5mm range, the UO Orthos compare favorably to the Tele Vue Plössls, Celestron Ultimas, and Orion Ultrascopics. The Orthos cost a bit less, provide slightly greater eye relief, and (some believe) slightly sharper and higher-contrast images.

UO actually sells three lines of Orthos. Their original line, the University Abbe Orthoscopics, cost $60 each and are available in 4, 5, 6, 7, 9, 12.5, 18, and 25mm focal lengths. The University H.D. Abbe Orthoscopics sell for $80 and have several minor physical enhancements, along with slightly better coatings. They're available in 5, 6, 7, 9, 12.5, and 18mm focal lengths. Finally, the 9, 12.5, and 18mm Orthoscopic Planetary Series (O.P.S) eyepieces sell for $120 and have specialized coatings optimized for Lunar and planetary observation.

The well-respected Japanese optical company Vixen, now a subsidiary of Tele Vue, manufactures two lines of eyepieces. The Lanthanum Super Wides are true premium eyepieces. The standard Lanthanum series eyepieces retain the 20mm eye relief of the premium line, but have apparent fields of view of 45° to 60°, rather than the 65° to 72° of the premium series. The Vixen Lanthanum eyepieces were formerly sold by Orion and Celestron under their own names, but are now available only from Vixen. The Lanthanum eyepieces are first-rate optically and mechanically, with the fit and finish we've come to expect from Japanese eyepieces. We think of them as Plössls that just happen to have 20mm of eye relief regardless of their focal lengths. The Lanthanums are sharp, contrasty, and have excellent edge performance, but they also have a reputation for being a bit dimmer than premium eyepieces, Plössls, and orthoscopics. That may be true, but if so, we've never been able to detect it.

These are by no means all of the premium and semi-premium eyepieces available. Zeiss, for example, makes world-class eyepieces, as does Takahashi. TMB makes very specialized (and expensive) eyepieces designed for planetary observing. But you won't find any of those in most astronomers' eyepiece cases, and for good reason. The models we've listed offer the best combination of features, performance, and price available.

If you wonder why we haven't talked about economy eyepieces, it's because we don't recommend buying them. Economy eyepieces generally fall into one of the following three categories:

Obsolete designs

With very few exceptions—such as the expensive TMB Monocentric eyepieces—a well-corrected eyepiece requires at least four lens elements. Older three-element designs, such as the Kellner (and its modern derivative, the Edmund RKE), the Modified Achromat (MA), and the Super Modified Achromat (SMA) are cheaper to produce than modern designs, but suffer from aberrations, short eye relief, narrow apparent fields of view, and other problems.

Older wide-field designs, including the König and Erfle, use from four to eight elements and are suitable for use in slow focal ratio scopes, but they generally provide very poor edge performance in faster scopes. (Owners of fast Dobs pronounce Erfle "awful.")

 Ironically, the excellent Antares Elite Plössl, Celestron Ultima, and Orion Ultrascopic eyepieces—which are marketed as "five-element Plössls" or "Super Plössls"—aren't Plössls at all. They are actually Erfles that use Heinrich Erfle's original 2-1-2 lens element design. But, because they restrict their apparent fields of view to about 50° rather than the 60° to 70° of typical Erfle eyepieces, they are usable without significant aberrations in scopes down to f/5 or so.

Chinese Plössls

Most inexpensive scopes are bundled with one or two Chinese Plössls, usually 25mm or 26mm and 9mm or 10mm. Those eyepieces are usable until you can get something better, but their polish and coatings are noticeably inferior to those of eyepieces that don't cost all that much more. For example, Chinese Plössls typically sell for $40 to $70, depending on focal length, while the much superior Antares Elite, Celestron Ultima, Orion Ultrascopic, and Tele Vue Plössls sell for $75 to $110. If you're on an extremely tight budget, Chinese Plössls offer a lot of bang for the buck. But if paying an extra $20 to $40 for an eyepiece is no big deal to you, you'll be much happier in the long run with the better-quality eyepieces.

Cheap knock-offs

Chinese optical factories have begun making knock-offs of some premium and semi-premium Japanese eyepiece lines. For example, when Orion introduced their Chinese Epic ED-2 line of eyepieces, a lot of people got excited. These $70 eyepieces promised 20mm of eye relief and 55° apparent fields of view. There was a lot of debate about whether the Epic ED-2 eyepieces were knock-offs of the $130 50° Vixen Lanthanums or the (then) $228 60° Tele Vue Radians. Everyone wants a bargain.

Then, as Epic ED-2 eyepieces made it into astronomers' hands, reality set in. The Epic ED-2 eyepieces weren't "Radian killers" nor even "Lanthanum killers." They were $70 Chinese eyepieces with mediocre fit and finish and optical performance well below that of more expensive eyepieces. They suffer from ghosting and flaring, and edge performance in fast scopes is poor. Unless you place your eye just so, they also suffer from blackouts, as the image simply disappears. Although some astronomers are happy with them for the price, no one mistakes them for a Radian or Lanthanum.

The moral is, when it comes to eyepieces, TANSTAAFL, There Ain't No Such Thing As A Free Lunch. If cheap eyepieces performed like expensive eyepieces, no one would buy the expensive eyepieces. Duh.

Choosing an Eyepiece Set

When it comes to choosing eyepieces, most beginning amateur astronomers follow a predictable path. They start with the eyepieces supplied with their scopes and soon decide they need more eyepieces. A full set of focal lengths! Wider fields! They buy a lot of cheap eyepieces, hoping against hope to find what they want without paying high prices for premium eyepieces. Then one night they make the mistake of looking through someone else's premium eyepiece. The scales fall from their eyes, and they realize just how much better premium eyepieces really are. So they buy their first premium eyepiece, and they're hooked. But they find themselves with one premium eyepiece and a case full of cheap eyepieces that are pretty much worthless.

We suggest you skip the first part and go straight to the second. Make your first eyepiece purchase a premium model. Yes, if you're on a budget, the thought of buying premium eyepieces is pretty intimidating. After all, a full set of premium eyepieces may cost $3,000 or more. But here's the secret: *you don't need a full set.*

Most experienced amateur astronomers use at most three or four eyepieces regularly, and if you have a good Barlow or Powermate, you can get by with fewer. For example, Figure 4-12 shows the set we use for 99% of our observing in our 10" f/5 Dob:

> 30mm Orion Ultrascopic ($120; 42X magnification; 1.25° true field)
> 14mm Pentax XL ($225; 90X magnification; 0.73° true field)
> 10mm Pentax XW ($310; 125X magnification; 0.56° true field)
> 14mm Pentax XL with 2X Barlow (180X magnification; 0.36° true field)
> 10mm Pentax XW with 2X Barlow (250X magnification; 0.28° true field)

Figure 4-12. Our standard eyepiece set: 30mm Orion Ultrascopic, 14mm Pentax XL, 10mm Pentax XW

Here's what our observing buddy Paul Jones uses regularly in his 8" f/10 SCT:

> 27mm Tele Vue Panoptic ($330; 75X magnification; 0.90° true field)
> 14mm Pentax XL ($225; 145X magnification; 0.45° true field)
> 7mm Tele Vue Nagler ($180; 290X magnification; 0.28° true field)

Here's what our observing buddy Steve Childers uses regularly in his 17.5" f/5 Dob:

> 40mm Pentax XL ($245; 56X magnification; 1.17° true field
> 27mm Tele Vue Panoptic ($330; 82X magnification; 0.83° true field)
> 14mm Pentax XL ($225; 159X magnification; 0.41° true field)
> 10.5mm Pentax XL ($225; 212X magnification; 0.31° true field)
> 8mm Tele Vue Radian ($240; 278X magnification; 0.22° true field)

 With the exception of the 40mm Pentax XL and the 27mm Tele Vue Panoptic, all of these eyepieces use 1.25" barrels, and so they can also be used in our various other scopes that have 1.25" focusers.

So, assuming you're starting with only the bundled eyepieces supplied with your scope, how do you begin building an optimum eyepiece collection? We suggest you use the following guidelines.

- If budget is not an obstacle, simply buy the eyepieces you need. If budget is an issue, your first acquisition should be a top-notch 2X Barlow **[Hack #46]**. A Barlow effectively doubles your eyepiece collection at minimal cost.

- If your budget permits, buy only premium eyepieces. If you are on a tight budget, buy semi-premium models. You're better off with only two or three premium or semi-premium eyepieces than with an entire case full of cheap eyepieces.

- Everyone needs a low-power eyepiece **[Hack #48]**, but, if you're going to make compromises, this is the place to do it. When you're getting started, the low-power eyepiece bundled with your scope will suffice until you've filled out your collection with higher priority eyepieces. If you have no low-power eyepiece, fill this slot with a $50 Chinese 32mm Plössl for the time being.

- Your first premium eyepiece should be a moderate-power eyepiece that can be Barlowed to provide a high-power eyepiece. This will be your work-horse eyepiece. For use with a 2X Barlow, choose an eyepiece that provides a magnification roughly half the aperture of your scope in millimeters, but no more than 135X to 150X. For a 6" (150mm) scope, we recommend an eyepiece that provides 60X to 90X; for an 8" (200mm) or larger scope, 90X to 125X. If you plan to make do with just one premium eyepiece and a 2X Barlow for a while, choose an eyepiece that provides a magnification in the upper part of the range, e.g., 90X for a 6" scope or 125X for an 8" or larger scope.

> For scopes under 8", aperture determines maximum useful magnification, which is roughly 1.5 times the scope aperture in millimeters. For scopes 8" and larger, atmospheric instability determines maximum useful magnification, which is typically 250X to 300X on steady nights in most locations. Even if you have a large scope, there is little point to having an eyepiece that provides more than 250X to 300X natively. On those rare nights when seeing is good enough to use magnifications higher than 300X, you can simply Barlow one of your standard eyepieces to achieve higher magnification.

- For your second premium eyepiece, buy one on the other end of the ranges just given. For example, if you have an 8" or larger scope and bought a 125X eyepiece as your first premium eyepiece, the second one should yield a native magnification on the other end of that range, 90X in this case. With the 2X Barlow, that gives you a very useful range of magnifications—90X, 125X, 180X, and 250X.

But not everyone can afford even one premium eyepiece, and there are even some people (most with driven scopes) who actually prefer the 50° apparent fields of Plössls and similar eyepieces. If you're in that situation, here's what we recommend:

- Buy a top-notch 2X Barlow, such as the Orion Ultrascopic, the Tele Vue, or the Celestron Ultima.

- Use your current low-power eyepiece as your finder eyepiece. If you don't have a low-power eyepiece, buy one **[Hack #48]**.

- If you must wear eyeglasses while observing, or if long eye relief is otherwise important to you, buy Vixen Lanthanum eyepieces. If you're not concerned about eye relief, buy the less expensive Tele Vue Plössls, Antares Elites, Celestron Ultimas, or Orion Ultrascopics.

> If you wear eyeglasses only to correct for near- or farsightedness, you needn't wear them while observing. Simply use the telescope focuser to accommodate your own vision. If your glasses correct for astigmatism, you'll need them when using an eyepiece that provides a large exit pupil, such as a 32mm finder eyepiece. However, many people find that their astigmatism is limited to the outer portion of their pupil. When you observe at moderate to high powers, which is to say using a 3mm or smaller exit pupil, you are using only the center of your pupil and may not need your glasses.
>
> We don't wear glasses to observe, but we've found, as have many others, that we much prefer the 20mm eye relief provided by the Lanthanums and many premium eyepieces. It's simply more comfortable to observe with an eyepiece that provides plenty of eye relief, and if you're comfortable, you tend to spend more time looking at the objects and teasing out finer detail.

- There's a good chance that the focal length you might otherwise choose for your moderate-power eyepiece is already provided by your low-power eyepiece when you use it with your Barlow. If so, choose a focal length somewhat shorter than the Barlowed combination. For example, if your low-power eyepiece is 25mm, that gives you 12.5mm Barlowed. Choose an eyepiece in the 10mm range, which also gives you the equivalent of a 5mm eyepiece when Barlowed, providing both a moderate magnification for general viewing and a higher magnification for planetary and Lunar observing.

Although many inexpensive scopes are bundled with a 9mm or 10mm Plössl, we suggest you not consider it when planning your eyepiece purchases. This eyepiece is generally of lower quality than the bundled 25mm or 26mm eyepiece, and it should be a priority to replace it as soon as possible.

- Fill in your range by adding one or two more eyepieces with focal lengths spaced appropriately to give a useful range of magnifications, with and without the Barlow, up to the maximum useful magnification of your scope. For example, for a scope with 1,200mm focal length—a typical 6", 8", or 10" Dob—we might add a 15mm 50° eyepiece, to give us the following options:

  25mm (48X; 1° true field)
  15mm (80X; 0.63° true field)
  25mm with 2X Barlow (96X; 0.52° true field)
  10mm (120X; 0.42° true field)
  15mm with 2X Barlow (160X; 0.31° true field)
  10mm with 2X Barlow (240X; 0.21° true field)

If we wanted to add another eyepiece to this set, we'd probably choose a 12mm model. That makes the 25mm with Barlow superfluous, but when Barlowed, it also adds 200X to your repertoire. 160X is a bit too little for planetary observing, and 240X may be a bit too much on some nights. Having a 200X option puts you right in the middle on nights when the atmosphere is just a bit too unstable to support 240X.

Finally, some people want to spend as little as possible on an eyepiece collection, either because they simply can't afford good eyepieces or because they're new to the hobby and aren't sure they're going to maintain their initial interest. If you're in that group, we recommend simply adding a good 2X Barlow and using the eyepieces supplied with your scope until you can afford something better or decide that you're hooked on the hobby. With the Barlow, you have a decent selection of magnifications and fields of view that will tide you over until you're ready to buy some eyepieces.

Chart Your Eyepiece Characteristics

Knowing the capabilities of your eyepieces makes it easier to choose the right one. Paste a reference chart to your scope to keep this information at your fingertips.

If you have only one scope and a few eyepieces, it's easy enough to remember the magnification, field of view, and other characteristics of each eyepiece. But if you have multiple scopes and many eyepieces, it can be difficult to remember all the details.

For example, assume you have three telescopes, an 80mm f/5 [Hack #9] short-tube refractor with a focal length of 400mm, a 10" f/5 Dob with a focal length of 1,255mm, and an 8" f/10 SCT with a focal length of 2,032mm. You also have a *very* nice selection of Tele Vue eyepieces—41mm, 27mm, and 19mm Panoptics; and Radians in 14mm, 12mm, 10mm, 8mm, and 6mm focal lengths—and a 2.5X Powermate, which effectively multiplies the power of any eyepiece it is used with by 2.5X.

Your first attempt at an eyepiece chart for the 400mm short-tube refractor might look something like Table 4-3.

Table 4-3. An eyepiece chart for a 400mm f/5 refractor

| Eyepiece | Alone | | | | w/ Powermate | | | | |
	Mag	Field (°)	Field (')	Pupil (mm)	EFL (mm)	Mag	Field (°)	Field (')	Pupil (mm)
41mm Panoptic	10X	6.6	395.3	8.2	16.4	24X	2.6	158.1	3.3
27mm Panoptic	15X	4.4	262.1	5.4	10.8	37X	1.7	104.9	2.2
19mm Panoptic	21X	3.1	183.1	3.8	7.6	53X	1.2	73.2	1.5
14mm Radian	29X	2.1	123.8	2.8	5.6	71X	0.8	49.5	1.1
12mm Radian	33X	1.8	108.3	2.4	4.8	83X	0.7	43.3	1.0
10mm Radian	40X	1.5	90.2	2.0	4.0	100X	0.6	36.1	0.8
8mm Radian	50X	1.2	71.3	1.6	3.2	125X	0.5	28.5	0.6
6mm Radian	67X	0.9	54.1	1.2	2.4	167X	0.4	21.7	0.5

This chart presents a lot of very useful information in a small space:

- Magnifications are useful to know for obvious reasons.
- The Field (°) columns list the field in degrees, which is useful as a rough indication of how suitable the eyepiece is for observing a given object. For example, if you're observing a large galaxy, nebula, or star cluster, you probably want a wide field of view. Conversely, if you're trying to tease out detail in a tiny globular cluster or planetary nebula, you probably want to put some serious power on and narrow the field of view.
- The Field (') columns list the field of view in arcminutes, which is useful for star hopping, estimating separation of close objects, and so on.
- The Pupil columns list the exit pupil provided by the various eyepieces and combinations, which are useful for several reasons. First, human visual acuity peaks between 2mm and 3mm, which is the ideal range for observing any object. Second, an exit pupil of 1.0mm to 0.7mm is the smallest suitable for general observing. Below that, diffraction and floaters (microscopic bits of material in your eye) begin reducing acuity significantly. Third, an exit pupil around 0.5mm is the smallest useful for specialized observing, such as splitting double stars. Finally, some interference filters work best within narrow ranges of exit pupil size.
- The EFL (effective focal length) column lists the equivalent focal length of each eyepiece when used with the Powermate.

Unfortunately, there are also a few things wrong with this chart. Most obviously, the 41mm and 27mm Panoptics have 2" barrels, while the Powermate accepts only 1.25" eyepieces. Oops. Less obviously, there is some overlap in the combinations. For example, you probably would choose to use the 8mm Radian alone (50X, 1.2° FoV) rather than the 19mm Panoptic with the Powermate (53X, 1.2° FoV). The Radian has much better eye relief than the Panoptic/Powermate combination, and you'll be looking through less glass. Removing the superfluous combinations gives us a chart that looks something like Table 4-4.

Table 4-4. An improved eyepiece chart for a 400mm f/5 refractor

	Alone				w/ Powermate				
Eyepiece	Mag	Field (°)	Field (')	Pupil (mm)	EFL (mm)	Mag	Field (°)	Field (')	Pupil (mm)
41mm Panoptic	10X	6.6	395.3	8.2					
27mm Panoptic	15X	4.4	262.1	5.4					
19mm Panoptic	21X	3.1	183.1	3.8					

Table 4-4. An improved eyepiece chart for a 400mm f/5 refractor (continued)

Eyepiece	Alone Mag	Field (°)	Field (')	Pupil (mm)	w/ Powermate EFL (mm)	Mag	Field (°)	Field (')	Pupil (mm)
14mm Radian	29X	2.1	123.8	2.8					
12mm Radian	33X	1.8	108.3	2.4	4.8	83X	0.7	43.3	1.0
10mm Radian	40X	1.5	90.2	2.0	4.0	100X	0.6	36.1	0.8
8mm Radian	50X	1.2	71.3	1.6	3.2	125X	0.5	28.5	0.6
6mm Radian	67X	0.9	54.1	1.2	2.4	167X	0.4	21.7	0.5

You can still use the 19mm Panoptic or the 14mm Radian with the 2.5X Powermate, of course. But having only optimum combinations listed makes it faster to choose the ideal combination.

Turning to the 1,255mm f/5 Dob, we once again run the calculations to reflect the higher magnifications and smaller fields with the longer scope. We also considered eliminating one or two combinations. The 6mm Radian with the 2.5X Powermate yields 523X magnification in our 1,255mm Dob. The atmospheric turbulence in our area makes that too much to be usable on any but the best nights. In fact, we can seldom push much higher than 300X, so we considered eliminating the 8mm Radian/Powermate combination as well. We kept both, though, because they are usable on some nights, particularly for Lunar observing. Table 4-5 shows our results for the 1,255mm Dob.

Table 4-5. An eyepiece chart for a 1,255mm f/5 Dob

Eyepiece	Alone Mag	Field (°)	Field (')	Pupil (mm)	w/ Powermate EFL (mm)	Mag	Field (°)	Field (')	Pupil (mm)
41mm Panoptic	31X	2.1	126	8.2					
27mm Panoptic	46X	1.4	83.5	5.4					
19mm Panoptic	66X	1	58.3	3.8					
14mm Radian	90X	0.7	39.4	2.8					
12mm Radian	105X	0.6	34.5	2.4	4.8	261X	0.2	13.8	1.0

Table 4-5. An eyepiece chart for a 1,255mm f/5 Dob (continued)

	Alone				w/ Powermate				
Eyepiece	Mag	Field (°)	Field (')	Pupil (mm)	EFL (mm)	Mag	Field (°)	Field (')	Pupil (mm)
10mm Radian	126X	0.5	28.8	2.0	4.0	314X	0.2	11.5	0.8
8mm Radian	157X	0.4	22.7	1.6	3.2	392X	0.2	9.1	0.6
6mm Radian	209X	0.3	17.3	1.2	2.4	523X	0.1	6.9	0.5

Finally, we recalculate the chart for a 2,032mm 8" f/10 SCT. Although we left the eyepiece/Powermate combinations for the 12mm and 10mm Radians, the truth is that those combinations might actually be useful only once or twice a year, unless you live in an area like southern Florida that has extremely stable air. In practical terms, the Powermate would probably never come out of our eyepiece case. Instead, we'd just use the eyepieces natively. They provide an excellent range of magnifications, fields of view, and exit pupil sizes all by themselves. Table 4-6 shows our results for a 2,032mm SCT.

Table 4-6. An eyepiece chart for a 2,032mm f/10 SCT

	Alone				w/ Powermate				
Eyepiece	Mag	Field (°)	Field (')	Pupil (mm)	EFL (mm)	Mag	Field (°)	Field (')	Pupil (mm)
41mm Panoptic	50X	1.3	77.8	4.1					
27mm Panoptic	75X	0.9	51.6	2.7					
19mm Panoptic	107X	0.6	36	1.9					
14mm Radian	145X	0.4	24.4	1.4					
12mm Radian	169X	0.4	21.3	1.2	4.8	423X	0.1	8.5	0.5
10mm Radian	203X	0.3	17.8	1.0	4.0	508X	0.1	7.1	0.4
8mm Radian	254X	0.2	14	0.8					
6mm Radian	339X	0.2	10.7	0.6					

You can calculate your own charts based on the measured or nominal focal length of your scope(s) and the fields of view of your eyepieces. Once you've created a chart for each of your telescopes, print it out, laminate it, and paste it on the side of the scope near the focuser. You'll always have these important eyepiece characteristics immediately available.

HACK #51 View Dim Objects in the Same Field as a Very Bright Object

Build an occulting eyepiece.

Almost any modern eyepiece provides reasonably good contrast and minimum glare. There are times, though, when even the best standard eyepiece isn't quite good enough to show an extremely dim object in close proximity to an extremely bright object. For those special situations, you need an *occulting eyepiece*, which you can make at no cost by temporarily modifying a standard eyepiece.

For example, you may be trying to observe Deimos and Phobos, the tiny moons of Mars. (Phobos is about 26×22 kilometers and Deimos only about 15×12 kilometers.) The problem is that these tiny moons are located within a few arcseconds of Mars and are 13 to 15 magnitudes—nearly 1,000,000 times—dimmer than the planet.

Imagine a 1,000,000 candlepower searchlight pointed at you from 10 miles away, with someone standing two feet from the searchlight holding a lit candle. Resolving the Martian moons without being swamped by the light from Mars is harder than resolving the candle flame without being blinded by the searchlight.

Similarly, you may be trying to view a very tight double star with a primary that is many magnitudes brighter than the companion. Even the best standard eyepiece isn't good enough to separate the dim companion from the glaring primary because the light from the primary star flares in your eyepiece, subsuming the tiny amount of light from the companion. Or, you may be trying to view a dim DSO near a bright star, for example the dim galaxy NGC 404 in Andromeda, which is located only seven arcminutes from the bright star Mirach.

In all of these situations, the problem is that there is a very bright object in the same field of view as the very dim object you want to see. Your first thought might be simply to put the bright object just outside the field of view of the eyepiece, leaving only the dim object in the field of view. There are two problems with that idea:

- All eyepieces provide inferior images of objects near the edges of their fields of view compared to an object located near the center of the field. (Premium eyepieces have excellent edge performance, but it is still inferior to the best they can do with the object centered in the field.)

- Many eyepieces exhibit severe ghosting and flaring of objects located just outside their fields of view.

The solution is to build an occulting eyepiece. An occulting eyepiece uses a small piece of opaque material—we use aluminum foil or thin plastic—at the field stop of the eyepiece to provide an occulting bar. Because the occulting bar is in the same plane as the field stop, it is sharply focused when the object is in focus. By moving the scope and rotating the occulting eyepiece as necessary, you can use the occulting bar to block the bright object, leaving the dim object visible, and in the center of the field.

It takes about five minutes to convert a standard eyepiece to an occulting eyepiece, costs nothing, and the process is easily reversible. What more could you want? There are a couple of caveats, though:

- Some eyepieces are unsuitable because their field stops are inside the optical assembly of the eyepiece. Although it's sometimes possible to modify such an eyepiece, it's generally a very bad idea to take apart the optical assembly of an eyepiece. Use a different eyepiece instead, one that has an accessible field stop.

- Whatever material you use for the occulting bar, make sure it can provide a sharp edge. Minor bumps or unevenness on the edge are OK, and, in fact, can be useful in getting the occulting bar arranged just so to cover the bright object. What you want to avoid are fuzzy edges, such as those produced with paper, cardboard, and similar materials.

- For this example, we used aluminum foil for visibility in the photographs. Although we have used aluminum foil in the past, its bright surfaces introduce reflection and flaring problems unless you darken them. We've used black spray paint or a felt-tip marker to darken the foil. Thin, flexible, opaque black plastic is generally easier to work with, although it can be difficult to get as sharp an edge with plastic as you can with foil.

Figure 4-13 shows the inexpensive Chinese 9mm Plössl that we used to build our occulting eyepiece.

Figure 4-13. The unmodified Plössl eyepiece that we started with

Plössl eyepieces are generally a good choice for modification. Their field stops are outside the lens assembly and are easily accessible. To begin, unscrew the barrel from the eyepiece to reveal the field lens assembly and field stop, visible at the center of the disassembled eyepiece in Figure 4-14.

Trim a small piece of material to form the occulting bar, making sure that at least one edge is cut sharply. Depending on your needs and the size of the field lens assembly, you can use a bar that occults only a small section of the field lens or, as shown in Figure 4-15, a bar that covers an entire hemisphere of the field lens. Press the occulting bar gently into place, securing it with a small dab of glue or a piece of tape. There's no mechanical stress on the occulting bar, so all you need is enough adhesion to keep it from falling out of the eyepiece under its own weight. We generally use a tiny dab of household glue—just enough to cover the tip of a toothpick—to secure each side of the occulting bar.

If you can determine the exact plane of the field stop, and you are using foil or a similarly flexible material, you can gently press down on the occulting bar to force it a bit deeper into the field lens assembly, to put it exactly in the

Figure 4-14. Remove the barrel from the Plössl eyepiece to reveal the field lens and field stop

plane of the field stop. That's what we did for this eyepiece, as you can see from the slight downward curve in the foil.

Once you have the occulting bar secured, simply screw the barrel back into the lens assembly, as shown in Figure 4-16. That's all there is to it. You can return the eyepiece to its original state simply by removing the barrel, peeling off the occulting bar, and reinstalling the barrel.

During the 2001 Mars close approach, we used this eyepiece in our attempt to bag Deimos and Phobos. It's a mediocre eyepiece, certainly. It doesn't have the best coatings, and the polish isn't up to the standard of world-class eyepieces of similar design like the Tele Vue Plössls, Orion Ultrascopics, or Celestron Ultimas. But it was what we had available (and, frankly, the only one of our eyepieces we were willing to disassemble). It yields about 280X magnification when used with our 2X Barlow in our 10" f/5 Dob, and about 420X when used with our 3X Barlow.

Figure 4-15. The occulting bar in position and glued down

And it worked! Although Phobos—orbiting so very close to Mars—eluded us, we were able to bag the dimmer, smaller Deimos, which orbits Mars at a greater distance. Next time, we'll be prepared with an eyepiece with better coatings, and perhaps we'll be able to bag Phobos as well.

> If you attempt to bag the Martian moons, understand that they are an extremely challenging target. Although it has been done with smaller scopes, we consider an 8" or 10" scope a realistic minimum. Everything must be perfect if you are to succeed. The atmosphere must be extremely transparent and stable, the site very dark, and your scope collimated perfectly **[Hack #40]**. It's also critical that your entire optical train be scrupulously clean **[Hack #34] [Hack #52]**. Even small smudges or dust particles scatter light, reducing contrast and contributing to flare.

Figure 4-16. The completed occulting eyepiece

HACK #52   Clean Your Eyepieces and Lenses Safely

Seven steps to better images.

All optical instruments use lenses in one form or another. Refractors have a single large objective lens, and binoculars have two. SCTs and other cata-dioptric scopes have a corrector plate, which is actually a large lens. Even Newtonian reflectors have lenses in their eyepieces.

Lenses inevitably become dirty with use. A dirty lens provides poor, low-contrast images, so it's necessary to clean your lenses from time to time. But lenses are fragile items—particularly their anti-reflection coatings—so it's very easy to do more harm than good if you clean them improperly.

The first rule of cleaning eyepieces and lenses is to do everything you can to avoid the necessity. Store your optical gear in a dust-free location or in sealed containers. Keep the lens caps on when you are not using the instrument. Handle the equipment carefully to avoid getting fingerprints (or nose-prints) on the lenses. Avoid cleaning lenses unnecessarily. A few specks of dust or a tiny smudge has no effect on image quality. More lenses have probably been ruined by unnecessary, careless cleaning than by any other cause.

Fortunately, the proper lens-cleaning procedure is straightforward. Use the following steps to clean your lenses safely and effectively.

1. Establish a clean, uncluttered, well-lit working environment. The kitchen table is usually a good choice. You'll also need a strong light, such as a flashlight, that you can direct at a grazing angle against the lens surface to reveal grit particles on the lens. If you are cleaning a large lens, such as a refractor objective, that you have removed from the scope, wear thin cotton gloves to avoid getting fingerprints on the glass.

Never clean a lens at the observing site if you can possibly avoid doing so. If you must clean a lens in the field, work inside your vehicle or another sheltered location where you can avoid the dust often present in outside air. Use a strong white light. It's impossible to clean a lens safely under red light because you can't see the grit particles that will scratch your lens.

2. Use an ear syringe or similar device to blow off any dust visible on the surface of the lens. Don't use your mouth, or you will probably spray saliva on the lens. (In an emergency, we have blown through an ordinary drinking straw to remove dust from an eyepiece.) Don't use canned air, either, because it may spray propellant onto the lens.

3. Examine the lens carefully under a strong light to make sure no dust or grit remains on the surface. You can remove a persistent particle by slightly moistening a tissue or cotton swab with Windex or a similar cleaning solution and blotting (not rubbing) the surface of the lens to pick up the particle. If necessary, use the blower again to remove any remaining dust.

Although you can buy lint-free lens cleaning tissue, we prefer to use ordinary facial tissue or toilet paper. If you do the same, make sure it's unscented and not treated with lotions or other adulterants. Toilet paper is particularly safe because you can simply unroll the outer layer or two to get a piece certain to be free of grit.

Some people prefer to use cotton swabs or cotton balls. If you do that, make sure they're pure cotton (often described as surgical cotton) rather than an artificial fiber. Cotton balls and swabs sold for medical use are usually pure, grit-free, sterile cotton. Those sold for fingernail polish removal are often made from artificial fibers and may contain grit.

4. Moisten a tissue or cotton swab with cleaning fluid. You want it damp, but not dripping wet. Begin at the center of the lens and wipe outward very gently, using a circular motion, as shown in Figure 4-17. If you've used too much cleaning fluid, it will spread over the lens surface. You want the swab or tissue to be just wet enough that you can watch a bead of cleaning fluid follow the movement of the swab or tissue. If you stop moving the swab or tissue, it should immediately reabsorb the bead of cleaning fluid. If you've used too much cleaning fluid, or if you move the swab or tissue too quickly, the bead breaks away from the swab or tissue and will dry on the lens, leaving spots.

Figure 4-17. Cleaning a binocular objective with a cotton swab

5. Replace the tissue or cotton swab frequently as you clean the lens. You may be able to clean an eyepiece or other small lens with one tissue or cotton swab. For larger lenses or smaller lenses that are badly smudged, using one tissue or swab too long merely spreads the gunk rather than picking it up.

6. The edge of the lens is the hardest part to get clean. As you approach the edge, replace your tissue or swab with a tissue folded into a sharp corner and moistened slightly with cleaning fluid. (Make sure not to touch the working area of the tissue with your finger to avoid contaminating it with skin oils.)

7. When you complete this procedure, the lens should appear clean when examined with a strong light at a grazing angle. If any lint fibers from the tissues or swabs remain on the lens, use your blower to remove them.

Some astronomers treat the choice of cleaning fluid as a religious issue, but the truth is that almost anything reasonable works fine. We have used commercial lens cleaning solutions, methanol (methyl alcohol), ethanol (ethyl alcohol), isopropanol (isopropyl alcohol), and acetone. (Be careful using acetone around painted or plastic surfaces; it may dissolve them.)

> Some astronomers clean their lenses with a LensPen (*http://www.lenspen.com*), but the LensPen scares us despite the reassurances of its maker. The LensPen uses a two-step cleaning process. You first use the brush end to remove particles. (That in itself makes us uncomfortable; we fear using a brush on a lens may scratch the coatings by dragging an abrasive particle across the surface.) Once the lens is free of dust, you rub the other end of the LensPen against the lens using a circular motion to wipe away fingerprints, smudges, film, and other contaminants. The LensPen is said to be good for 500 or more cleanings, depending on the surface area of the lens and how dirty it is. The idea of reusing a cleaning surface once, let alone 500 times, scares us to death.
>
> Perhaps we're just timid. Thousands of photographers and astronomers use these devices to clean their lenses, and we've never heard a report of a lens damaged by a LensPen. Even so, the traditional method is cheap, safe, and effective—albeit less convenient than using a LensPen—so that's what we continue to use.

Although some astronomers insist that only reagent-grade solvents are acceptable—and we suspect our organic chemist friend Paul Jones of using purer still spectroscopic grade solvents on his lenses—we've never had a problem using drugstore-grade USP solvents. Even the alcohols and acetone sold in metal gallon cans at the hardware store are fine. The important thing is to use a pure solvent rather than a mixture. For example, the acetone sold as fingernail polish remover often contains scents and oils that leave a film on the lens. Similarly, some drugstore alcohol contains scents, rubifacients, or other adulterants.

Our first choice of cleaning solution is isopropanol. We prefer 91% isopropanol, but that may be difficult to find. You can buy 70% isopropanol at any drugstore for about $1/pint, and it also works well. For water-soluble spotting, such as that caused by dried dew, commercial glass cleaners like Windex or Glass Plus work fine. (Just don't spray them on the lens as you would on a window.)

For particularly filthy lenses, you may need to use heroic measures. Public star parties are notoriously hard on eyepieces, which is why most amateur astronomers use only "throw-away" eyepieces at public observations. After such events, we have seen eyepieces contaminated with an incredible array of substances. Fingerprints and noseprints are common, of course, as is mascara, but you can expect your eyepieces to be attacked by anything up to and including bubblegum at public events. For cleaning such eyepieces, we *start* with acetone, which is usually our last resort.

HACK #53 Install a Unit-Power Finder

Find deep-sky objects quickly and easily.

Installing a *unit-power finder*, often mistakenly called a zero-power finder, is the best single upgrade any Deep-Sky Object (DSO) observer can make to a scope. A unit-power finder provides no magnification. Instead, it allows you to view the night sky naked eye with a superimposed dim red target-locating pattern. By orienting that pattern relative to the background stars, you can locate most deep-sky objects in a fraction of the time needed if you use an optical finder alone.

A unit-power finder is simplicity itself. No computers, no fancy optics—just a dim red bulls-eye pattern superimposed on the night sky. Figure 4-18 shows a simulation of using a Telrad unit-power finder to point the scope at the Great Orion Nebula, M42.

Figure 4-18. Using a Telrad unit-power finder to put M42 in the eyepiece

Red-Dot Finders

Rather than a bulls-eye pattern, a red-dot finder superimposes a simple (you guessed it) red dot. Red-dot finders are available in a wide range of prices, from the $21 StellarVue model (*http://www.stellarvue.com*) to the exquisite $205 Tele Vue Starbeam (*http://www.televue.com*). Some astronomers even mount the inexpensive red-dot sight made for Daisy BB guns as a telescope finder.

Any red-dot finder allows you to point your scope quickly and accurately. The problem is, that's all they do. Because they don't provide a bulls-eye pattern, they offer no help in locating objects geometrically. We suggest you avoid red-dot finders.

Bulls-Eye Finders

The real competition in unit-power finders is between the $40 Rigel Quik-Finder (*http://www.rigelsys.com*) and the $40 Telrad (no web site). Both products are extremely popular and have strong advocates in the amateur astronomy community. Any Telrad versus QuikFinder discussion soon degenerates into a religious debate. We have our own strong preference, but we'll try to present both products fairly.

The Rigel QuikFinder has the following advantages versus the Telrad:

- The QuikFinder is small, which makes it easier than the Telrad to fit between the focuser and optical finder, particularly on smaller scopes.

- The QuikFinder is much lighter than the Telrad, which may be an important consideration if your Dobsonian scope has balance problems. (The QuikFinder appears to be taller than the Telrad, which would be an advantage if true, but is actually an optical illusion caused by the smaller size of the QuikFinder.)

- The QuikFinder includes two bases, one intended for the sharply curved tubes of refractors and other small scopes and the other for the less-curved tubes of larger reflectors and SCTs. Additional bases are available for the Telrad, but you must buy them separately.

- The QuikFinder is easier to mount on some scopes than the Telrad because its base plate is shorter. For example, when our friend and colleague Paul Jones bought a Telrad to mount on his 8" SCT, he found that there was nowhere on the tube he could affix the long Telrad baseplate that did not interfere with his mounting rings. Similarly, on some small Dobs and many refractors, there's simply no convenient location to mount a Telrad. The QuikFinder fits in much smaller places than the Telrad, and so it may be your only realistic choice even if you prefer the features of the Telrad.

- The QuikFinder includes built-in blinking as a standard feature, which is very useful when you need to orient the bulls-eye against a field of dim stars. You can add blinking to the Telrad with about $5 worth of parts if you're comfortable building electronic projects. You can also buy a $20 third-party Telrad blinker if you want to avoid soldering.

The Telrad also has several advantages versus the QuikFinder:

- The Telrad provides a three-circle, bulls-eye pattern—0.5°, 2°, and 4°—versus only the 0.5° and 2° circles with the QuikFinder. The usefulness of that extra 4° circle is difficult to overstate, particularly when you are trying to locate objects that are far from a bright star.

- The Telrad projects its virtual bulls-eye pattern at infinity, which means it has zero parallax. If you center a star or planet in the Telrad, it remains centered regardless of your eye position because the object and the bulls-eye pattern are both at infinity. The QuikFinder has moderately severe parallax because its virtual bulls-eye pattern has an apparent distance of only a few yards. That means moving your eye relative to the QuikFinder also changes the relative position of the nearby bulls-eye pattern relative to the infinitely distant object. Unless you place your eye in exactly the same position relative to the QuikFinder every time, you'll find that the position of the object in your eyepiece is unpredictable. When you use high magnification for planetary observing, the object may be outside your eyepiece field of view. The parallax problem with the QuikFinder is less important for DSO observing, or if you also have a crosshair optical finder installed.

> To see the effect of parallax, hold up a finger at arm's length. Close one eye and align the finger with a distant object. Move your eye an inch or two from side to side. The relative positions of your finger and the distant object change, but not by much. Now repeat the experiment with the finger only a couple inches from your eye. The same amount of head movement causes a much larger displacement between your finger and the distant object. If your arm was long enough to put your finger on the distant object, it wouldn't matter where your eye was positioned: the finger would remain exactly aligned with the object. That's how the Telrad works, because its virtual bulls-eye is at infinity, the same as the object. The QuikFinder exhibits parallax because its virtual bulls-eye, like your finger, is at an intermediate distance between your eye and the object.

- The Telrad is much more heavily constructed and robust than the Quik-Finder. Both are constructed of plastic, but the Telrad is built like a tank. We leave our Telrads attached when our scopes are broken down for transport, and we've never had a problem. We wouldn't risk that with a QuikFinder, which is much more fragile.

- The Telrad dims the view less than the QuikFinder. Both products use ordinary glass rather than coated optical glass, and both dim the view relative to an unobstructed naked-eye view. But the Telrad dims the view noticeably less to our eyes.

> You can use either finder with both eyes open to see dimmer stars than are visible looking only through the finder because your brain merges the view from both eyes. The only problem with this method is that it can be physically awkward, depending on the type of scope you use and the location of the object you are trying to find. With a Dobsonian scope, for example, you may find yourself squatting or lying behind the Dob and looking up the full length of the tube.

- The Telrad uses standard alkaline batteries rather than the lithium batteries used by the QuikFinder. Both devices seem to live forever on one set of batteries; the dim red LED consumes next to no power. However, if your batteries do fail in the field and you've forgotten to bring spares, it's a lot easier to borrow alkalines for the Telrad than lithiums for the QuikFinder.

- The Telrad is more widely supported. More third-party accessories such as dew shields, dew heaters, flip-down mirrors, and so on are available for the Telrad. Also, many field guides, such as Harvard Pennington's excellent *The Year-Round Messier Marathon*, are based on 4° Telrad circles rather than 2° QuikFinder circles.

We much prefer the Telrad, but we have no argument with those who choose the QuikFinder. But do install one or the other if you have any interest at all in DSO observing. Until you try it, you won't believe how much easier it is to find objects using a bulls-eye finder.

Mounting a Unit-Power Finder

The Telrad and QuikFinder both use two-part assemblies. The finder unit mounts to a separate base, which is affixed to the scope using double-sided foam tape. This method has several advantages. First, it allows you to use the finder on more than one telescope simply by installing additional bases on your other scopes and moving the finder between scopes as needed. Second, it makes it quick and easy to remove the finder when you store or transport your scope, thereby protecting the finder from damage. Third, because the base is permanently affixed to the scope, the finder keeps its alignment quite well. You can usually remove and replace the finder without having to do more than a quick tweak to its alignment. Finally, the double-sided tape, while it provides a reliable mounting, also serves as a break-away failure point. If you do happen to bang your finder on a doorway, the tape usually gives way, preventing damage to the finder itself.

The first step in mounting the bulls-eye finder is deciding where to place it. **Figure 4-19** shows what we consider to be the ideal position on a Dobsonian scope, which is between the focuser and the optical finder. Positioning the bulls-eye finder here allows you to use the eyepiece, bulls-eye finder, and optical finder without having to move your head much.

Figure 4-19. A Telrad mounted on a 10" Dobsonian scope

If you're uncertain about the best mounting location, you can use duct tape or heavy rubber bands to mount the finder temporarily. Once you determine which location best fits your preferences, use the mounting tape to affix the base permanently. And we do mean permanently. When we first used it, we thought double-sided foam tape would be a precarious mounting method, but if you affix the foam tape properly it's about as reliable as using bolts. We mounted a Telrad on our 10" Dob with foam tape. After four years of using the scope in everything from intense heat to intense cold, the Telrad base is stuck to the tube as firmly as ever.

Once you have chosen the location for your finder, clean and polish that part of the tube. The finder base attaches with double-sided tape, so you want to make sure the surface is smooth and clean before you press the base into position. Otherwise, the tape may fail to adhere properly and the finder may fall off when you least expect it.

The ideal time to mount the finder is a night with a full moon. (Yes, we know that sounds strange, but bear with us....) To mount the finder, follow these steps:

1. Make sure all of the alignment screws are set to the middle of their travel. If necessary, count the number of turns from one extreme to the other and then back off half that number of turns. The goal is to leave as much room as possible for adjustment on either side of center.

2. Connect the finder to the base, and peel off the glossy paper or plastic that protects the sticky surface of the tape. Turn on the finder and adjust the bulls-eye pattern to moderate brightness.

3. Get the moon centered in the field of view of your telescope, using a medium- to high-power eyepiece. Ideally, use an eyepiece that provides a 0.5° field of view, which corresponds almost exactly to the apparent size of the Lunar disk and to the inner circle of the bulls-eye finder. If your scope tracks, turn on the motors. If you have a Dobsonian or other non-tracking scope, you'll need to work quickly to get the finder attached before Luna drifts very far.

4. With Luna centered in the telescope field of view, position the bulls-eye finder close to the telescope tube, but not quite touching it. Center Luna in the smallest, inner circle of the bulls-eye finder, which it fits almost exactly. Keeping Luna centered, gently press the bulls-eye finder against the tube. The tape should adhere instantly. Press down sufficiently to make sure the tape is fully adhered.

> You can, of course, accomplish the same thing by centering a bright star or planet in the bulls-eye finder, or for that matter a distant church steeple or radio tower. The Telrad and QuikFinder both have more than enough adjustment movement to correct any minor misalignment. But we like the precision of matching the 0.5° finder circle to a 0.5° moon.

Be careful if you mount a Telrad to a refractor, small reflector, or other scope with a small tube diameter. The Telrad mounting bracket is only slightly curved, to fit the curve of larger diameter tubes. As we found out by experience, the curve is insufficient to allow the mounting bracket and foam tape to make firm contact with a 90mm refractor tube. When we mounted it, the bracket appeared to fit and the foam tape seemed to have a solid grip.

Alas, we found out differently as we were setting up for a public observation one night. When we unzipped the refractor carrying case, we found the Telrad lying loose at the bottom of the case. Fortunately, there's an easy way to solve the problem. Simply add a second layer of double-sided foam tape, which you can buy at any hardware store. The additional thickness of the second layer allows the mounting bracket to make firm contact with the tube, as shown in Figure 4-20.

Figure 4-20. Mounting a Telrad on a refractor

 If you ever need to remove foam tape, as we did when we replaced the Telrad on our 90mm refractor, you'll find the adhesive is incredibly tenacious. We ended up with part of the tape stuck to the Telrad base and part to the refractor tube, and none of it wanted to come off. We tried peeling it with our fingernails, but it came away only in tiny chunks of foam. The best way we found to remove it was to use dental floss to "saw" under the foam. Once the foam has been peeled away and the adhesive exposed, use WD-40 or commercial adhesive remover to remove the remaining adhesive.

HACK #54 Upgrade Your Optical Finder

Match your optical finder to your type of telescope and your observing habits.

An optical finder is a low-power, wide-field refractor telescope with a crosshair eyepiece. The finder has a much wider field of view than the main telescope—typically 4° to 8° versus perhaps 1°—so it is much easier to acquire an object in the finder than with the main scope. If the finder is properly aligned with the main telescope, centering an object in the finder crosshairs also centers the object in the field of view of the main telescope.

Most inexpensive and midrange telescopes come with some sort of optical finder. The type, size, and quality of the finder varies with the size and price of the scope.

- Cheap small scopes include a cheap, tiny finder, usually 5 power with a 24mm single-element plastic objective (5X24). These so-called finders are worse than useless and should be replaced immediately with something better.

- Mid-size and mid-price scopes usually provide a 6X30 achromatic finder with coated glass lenses. A 6X30 finder may be usable until you can replace it with something better, but it provides insufficient magnification and light gathering. A 6X30 finder shows stars not much below 8th magnitude, which is not deep enough for serious use. Unless you observe Luna and the planets only, make it a high priority to replace the 6X30 finder with a suitable 50mm finder.

- Larger and more expensive scopes may include a 7, 8, or 9X50 achromatic finder with multicoated glass lenses. The larger objective and (usually) better coatings of a 50mm finder allow it to show stars down to about magnitude 10, which makes it suitable for all but the largest main scopes. Many bright DSOs, including most of the Messier objects, are at least dimly visible in a 50mm finder from a dark site. (Interestingly, objects are often visible in an optical finder that are invisible in a binocular of identical magnification and aperture. This is because the finder is mounted on the scope and so provides a stable view.) But even if your scope came with a 50mm finder, it may make sense to replace it with another 50mm finder of a different design.

We consider a 50mm finder the best choice for any scope from the smallest to perhaps 12.5" or 15". For scopes larger than that, an 80mm finder is usually a better choice.

Finders are made in the following four designs:

Right-angle, correct-image

A right-angle, correct-image (RACI) finder, shown in **Figure 4-21**, uses an Amici prism diagonal to provide an uninverted image that is correct left-to-right, as shown in **Figure 4-22**. A RACI finder has two primary advantages. First, it provides a correct-image image view of the night sky that corresponds to your star charts and to what you see with your eyes, your binocular, and your unit-power finder. You needn't do any mental or physical gymnastics to match image orientations from one source to another. Second, because the RACI finder positions the eyepiece at 90° to the optical axis of the main scope, you can switch between using the RACI finder, the unit-power finder, and the main scope eyepiece by moving only your head. Depending on the type of scope and where it is pointing, with a straight-through finder you may have to physically contort yourself to look through it.

Figure 4-21. A right-angle, correct image (RACI) finder

Figure 4-22. The correct image provided by a RACI finder

RACI finders have two disadvantages. First, because the eyepiece of the finder is at 90° relative to the optical axis of the main scope, you cannot use the RACI finder to acquire objects. That means you'll need a second finder of some sort, such as a Telrad unit-power finder. Second, the Amici prism absorbs some light, so the view through a RACI finder is marginally dimmer than that through a straight-through finder or one that uses a mirror diagonal.

We consider a RACI finder the best choice for Newtonian reflectors, including Dobs, and for any other scope that places the eyepiece at the front of the optical tube assembly (OTA). A RACI finder is also a good choice for a refractor, SCT, or other scope that places the eyepiece at the rear of the OTA, for the same reason that star diagonals are usually used in such scopes: unless the scope is mounted very high and you observe standing up [Hack #60], the eyepiece placement of a RACI finder makes it much more convenient to use than a standard straight-through finder.

Straight-through

A straight-through finder is simply a standard refractor with the eye-piece in-line with the optical axis of the scope. This is the traditional finder style, and it is the type of finder supplied with most scopes. A straight-through finder produces an image that is inverted but correct left-to-right, as shown in Figure 4-23. That means you can use your star charts simply by inverting them top-for-bottom.

Figure 4-23. The inverted image provided by a straight-through finder

A straight-through finder has several advantages. First, it is the least expensive design to produce, so at a given price point a straight-through finder will generally be of higher optical and mechanical quality than any of the other designs. Second, all other things being equal, a straight-through finder produces the brightest image because it has none of the additional optical elements needed to erect the image or bend it at a 90° angle. Third, because the optical axis of the finder is parallel to the optical axis of the main scope, a straight-through finder can be used directly to point the scope. Finally, again because the optical axes are parallel, you can use the "both eyes open" method to view the night sky directly with one eye while using the other eye to view through the finder. Your brain superimposes the magnified, brighter finder view on the naked-eye view, allowing you to place the finder precisely where it needs to be to put an object in the eyepiece of the main scope.

A straight-through finder has some disadvantages as well. First and fore-most, the eyepiece position may make it awkward to use the finder, par-ticularly on a Newtonian reflector that is pointed near zenith. Second, because it inverts the image, you must mentally translate between the finder image and the view of the night sky provided by your eyes, your binocular, and your unit-power finder. Finally, if you use the both-eyes-open method, you are superimposing an inverted finder image against the correct image of the night sky provided by your other eye.

We consider a straight-through finder a usable choice for a refractor, SCT, or other scope that places the eyepiece at the rear of the OTA. In our opinion, a straight-through finder is not the best choice for a Newtonian reflector or other scope that places the eyepiece at the front of the OTA.

> In fairness, we should point out that many experienced observers actually prefer a straight-through finder on a Newtonian reflector because it allows them to use the both-eyes-open method. Choose according to your own preferences and practices.

Correct-image, straight-through

A correct-image, straight-through (CIST) finder is, in effect, half of a binocular. Like a binocular, a CIST finder uses a Porro prism to provide an image that is uninverted and correct left-to-right. (In fact, some astronomers literally cut a binocular in half and use each half as a finder scope.) A CIST finder combines the advantages and disadvantages of RACI and straight-through finders. The CIST design has never really caught on for mainstream use, although some inexpensive Chinese short-tube refractors include a 5X24 or 6X30 CIST finder.

Right-angle

A right-angle finder uses a standard mirror diagonal and produces an image that is right-side-up but mirror-reversed (flipped left-to-right), as shown in Figure 4-24. Although a right-angle finder provides the same easy eyepiece accessibility as a RACI finder, we consider it a poor choice for any type of scope because of its mirror-reversed image. The flipped image means the only way to make the view in a right-angle finder correspond to your charts is to turn the chart face down and shine a light through the chart. Obviously that doesn't work with charts printed on both sides.

Figure 4-24. The inverted image provided by a right-angle finder

On balance, we believe a 50mm RACI finder is the best choice for nearly any scope. There are many RACI models to choose among. Fortunately, a finder need not be of particularly high optical quality because you are using it only to locate objects, not to observe them. That means finders, even reasonably good ones, are relatively inexpensive accessories.

Considering only the finder itself, we think Antares RACI models are the best choice for a replacement finder. (Antares also makes right-angle non-correct image finders, which we do not recommend.) Antares 7X50 RACI finders cost $90 ($105 with the optional mount), are available in various colors, and provide a removable eyepiece with diopter correction that allows you to focus the crosshairs independently from the focus of the finder itself. Unfortunately, the Antares finder mount uses the clumsy, old-fashioned, six-screw adjustment method.

Before you order a replacement finder, either verify that the new finder fits your current finder mount or replace the finder mount as well. If you replace the finder mount, verify either that the new finder mount comes with its own mounting plate or that it fits your current mounting plate.

We also recommend the 9X50 RACI finder sold by Orion for $65 including mount, which is what we use on our own scopes. The Orion RACI finder has the usual mediocre fit and finish of Chinese products, and its coatings are inferior to those of the Antares finders. The eyepiece and prism are not removable, nor does the eyepiece have a diopter adjustment. The crosshairs are rather thick and a bit rough. All of that said, the Orion RACI finder is more than good enough to do the job, and you can't beat the price. Finally, the finder mount supplied with the Orion RACI finder uses the modern two-screw adjustment method, which allows you to align the finder in seconds and holds alignment well.

HACK #55 Align Your Finder
Make sure you know where you're pointing.

We're always surprised by how many astronomers are unfamiliar with as basic a procedure as aligning the finder scope. At one public observation, we spoke with a visitor who had had his scope for months and was ready to give up the hobby. Why? Because he couldn't find anything.

At first, we assumed that this gentleman was merely undergoing the "newbie blues," unable to locate objects because he hadn't yet learned to star hop **[Hack #21]** or to use geometric location methods **[Hack #20]**. So we decided to show him how to find the Messier open clusters in Auriga, M36, M37, and M38. We pointed the finder where M37 should have been, expecting to see it in the finder. It wasn't visible, which we wrote off to the small aperture of his 6X30 finder. But when we looked in a wide-field eyepiece, M37 was nowhere to be seen. Hmmm.

We swung the scope down to get a distant antenna beacon light centered in the wide-field eyepiece, and then looked in the finder, expecting it to be off a bit. Holy cow! Not only wasn't the beacon centered in the crosshairs of the finder, it wasn't even within the field of view at all. No wonder this guy couldn't find anything. He had no idea where his scope was pointing.

When we asked him how he aligned his finder, we learned that he'd never aligned it and didn't realize that finder alignment was a necessary step. He'd unpacked his scope from the shipping carton, attached the finder, and let it go at that. The mechanics of the finder mount and bracket ensured that the finder was pointed generally in the right direction, but that was about it. So we showed him how to align his finder scope, and explained that he should check the alignment every time he set up his scope.

Aligning a finder is easy enough to do. It may take a few minutes the first time you do it, but after that it should take only a few seconds to check the alignment when you set up your scope and tweak it if necessary.

To begin aligning your finder, select a distant object as an alignment target. During daylight, use a distant telephone pole, church steeple, radio tower, or similar object. For accurate alignment, the object should be at a distance of at least 500 times the focal length of the scope. For typical amateur scopes, that translates to between 1/8 mile and 5/8 mile.

> Make sure the object you choose is unique and easily identi-
> fied. We once (mis)aligned our finder by pointing the finder
> at one telephone pole while, unbeknownst to us, the tele-
> scope was pointing at a different telephone pole. The best
> objects are therefore those that are easily and unambigu-
> ously identifiable, such as a church steeple on a distant ridge
> or a radio tower with a blinking light.

At night, you may have no alternative but to align on a bright star or planet. The advantage to using a celestial object is that there is no question that your target is sufficiently distant. The disadvantages are that it can be very hard to be certain that you have the same star in the finder and telescope eyepiece, and that most stars drift across the field of view because of the earth's rotation, so you're trying to align on a moving object. The easy solution, at least if you are in the Northern hemisphere, is to use Polaris as your alignment star. Because it is located only about 0.75° from the north celestial pole, Polaris has little apparent motion.

After you've selected an alignment target, center the object in your telescope's eyepiece, as shown in Figure 4-25. Use your lowest power eyepiece to make it easier to locate the object, and simply pan the scope around until the object appears in the field of view. If your finder is already roughly aligned, you can use it to get you approximately on target. Otherwise, you may have to scan around for quite a while to get the object in your eyepiece field of view.

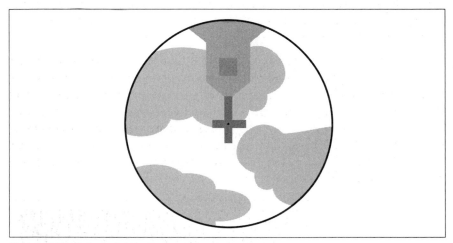

Figure 4-25. Center the alignment target in your low-power eyepiece

Depending on the type of telescope you have, the image in your eyepiece may be erect or inverted, and correctly oriented or flipped left-to-right. That part isn't important. All that matters is that you have the alignment object centered in the field of view of the telescope eyepiece. Once you have done that, without moving the telescope, look through your finder. You'll probably see something like Figure 4-26.

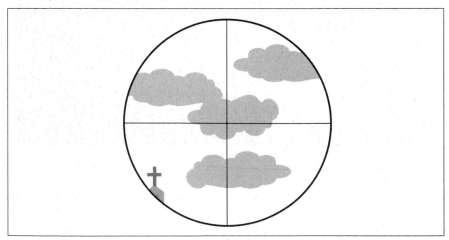

Figure 4-26. The view through the finder before it is aligned

Unless the finder is badly misaligned, the alignment target should be visible somewhere in the field of view of the finder. If it is not, loosen the screws that secure the finder mounting bracket to the scope, move the finder until the object is within its field of view, and then retighten the mounting screws. You'll almost certainly have moved the scope enough during this procedure to decenter the alignment object in the scope eyepiece, so recenter the alignment target in the scope's eyepiece before you proceed.

Finder mounts differ in the provisions they make for aligning the finder scope. The best type of finder mount, shown in Figure 4-27, supports the finder within a collar, with a rubber O-ring seal at the front. At the rear of the collar are two adjustment screws and a third, spring-loaded post that bears against the finder. Aligning the finder with such a mount usually takes only seconds. Also, the finder tends to stay in alignment because the screws are a fixed reference point and the spring-mounted post always forces the finder into the correct position against these screws.

Figure 4-27. Use the finder adjustment screws to center the crosshairs on the object

Another type of finder mount, which is despicable and unfortunately common, uses two rings, each with three adjustment screws spaced 120° apart. To align the finder with this type of finder mount, you have to back out some screws as you tighten down other screws. Because the front rings and back rings define two planes of alignment, getting the finder aligned in such a mount is often an exercise in frustration. Also, such mounts tend not to hold alignment well, so it's often necessary to realign the finder from scratch each time you set up the scope. Conversely, the collar-based, two-screw mount described earlier usually requires minor tweaking at most.

If your scope has one of the hateful two-ring finder mounts, we recommend replacing it with a collar-based finder mount. You can buy such mounts from Orion and other astronomy specialty vendors.

If you have a refractor, your finder mount probably has a dovetail base that mates to a corresponding slot on the left side of the focuser body. You can thank Vixen, a Japanese optical company with otherwise sane engineers, for that stupid arrangement. They originated that design and nearly every other refractor maker copied it.

If you are right-eyed or ambivisual, you probably don't understand the problem. But if you are left-eyed, the problem is evident the first time you try to use the finder. The scope body and focuser tube are so close to the finder eyepiece that there's insufficient room to use your left eye. You're forced willy-nilly to use your right eye, whether you can focus through the finder with it or not.

Some finderscopes have adjustable focus, so you may be able to focus the finder to suit your right eye. If the finderscope is not focusable or the range of adjustment is insufficient to accommodate your vision, the best solution is to substitute a finder mount with a much longer stalk. That replacement may or may not have a dovetail foot that fits that slot on the finder base. If not, you'll have to remove the focuser and objective lens cell and drill mounting holes in the optical tube itself. Be certain there are no bits of metal left in the tube before you reinstall the focuser and objective lens cell.

If you're going to replace the finder mount, consider replacing the finderscope itself at the same time. Most refractors are supplied with inadequate, small finders, often 6X30 or less. A proper optical finder, even for a small refractor, has at least a 50mm objective.

Once you have the alignment target within the finder field of view, use the finder adjustment screws to center the finder crosshairs on the alignment target, making sure not to jiggle the scope. The goal is to have the intersection of the crosshairs covering the center of the alignment target exactly, as shown in Figure 4-28.

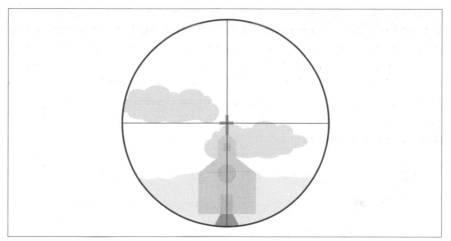

Figure 4-28. The view through the finder when it is properly aligned

Ordinarily, it's sufficient to align the finder with a low-power eyepiece and "eyeball it" to decide when the object is centered in the telescope eyepiece. If you have a crosshair eyepiece for your telescope, by all means use it for aligning the finder, but it's ordinarily good enough just to judge by eye when the object is properly centered.

For very critical work, it may be desirable to fine-tune finder alignment using a high-power eyepiece in the scope. To do that, first do a rough alignment using the low-power eyepiece. Once the finder crosshairs are as close to being exactly aligned as you can judge with the low-power eyepiece, simply replace the eyepiece with a higher power eyepiece and repeat the process.

HACK #56 Determine Your Optical Finder's Field of View

For simple pointing, the field of view doesn't matter, as long as it's wide enough for you to locate the object easily. For star hopping, it's important to know the field of view of your finder with reasonable accuracy.

An optical finder serves two purposes. The obvious one is to help you point your scope at bright stars and other objects. The less obvious purpose is to provide a known field width when you are star hopping **[Hack #21]**. There are three ways to determine the field of view of your optical finder:

- Look it up. Most manufacturers publish the field of view. Unfortunately, the stated field is not always accurate, particularly for inexpensive Chinese finders. For example, a finder with a stated 5.5° field may have an actual field of 5° (or 6°).

- Drift test the finder [Hack #57]. Although drift testing gives very accurate results, it is time consuming. If you are using a new or borrowed finder, you may not have time to do a drift test.

- Use the finder to look at pairs of bright stars with known separations. For example, if the published field of view of the finder is 5.5°, two stars separated by exactly 5.5° should just barely fit in the finder's field of view. If they fit with room to spare, estimate how much "extra" field you have. If it's about 5%, the actual field of the finder is 5.5 × 1.05, or about 5.8°. If they don't quite fit, you can use another pair of stars with a somewhat smaller separation and estimate the actual field quite closely.

Most optical finders have fields of view between 4.5° and 8°. Fortunately, there are two prominent groups of stars that provide several bright star pairs with separations suitable for estimating your finder's field very accurately. The Big Dipper, shown in Figure 4-29, provides star pairs with separations from 4.4° to 7.9°.

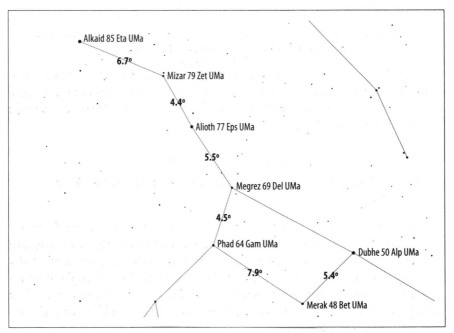

Figure 4-29. Use the Big Dipper to determine your finder field of view

Cassiopeia, shown in Figure 4-30 provides bright star pairs with separations from 4.3° to 7.3°.

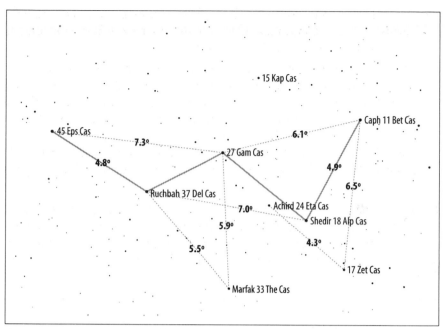

Figure 4-30. Use Cassiopeia to determine your finder field of view

The Big Dipper and Cassiopeia are both circumpolar, but on opposite sides of the pole. That means that when the Big Dipper is low, Cassiopeia is high, and vice versa. One or the other is always high enough to use to measure the field of view of your finder.

HACK #57 Determine Your True Field of View

A drift test lets you figure out just how much of the sky you can keep in your sights.

Any combination of telescope and eyepiece has a specific *true field of view* (*TFoV*), which is determined solely by the focal length of the telescope and the *field stop* diameter of the eyepiece. TFoV quantifies the amount of sky visible in a particular scope with a particular eyepiece. For example, if a particular telescope/eyepiece combination provides a 1° TFoV, two stars that are separated by exactly 1° will just fit into the eyepiece field, with each star on opposite edges of the field.

It is important to know the TFoV of your eyepieces in your scope and the TFoV of your finder because you use the true field to match the stars that are visible in the eyepiece to those on your charts and to plan and execute star hops **[Hack #21]**. There are several methods, of varying accuracy, to determine TFoV.

The Apparent Field of View (AFoV) method

The quickest way to determine TFoV is to divide the *Apparent Field of View (AFoV)* of the eyepiece by the magnification provided by that eyepiece in a given scope. For example, if your scope has a focal length of 1,200mm and you use a 25mm Plössl eyepiece with an AFoV of 50°, you can calculate the TFoV as follows:

- Divide the focal length of the scope, 1,200mm, by the focal length of the eyepiece, 25mm, to determine that that combination provides 48X magnification.

- Divide the AFoV, 50°, by the magnification, 48X, to yield a TFoV of 50/48, or about 1.0417°. You can multiply the result in degrees by 60 to convert it to arcminutes, which are used more commonly to refer to an eyepiece field of view. A 1.0417° field is 62.5 arcminutes, which is abbreviated 62.5'.

There are several problems with using this method:

- The actual focal lengths of both the telescope and eyepiece often differ 5% or more from nominal stated value because of normal manufacturing variances. For example, a telescope with a stated focal length of 1,200mm may actually be 1,173mm (or 1,234mm), and an eyepiece labeled 10mm may in fact be 9.7mm (or 10.4mm).

- The specified AFoV is often an approximation. An eyepiece with a nominal 55° AFoV may actually provide as little as 50° or as much as 60°. The nominal AFoV of premium eyepieces tends to be reasonably accurate; that of cheap Chinese eyepieces tends to be inaccurate and very optimistic. Also, the true AFoV of different focal length eyepieces in a series with nominally identical AFoVs may differ significantly from one focal length to another.

- Because they represent a spherical surface (the sky) as a plane (the view in the eyepiece), all eyepieces must introduce some distortion. That means that even if the AFoV is specified accurately, it cannot easily be translated to an accurate TFoV. For example, Tele Vue Panoptic eyepieces have a nominal AFoV of 68°, which is accurate insofar as it goes. But the distortion present in a Panoptic eyepiece means that if you measure the TFoV accurately you'll find that it corresponds to an undistorted AFoV of less than 65°. (We're not knocking Tele Vue or Panoptics here; Tele Vue is a first-rate optical maker, and Panoptics are superb, world-class eyepieces. Distortion is simply an optical fact of life.)

The AFoV method suffices for a quick-and-dirty calculation of TFoV and is good enough for most purposes, but it is not accurate enough for critical work.

The field-stop diameter method

Because the actual TFoV is determined solely by the focal length of the telescope and the field-stop diameter of the eyepiece, it is possible to calculate very accurate TFoV figures if you have an accurate focal length for your scope and have the field-stop diameter for the eyepiece in question. Unfortunately, Tele Vue is the only eyepiece maker that routinely publishes field-stop diameters for its eyepieces. For some eyepieces, you can use calipers to measure the field-stop diameter yourself. However, the field stop of many eyepieces is inside the eyepiece rather than exposed, which means you have to disassemble the eyepiece to measure its field stop, which is not a recommended procedure.

If you have or can get accurate values for your scope's focal length and your eyepieces' field-stop diameters, you can calculate an accurate TFoV. To do so, divide the eyepiece field-stop diameter in mm by the focal length of the scope and multiply the result by $(360/2\pi)$ or about 57.2958. (There are 2π radians in a full circle of 360°. Dividing 360° by 2π yields about 57.2958° per radian.) For example, if you use a 27mm Tele Vue Panoptic eyepiece with a 30.5mm field stop in a scope of 1,255mm focal length, the TFoV is $(30.5/1,255)*(360/2\pi)$, or about 1.39°. Multiplying that result by 60 yields the TFoV in arcminutes, about 83.5'.

It's interesting to compare the results of this method with the AFoV method. Assuming a telescope focal length of 1,255mm and a 27mm Panoptic with a 68° AFoV, the AFoV method yields a true field of about 1.46°. Calculating based on the 27mm Panoptic field stop diameter of 30.5mm yields a true field of view of only 1.39°, which specifies accurately the amount of sky visible with that telescope/eyepiece combination. Back-converting the accurate TFoV to determine a "real" AFoV gives us about 64.72° rather than the nominal 68°. Again, this is not a knock on Tele Vue or Panoptic eyepieces; the AFoV in Panoptics simply "looks" wider than it really is.

The drift-testing method

Sometimes the best way to be sure of something is just to go and look at it. That's the basis of the drift-testing method of determining TFoV. Drift testing is simple in concept. You place a star on one edge of the field of view of an eyepiece and then use a stopwatch to measure how long it takes to drift all the way to the opposite edge of the field of view. Because stellar motion (actually, Earth's motion) is known very precisely, it's easy to convert the elapsed time to an accurate TFoV.

There are two caveats with this method:

- You must be certain that the star drifts across the actual full diameter of the field of view, rather than across a chord.

- For perfectly accurate results, the star you use must be located exactly on the celestial equator [Hack #17], at declination 0°0'0". This is true because a star at other than exactly 0 declination has a different apparent drift rate. In practice, any star within ±5° or so of declination 0 yields sufficiently accurate results for all but the most critical work. If Orion is visible, use Mintaka (the belt star on the Bellatrix side), which is at declination about 0°18'. If Orion isn't up, other good choices are Porrima (γ-Virginis) at about 1°29', Sadalmelik (α-Aquarii) at about -0°17', and σ-Serpens at about 1°00'. (You can use any star for precise drift testing if you apply a declination correction factor, described in the next section.)

Drift Testing in Detail

These are the usual instructions for drift testing. Place the chosen star just outside of the field of view of the eyepiece. When the star appears at the edge of the field of view, start your stopwatch. Just as the star disappears on the other side of the field of view, stop the stop watch and record the elapsed time. Calculate the TFoV in arcminutes using the formula:

```
TFoV = T * 0.2507 * Cos(Dec)
```

where T is the elapsed time in seconds and Cos(Dec) is the cosine of the declination in decimal degrees of the star you used for testing. That's fine as far as it goes, but it does require a bit more explanation. Let's look at each of these terms:

T

T is the time in seconds required for the star to drift across the entire diameter of the eyepiece field.

0.2507

The factor 0.2507 is the constant that converts seconds of time to the TFoV in arcseconds. The earth rotates on its axis 360° in 24 hours, which translates to 15° per hour, 1° every four minutes, 1/4° per minute, and 1/240° per second. There are 60 arcminutes per degree, so Earth rotates at 60/240, or 0.25 arcminutes per second. (This is known as the *Solar Time Constant*.) So why the extra decimal places?

Up to this point, we've considered only Earth's rotation on its axis, which is the basis of *Solar time*. But to calculate stellar motion properly, we have to use *sidereal time* (star time), which is slightly different from

Solar time. The difference exists because as Earth rotates on its axis, it also orbits Sol. Over the course of one year, Earth makes one full additional rotation in its orbit, and this must be taken into account.

The Solar year is 365.2425 days long, which is 525,949.2 minutes or 31,556,952 seconds. A full circle is 360°, or 21,600 arcminutes. The correction for sidereal time therefore adds (21,600 arcminutes / 31,556,952 seconds) = 0.00068447675+ arcminutes/second to the apparent drift speed of a star located on the celestial equator. (This figure is only a very close approximation because Earth's orbital speed around the Sun varies according to its orbital position; when Earth is near Sol, it moves faster than when it is more distant.) Adding the ~0.0006845 correction to the Solar time constant 0.25 and rounding the result gives us 0.2507.

Cos(Dec)

If the star you use for testing is exactly on the celestial equator (declination 0), this factor drops out because the cosine of 0 is 1. If the star is not on the celestial equator, its apparent motion is slower. The larger the absolute value of the declination, the slower the apparent motion. A star located at exactly +90° or -90° declination has no apparent motion at all because it is exactly on the pole of the axis around which Earth rotates. Polaris, for example, is located at declination +89°15', and it requires 24 hours to trace a 1.5° circle around the pole.

Using a "slower" star without a correction factor would overstate the TFoV of an eyepiece. (The star takes longer to drift across the field, so the TFoV appears larger than it really is.) But you can use any star for drift testing if you calculate its declination correction factor. To do so, convert the star's declination to a decimal value and determine the cosine of that value. For example, if you use Antares, convert its declination of -25°26' to -25.433°. The cosine of -25.433° is about 0.9031.

Let's assume that you've drift tested using Antares, which took exactly 4:43 or 283 seconds to drift across the diameter of the eyepiece field. What is the TFoV of that eyepiece in arcminutes?

```
TFoV = 283s * 0.2507 arcminutes/s * 0.9031 = 64.1 arcminutes
```

The TFoV of this eyepiece in this scope is 64.1 arcminutes, or about 1.07°.

But the timing and calculations are the easy part. The first time you actually drift test an eyepiece, you'll learn that the hard part is getting the star to drift across the full diameter of the eyepiece field instead of a chord. If you try to do this by trial and error, you'll waste a lot of time and become quite frustrated. We know. We've watched people do it. However, there are some easy solutions:

- If you have a polar-aligned equatorial mount, center the chosen star in the eyepiece field, turn off your drive motor(s), and use your RA slow-motion control to put the star just outside the eastern edge of the eyepiece field. As the star appears in the field, start your stopwatch. Stop it just as the star exits the eyepiece field.

- If you have a Dobsonian or other alt-az mount, simply center the star in the eyepiece field and start your stopwatch. The human eye is very, very good at centering objects. If you start with the star centered in the eyepiece field, by definition it will cross a half-diameter of the eyepiece field as it drifts to the edge. You can use the same calculation described previously and simply double the result to determine the TFoV of the eyepiece.

- If you have an alt-az scope but prefer to use a full-diameter drift test, first do a half-diameter drift test described in the preceding bullet point. Watch where the star leaves the field of view, and then restart the test with the star entering the view at the opposite point on the clock face. (We prefer to use half-diameter drift tests for alt-az scopes; we can do twice as many tests in the same time and average the results.)

Once you determine an accurate TFoV for a particular eyepiece in one scope, you can back-convert that TFoV to a field stop diameter for that eyepiece by multiplying the TFoV in decimal degrees by the focal length of the scope in mm and then dividing by $(180/\pi)$ or about 57.2958. For example, to back-convert the 64.1 arcminute field we determined using a scope with a focal length of 1,255mm, convert the 64.1 arcminute field to 1.07° and use the formula 1.07 * (1,255/57.2958) = 23.44mm.

If you know the focal length of your other scopes, you can determine their true fields of view without drift testing simply by using the field-stop diameter calculation described in the preceding section.

Drift Testing Your Finder

Most astronomers use drift testing only for eyepieces, but it's just as accurate a means of determining the true field of view of optical finder scopes. Of course, finders have much wider fields of view than eyepieces, so it may take half an hour or more to do a full-diameter drift test on a finder, which you really don't need to do. Finders have nicely centered crosshairs, so it's trivially easy to do an accurate half-diameter drift test. If you center the star in the crosshairs, there's no question that the star crossed the exact center of the field.

When we test a new finder, we simply point it at a bright star when we first set up the scope and then go about our business. If the star is anywhere near the celestial equator, we know about how long the half-diameter drift test should take, based on the published FOV of the finder. For example, if the finder supposedly has a 6° field, we know that our half-diameter drift test should take about 12 minutes (the 3° half-diameter at about four minutes per degree). We center the star in the crosshairs and start our stopwatch. About 10 minutes later, we wander back to the scope and watch as the star drifts out of the field of view. We jot down the elapsed time and which star we used, and do the calculations later at our convenience.

It's surprising how much the actual FOV of a finder may differ from published specifications. We saw one finder with a nominal FOV of 6.5° that turned out to be more like 5.8°. Conversely, we remember another finder with a supposed 5° FOV that actually had a 5.5° FOV. That amount of difference can be significant, particularly if you are a dedicated star hopper, so it's worth testing your own finders to determine their actual fields of view.

HACK #58 Enhance Lunar and Planetary Contrast and Detail

Choose a basic set of filters to improve the view.

Just as photographers use filters to alter the images captured by their cameras, astronomers use filters to alter the images visible in their eyepieces. A filter, by definition, can add nothing to an image; it can only take away. But, like the sculptor who creates an elephant from a block of stone by cutting away everything that doesn't look like an elephant, an astronomer may use filters to create more detailed views of some objects by removing extraneous light to allow the subtle details of the object to show through.

Astronomical filters are available in several sizes. The most common are those that fit standard threads on 1.25" and 2" eyepieces and the visual backs of Schmidt-Cassegrain Telescopes (SCTs). Meade goes its own way, using a non-standard thread on their 1.25" and 2" eyepieces and filters, and Questar uses still a different thread. A few filters are available for 0.96" eyepieces, which are common in Japan and with some old telescopes but rare elsewhere. Filters for 1.25" eyepieces are available in by far the widest variety. The selection for 2" eyepieces is somewhat more limited, and those for SCT visual backs are even more limited.

If you use both 1.25" and 2" eyepieces, consider buying only 2" filters. Although a 2" filter cannot be attached directly to a 1.25" eyepiece, there are two common workarounds. First, many 1.25"/2" focuser adapters provide threads for mounting a 2" filter, which can then be used with any 1.25" eyepiece you insert into the adapter. Second, if you have a Newtonian (or Dobsonian) reflector, you can install a filter slide, which attaches to the inside of the tube between the bottom of the focuser and the secondary mirror. With a filter slide, you can quickly position any of several filters under the focuser simply by moving the slide. Finally, of course, you can simply hold a 2" filter between your eye and the eye lens of the eyepiece, assuming the eyepiece provides sufficient eye relief.

Enhancing Color Contrast

Color filters, sometimes called *planetary filters* or *Lunar/planetary filters*, are simply discs of colored glass, ground optically flat on both sides and with antireflection coatings applied to both surfaces. By selectively transmitting parts of the visible spectrum and blocking other parts, color filters can enhance visual contrast and detail in bright objects like Luna and the planets. All color filters use the century-old Kodak Wratten numeric designations familiar to photographers. Any two filters that have the same Wratten designation, regardless of manufacturer, have similar or identical transmission characteristics.

Transmission curves are available for all Wratten color filters. These curves graph the transmission (or its inverse, the density) of a filter across the visible spectrum. Figure 4-31 shows an example transmission curve, this one for a #58 Green filter. The same data may be presented in tabular form, as shown in Table 4-7. These data tell us that the #58 Green filter passes most of the green light, with maximum transmission centered on about 530 nanometers (nm). Blue-green wavelengths are attenuated, and everything shorter than about 465 nm—blue through ultraviolet—is blocked completely, as are yellow through red wavelengths from about 620 nm (orange) through 700 nm (deep red). At about 700 nm, the #58 Green filter begins providing higher transmission, reaching nearly 100% at wavelengths of 800 nm (infrared) and higher.

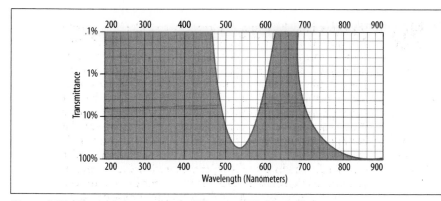

Figure 4-31. Transmission curve of a Wratten #58 Green filter

Table 4-7. Transmission data for Wratten #58 Green filter

Wavelength	Percent transmittance
400	--
10	--
20	--
30	--
40	--
50	--
60	--
70	0.23
80	1.38
90	4.90
500	17.70
10	38.80
20	52.20
30	53.60
40	47.60
50	38.40
60	27.80
70	17.40
80	9.00
90	3.50
600	1.50
10	0.41
20	--
30	--
40	--
50	--
60	--
70	--
80	--
90	--
700	0.53

Some experienced Lunar/planetary observers swear by color filters, while others (including us) find them of very limited use. Although a filter changes the appearance of the object dramatically—a #25 Red filter, for example, turns everything bright red—the degree of contrast and detail enhancement is usually quite subtle. Color filters are not a magic bullet for enhancing Lunar/planetary observing, but they are relatively inexpensive and you may find the enhancements they provide worth their small cost.

> Consider transmission percentage when you choose color filters. Small telescopes provide dim images at Lunar/planetary powers even without a filter; if you add a dense filter, the image may darken unacceptably. At 25X/inch to 30X/inch magnification, which is to say typical Lunar/planetary powers for mid-size scopes, we think you'll find any filter with transmission less than 15% or so to be too dark to use. At 40X/inch to 50X/inch, typical Lunar/planetary powers for small refractors and similar instruments, any filter with transmission less than 40% to 50% is likely to dim the image unacceptably.

We think the four color filters listed in Table 4-8 are most useful for general Lunar/planetary observing with mid-size to large telescopes. You can buy them individually for $10 or $15 each in 1.25" or $20 to $25 each in 2" versions. Some astronomy specialty vendors sell this or a similar selection as a set, often at a significantly lower price versus buying them individually.

If you want to start with just one color filter, we think the #80A Medium Blue is the best choice for mid-size or larger scopes. Although it is not necessarily the best choice for any one purpose, it is helpful for many objects, from Luna to Venus to the superior planets. The #80A even serves as a "poor man's" light pollution filter, somewhat reducing the yellow-pink sky-glow caused by low-pressure sodium vapor lights. Unfortunately, the #80A is often a bit too dark for 4.5" and smaller scopes, particularly at high power. For such scopes, we recommend the #15 Deep Yellow as the best general-purpose filter. The #82A Pale Blue is a jack of all trades, and it would be our second or third choice for any size scope.

Table 4-8. Basic color filters

Wratten	Color	Transmission	Improves
#15	Deep Yellow	67%	Lunar feature contrast/terminator; Venus clouds; Mars clouds/ice caps; Mars, Jupiter, Saturn feature contrast; Uranus/Neptune detail
#58	Green	24%	Venus clouds; Mars ice caps; Jupiter GRS; Saturn clouds/belts/polar regions

Table 4-8. Basic color filters (continued)

Wratten	Color	Transmission	Improves
#80A	Medium Blue	30%	Lunar surface detail & feature contrast; Venus clouds; Mars clouds/ice caps; Jupiter belts/rilles/festoons/ GRS; Saturn belts/polar regions; reduces low-pressure sodium-vapor light pollution; most generally useful color filter
#82A	Pale Blue	73%	Luna/Mars/Jupiter/Saturn low-contrast features/detail; Jupiter/Saturn cloud belts; comet tails; reduce refractor false color; structure/detail in bright galaxies

Table 4-9 lists the four filters we consider most generally useful after the basic set. The #47 Violet, although it is too dense for use in small scopes, is probably the best overall choice for viewing the inferior planets, particularly Venus. The similarly dense #25 Red lightens warms colors and darkens cool colors dramatically, and it is the best choice when you need a deep-cutting contrast filter. The #11 Yellow-Green and the #21 Orange straddle the #15 Deep Yellow in both density and effects.

Table 4-9. Supplemental color filters

Wratten	Color	Transmission	Improves
#11	Yellow-Green	78%	Venus clouds; Mars maria/ice caps; Jupiter/Saturn clouds/feature contrast; Saturn Cassini Division; Uranus/Neptune detail/contrast
#21	Orange	46%	Mercury/Venus contrast against daytime sky; Mars maria; Jupiter/Saturn belts/festoons/polar regions; color correction with Mylar Solar filters; similar to #15 but with lower transmission and higher contrast
#25	Red	14%	Mercury features/contrast; Venus clouds/contrast/terminator; Mars maria/ice caps/surface detail; Jupiter clouds/belts/transits; Saturn clouds
#47	Violet	13%	Lunar detail; Mercury/Venus clouds/contrast/detail (many observers' first choice for Venus); Mars ice caps; Saturn ring structures

If you have the Basic and Supplemental sets and still feel the need for more color filters, consider one or more of those listed in Table 4-10. In general, filters in this group simply fine-tune the effects of the more commonly used filters. We consider all of these filters highly optional, so much so that we no longer own any of them.

Table 4-10. Less frequently used color filters

Wratten	Color	Transmission	Improves
#8	Light Yellow	83%	Moon features/contrast; Mars maria; Jupiter/Saturn belts; Uranus/Neptune detail; comet tail/coma detail; similar to #12 but with higher transmission and less pronounced effects
#12	Yellow	74%	Moon features/contrast; Mars maria; Jupiter/Saturn belts; Uranus/Neptune detail; intermediate between #8 and #15 in transmission and effects
#23A	Light Red	25%	Mercury/Venus sky contrast in daylight/twilight; Mercury features/contrast; Venus clouds/contrast/terminator; Mars maria/ice caps/surface detail; Jupiter clouds/belts/transits; Saturn clouds; similar to #25 with higher transmission and less pronounced effects
#29	Deep Red	8%	Mercury/Venus sky contrast in daylight/twilight; Mercury features/contrast; Venus clouds/contrast/terminator; Mars maria/ice caps/surface detail; Jupiter clouds/belts/transits; Saturn clouds; similar to #25, but much darker and with more pronounced effects
#38A	Deep Blue	17%	Venus clouds; Mars dust storms; Jupiter belts/GRS; Saturn rings/belts/clouds; bright comet tails
#56	Light Green	53%	Lunar detail; Mars dust storms/ice caps/clouds; Jupiter clouds/low-contrast detail; Saturn cloud/surface detail

If you want to try different filters before plunking down $10 or $20 each for mounted glass filters, you can use thin-film acetate filters to judge the approximate effect of different Wratten glass filters. A set of 100 acetate color filters is available for under $10 from Scientifics Online (*http://scientificsonline.com*), formerly known as Edmund Scientific.

Dimming the Image

Neutral-density (ND) filters and *Polarizing filters*, sometimes called *moon filters*, are used when you need to dim the image, for example, as if you are viewing Luna at low power. ND filters are available in various fixed densities from quite light to very dark, and are simply dyed glass filters with a neutral gray shade. ND filters may be labeled by their percentage transmission (e.g., an ND50 filter has 50% transmission and an ND25 25%) or logarithmically (e.g., an ND 0.3 filter has 50% transmission, an ND 0.6 25%, and an ND 1.0 10%).

Some filters marketed as "moon filters" combine neutral density with color filtration. For example, we have one "moon filter" of unknown provenance that appears to provide the combined effects of an ND50 filter with a #58 Green or #56 Light Green Wratten filter.

Polarizing filters are, in effect, variable neutral-density filters. They comprise two layers of Polarizing material in a mount that allows the layers to be rotated relative to each other. As you change the relative position of the Polarizing layers, the visible neutral density varies from moderate to high.

We consider neutral-density and Polarizing filters to be useless except in vary large instruments. Part of that, we suppose, is because the primary purpose of a telescope is to gather light, and most astronomers consider "wasting light" to be stupid, if not downright sinful. But the real reason is that there are better alternatives. If, for example, Luna is too bright in your eyepiece, that's nature's way of telling you to use more magnification, use a contrast filter, or both.

HACK Enhance Nebular Contrast and Detail
#59 Make the faint fuzzies a bit fainter but a lot less fuzzy.

As useful as color filters can be for Lunar and planetary observing, they're no help at all for observing nebulae. Nebulae are so dim that only the very brightest of them show even a hint of color, so the only thing a standard color filter does is dim them further.

Wouldn't it be nice if there were filters that selectively passed the light emitted by nebulae while stopping light pollution and other "bad" light? As it happens, such *nebula filters* do exist. Like any filter, nebula filters can't add anything; they can only take away. But, by selectively removing undesirable light while passing nearly all of the light emitted by nebulae, nebula filters enhance the level of contrast and detail visible in nebulae, even from a dark-sky site.

 For an in-depth discussion of the technical aspects of nebula filters, read *Choosing a Nebula Filter* by Greg A. Perry, Ph.D. (*http://members.cox.net/greg-perry/filters.html*).

Dimming Background Skyglow

Broadband filters, often called *Light Pollution Reduction filters* or *LPR filters*, are one of a class known as *interference filters*. Unlike color filters, which are

monolithic slabs of dyed glass, interference filters use multiple microscopi-
cally thin interference coatings to selectively block or pass specific wavelengths. Depending on the number and type of interference coatings
applied, the passbands of the filter may be broad or narrow, and contiguous
or separated. Figure 4-32, for example, shows the transmission curve of the
Orion SkyGlow, a typical broadband filter. Similar broadband filters are
available under various tradenames from Parks/Lumicon, Thousand Oaks,
and other astronomy vendors.

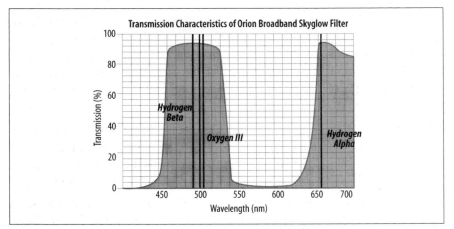

Figure 4-32. Transmission curve of the Orion Skyglow broadband filter

While stars have continuous emission spectra, various astronomical objects,
such as planetary nebulae and emission nebulae, emit light only at certain
wavelengths, which include:

- The bluish hydrogen-beta (H-beta or H-β) line at 486 nm
- The blue-greenish doubly ionized oxygen (Oxygen-III or O-III) lines at
  496 and 501 nm
- The blue-greenish cyanogen (CN) lines at 511 and 514 nm
- The red hydrogen-alpha (H-alpha or H-α) line at 656 nm

Broadband filters have relatively wide passbands in the blue/blue-green and
red parts of the spectrum to allow these astronomically important emission
lines to pass with little or no attenuation.

Conversely, broadband filters are designed to block portions of the spectrum lower than about 445 nm and between about 540 nm and 640 nm
because those portions of the spectrum include the emission lines of major
sources of light pollution, including:

- The purplish mercury (Hg) lines at 405 and 436 nm produced by mercury-vapor lights
- The yellow/yellow-orange mercury (Hg) lines at 546 and 579 nm produced by mercury-vapor lights
- The yellow line at 558 nm produced by natural air glow
- The yellow/orange sodium (Na) lines at 570, 579, 583, and 617 nm produced by low-pressure sodium-vapor lights

It is these "bad" emission lines that cause the pinkish-yellow skyglow in heavily light-polluted locations. In theory, then, a broadband filter should stop all of the "bad" light and pass all of the "good" light, allowing you to observe under even relatively bright suburban skies. And, in fact, broadband filters are usually marketed as "light pollution reduction" filters, suitable for use under moderately light-polluted conditions. Alas, the reality is different.

The real problem is that line-spectrum sources of light pollution, such as mercury-vapor and low-pressure sodium-vapor lights, are rapidly being replaced by continuous-spectrum lighting, primarily high-pressure sodium-vapor lights. This is occurring for both esthetic and economic reasons.

Esthetically, most people dislike the stark yellow brilliance of low-pressure sodium-vapor lights and the purple glow of mercury-vapor lights. (Also, mercury-vapor lights use the environmentally unsafe mercury, which presents disposal problems.) Economically, high-pressure sodium-vapor lights are more efficient than the older low-pressure sodium-vapor and mercury-vapor lights, which means they use less electricity and cost less to run for a given light output.

As light pollution sources increasingly shift to continuous-spectrum output, the value of broadband filters continues to shrink. Although they may provide minor image enhancement under bright urban and suburban skies by blocking most of the wavelengths emitted by sodium and mercury lights, few people consider the marginal improvement worth the relatively high cost of these filters, typically $60 to $100 in 1.25" versions and as much as $200 in 2" versions.

A broadband filter is most helpful with objects such as emission nebulae and planetary nebulae that emit most or all of their light at specific wavelengths. This is true because the broadband filter selectively dims unwanted wavelengths while allowing desirable wavelengths to pass with little or no diminution, thereby increasing the contrast of the object against the sky.

Continuous emitters, objects that emit across the entire visible spectrum—including stars, galaxies, and reflection nebulae—show less improvement with a broadband filter because the filter also blocks part of the light they emit. Still, because a broadband filter blocks "bad" wavelengths selectively, it does enhance contrast for galaxies and reflection nebulae, although the improvement is usually quite subtle.

Under typical urban observing conditions, light pollution is so severe that it effectively "swamps" the ability of a broadband filter to selectively block undesirable wavelengths. Although the broadband filter still eliminates sodium and mercury wavelengths, the sky remains so brightly lit by continuous-spectrum sources that it is difficult to locate faint fuzzies, let alone see them. The filtered view is marginally superior to the unfiltered view, but the improvement is so subtle that most observers won't consider a broadband filter very helpful for urban observing.

A broadband filter is more useful at darker site. Because it selectively blocks natural skyglow and the wavelengths emitted by sodium and mercury lights, a broadband filter can darken the sky background, increasing contrast and providing a more esthetically pleasing view. Although the effect of a broadband filter is subtle, it can improve contrast and reveal additional detail in continuous emitters like galaxies and reflection nebulae from darker sites.

A broadband filter is no substitute for a dark observing site. If you buy a broadband filter expecting it to magically slay light pollution and allow you to observe dim galaxies from beneath a streetlight, you'll be disappointed. But if you accept what it offers—selective blocking of most undesirable wavelengths—and use it appropriately, you'll find that a broadband filter provides marginal improvement for many objects.

> There is one activity for which a broadband filter is nearly indispensable. For film imaging at a site with slight light pollution, using a broadband filter extends the exposure time you can use while maintaining a dark-sky background, thereby allowing your images to reveal dimmer objects and more detail in brighter objects.

Enhancing Nebular Contrast

Narrowband filters, such as the Orion UltraBlock and the Parks/Lumicon Ultra-High Contrast (UHC), use the same interference-layer coating technology as broadband filters. They differ, as you might expect from their name, in the width and location of their passbands. Figure 4-33 shows the pass-band for an Orion UltraBlock narrowband filter. (Compare this figure to the preceding broadband transmission curve.)

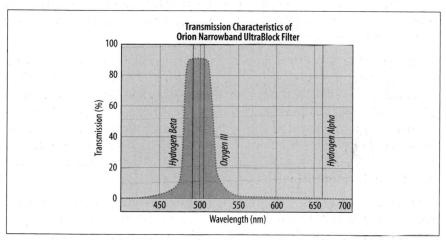

Figure 4-33. Transmission curve of the Orion UltraBlock broadband filter

Narrowband filters have very high transmission at the critical wavelengths of 486 nm (H-β) and 496/501 nm (O-III) and almost zero transmission outside that narrow band of spectrum. (Some narrowband filters also pass the 656 nm H-α line, but that wavelength seldom contributes much to the view unless you are using a large scope.) Because planetary nebulae and most emission nebulae emit most or all of their light in the passband of narrowband filters, viewing these objects using a narrowband filter selectively increases their contrast against the background sky, often dramatically.

With a narrowband filter, the background sky is darkened dramatically. You can see more of the extent of the object because dimmer portions of it that would otherwise be obscured by natural or artificial skyglow are visible with the filter blocking those parts of the spectrum. More fine, low-contrast detail is visible because you are looking only at the light emitted by the object, unsullied by light from other sources. In addition to improving the view of many objects, a narrowband filter also reveals objects that are entirely invisible without it. For example, from a reasonably dark site, we can glimpse the Rosette Nebula in Monoceros naked eye by looking through a narrowband filter.

The degree of improvement provided by a narrowband filter is strongly influenced by observing conditions. If you are not fully dark adapted **[Hack #11]**, a narrowband filter may show no improvement at all because much of the enhancement occurs in the fainter portions of the nebulosity.

Exit pupil size also matters. Narrowband filters provide the most enhancement with an exit pupil in the range from 7mm down to 2mm, which translates to 3.5X to 12.5X magnification per inch of aperture. Best results are in the 3.5X to 8X per inch range.

Because narrowband filters have such narrow passbands, they significantly darken not just skyglow but any continuous-spectrum light source, including stars, galaxies, and reflection nebulae (which glow by reflected starlight). Accordingly, they truly are "light pollution reduction" filters, but the cure may kill the patient.

That is, although a narrowband filter can eliminate or greatly attenuate light pollution even under bright urban skies, it does so by dimming the view dramatically. So much so that it may be difficult to see anything *except* line-emitting objects like emission nebulae, including the very stars that you need to see to guide you to the object. With a narrowband filter, bright stars appear dim, and dim stars are invisible.

Because a narrowband filter makes it difficult to see any but the brightest stars, the proper technique is to get the object in the field of view first, and then use the filter to view the object, either by screwing the filter onto the eyepiece or by "blinking" the object, which is to say holding the filter between the eyepiece and your eye and moving it in and out of your line of view.

Regardless, the benefits of a narrowband filter are indisputable. At light-polluted sites, a narrowband filter allows you to view planetaries and emission nebulae against a pleasingly dark sky background. At dark sites, the narrowband provides images significantly better than unfiltered views. We think a narrowband filter belongs in every DSO observer's eyepiece case.

We suggest you not dither about which brand to buy. The two market leaders are the Orion UltraBlock and the Parks/Lumicon UHC. Both are excellent products, and their similarities greatly outweigh their differences. We have done detailed A–B comparisons of the two, attaching them to two identical 14mm Pentax XL eyepieces and then swapping the eyepieces in and out of the same 10" scope. Our colleagues agree with us: although there are minor differences in the views, they're not worth worrying about. Buy whichever happens to be cheaper or on sale at the time. You'll be happy with either.

> One night, Robert and our friend Paul Jones decided to do an A–B test of the Ul' ıBlock against the UHC. Robert mounted his UltraBloc' ı his 14mm Pentax XL eyepiece. Paul mounted his Uŀ on his 14mm Pentax XL. We pointed Robert's 10" Dcᴗ at M42, the Great Orion Nebula, and began swapping the eyepieces in and out. Robert *slightly* preferred the view in the UltraBlock and Paul *slightly* preferred the view with the UHC—probably pride of ownership—but both agreed that the differences were almost unnoticeable. Barbara took a quick look at each and announced she much preferred our UltraBlock.
>
> So Paul decided to call her bluff. He used his body to conceal which eyepiece/filter he was putting into the focuser. He then called Barbara over and asked her which filter was in place. "That's ours!", Barbara announced emphatically. Paul and Robert just looked at each other, thinking "how could she possibly tell?" So we asked her. Barbara informed us that the fourth star in the Trapezium was easily visible with our UltraBlock, but very faint with Paul's UHC. We checked, and sure enough it was true. Barbara *notices* things.

Blocking All But a Specified Wavelength

Line filters can be thought of as narrowband filters on steroids. While a typical broadband filter has a 100 nm passband, and a typical narrowband filter has a 25 nm passband, a line filter may have a passband of only 8 nm or so. Line filters isolate one particular line (or a closely grouped pair of lines), transmitting nearly all of the light at that wavelength and blocking all other light. Accordingly, the view through a line filter is quite dim.

There are two line filters in common use among amateur astronomers:

Oxygen-III

> The *Oxygen-III filter*, also called an *O-III filter* or *O-3 filter*, is useful for many emission nebulae, but its most common use is to enhance the view of planetary nebulae, many of which emit light almost exclusively on the 496 nm and 501 nm lines of doubly ionized oxygen. Although a standard narrowband filter enhances many planetary nebulae, the O-III filter does a much better job on some planetaries, sometimes dramatically so. An O-III filter is usually the second choice of dedicated DSO observers, after the indispensable narrowband filter.

There are no hard-and-fast rules as to which is the best filter for observing a particular object or class of objects. Some planetaries, for example, reveal more detail with a narrowband filter than with an O-III filter. Conversely, some emission nebulae are shown to better advantage with an O-III filter than with a narrowband filter.

Hydrogen-beta

The *Hydrogen-beta filter*, also called an *H-beta filter*, *H-β filter*, or *Hydrogen-b filter*, is often referred to as the *Horsehead Nebula filter*, only partly in jest. The H-beta filter is extremely specialized. It is useful for only a handful of objects, of which the Horsehead Nebula is by far the most famous. But for those few objects, the H-beta filter is nearly essential. It's not overstating the case much to say that without an H-beta filter the Horsehead is elusive visually even in large scopes, while with an H-Beta filter the Horsehead becomes possible—although challenging—from a dark site with scopes as small as 8" to 12", and relatively easy in large scopes.

We consider the O-III filter the second most useful nebular filter overall, and one that any serious DSO observer will want to own. The H-beta filter is a different matter. If you are on a quest to view the Horsehead Nebula, by all means buy an H-beta filter of your own. Otherwise, there are better things to do with your money. Just borrow an H-beta filter from time to time when you need it.

Some interference filters, including those sold by Orion and Lumicon, sandwich their interference coatings between two layers of glass. These filters are as durable as standard dyed-glass color filters and can be cleaned safely just as you'd clean any other optical surface. Other interference filters, including those sold by Sirius and DGM optics, put their fragile coatings on an outside surface. These filters are extremely delicate and can be ruined even by careful cleaning. In fact, we have heard reports of such filters being damaged by dew.

Choosing Nebula Filters

There's no question that serious DSO observers should own and use nebula filters, but deciding how many you need, which types, and which brands is not easy. Fortunately, unless you are restricted by your budget, it's not an either-or situation. Buy more than one nebula filter, and use whichever provides the best view of the object you happen to be observing. We suggest the following guidelines:

- If you're going to buy only one nebula filter, choose a narrowband filter such as the Lumicon UHC or the Orion UltraBlock. A narrowband filter is the most generally useful of the bunch. It is the best filter for most objects that benefit from using a nebular filter, and it shows a greater improvement on more objects than any other type of nebula filter. A narrowband filter provides at least some improvement on nearly all of the objects that benefit from any type of nebular filter. In short, a narrowband filter is most likely to help and least likely to hurt the view of any nebular object. If you observe DSOs, you should make it a high priority to acquire a narrowband filter.

- If you buy a second nebula filter, choose an O-III filter. An O-III filter is the best choice for most planetary nebulae, and it improves many emission nebulae as well.

Although most emission nebulae are more enhanced by the narrowband filter than the O-III filter, there are reflection nebulae that appear better with the O-III. Similarly, although the O-III is the best choice for most planetary nebulae, some planetaries are better served by the narrowband filter.

To some extent, "best" is a matter of opinion. For example, when we tested the UltraBlock and UHC filters on M42 against the OIII, we were hard-pressed to say which provided the "best" view. They were different views, certainly. The extent of the nebula was larger with the narrowband filters, and the image was brighter, as you might expect. On the other hand, the OIII filter revealed fine detail that was not visible in either narrowband filter.

If you can afford only one nebular filter, the narrowband is unquestionably the better choice (unless you happen to specialize in planetary nebulae). Otherwise, buy one of each.

- Most DSO observers are content with just a narrowband filter, or with narrowband and O-III filters. But if there's room left in the budget for one more filter, add a broadband filter such as the Lumicon Deep-Sky or the Orion SkyGlow. There are some objects for which a broadband filter improves the view more than a narrowband or O-III filter. If the cost doesn't matter, buy a broadband filter and try it on various objects. But we suggest you consider the broadband filter a low-priority purchase.

- Finally, there's the H-beta filter. For the California Nebula, the Cocoon Nebula, the Horsehead Nebula, and a dozen or so other objects, the H-beta filter is almost a magic bullet. But the H-beta filter is of very limited help on most emission nebulae, and actually degrades the view for most planetary nebulae. The H-beta is a very specialized filter, and we suggest that you consider acquiring one a very low priority.

Specialty Filters

Various specialty filters exist that have even more narrowly defined uses than the H-Beta filter. For example, the Parks/Lumicon Swan-Band Comet filter is for observing comets, period. Its passband includes the cyanogen lines and the 501 nm O-III line, but not the 496 nm O-III line. This filter is useless unless you observe comets; if you do, it's very useful indeed.

Dedicated amateur astronomer Al Misiuk also happens to be an expert on interference coatings. He formed a company named Sirius Optics (*http://www.siriusoptics.com*), which produces various specialized interference filters. The first project Al took on was making a filter to improve the false color generated by achromatic refractors when viewing bright objects. Traditionally, astronomers used a yellow filter to reduce this blue-violet haze, but the transmission characteristics of Wratten color filters are not ideal for this purpose. The Sirius Optics MV-1 (minus violet) interference filter eliminates most false color while imparting only a mild greenish-yellow cast. The enhanced Neodymium Eyepiece Filter combines MV-1 interference coatings with neodymium optical glass to provide an almost completely neutral image. The MARS 2003 filter is an interference filter optimized for (you guessed it) Mars, although it is also useful on some other objects. The NIR1 Near Infrared Blocking filter is useful for astrophotographers.

In addition to specialty filters, Sirius Optics also produces more general-purpose interference filters, including the CE1 Contrast Enhancement filter (enhanced broadband), the NEB1 Nebula Eyepiece Filter (narrowband), and the PC1 Planetary Contrast Eyepiece Filter.

HACK #60 Please Be Seated

Use a good observing chair and add 2" of aperture for free.

We're always amazed at how many astronomers stand at their telescopes instead of sitting. An old rule of thumb says that observing while comfortably seated is the equivalent of adding a couple extra inches of aperture. It's true. When you observe seated you are more relaxed and less shaky, and that pays off in terms of being able to see more detail. A good observing chair is probably the single most helpful accessory you can add.

The type of observing chair you need is determined by the type of scope you use. That's true because the eyepiece position of some types of scopes moves dramatically with changes in the altitude of the object you are observing, while for other types of scopes the change in eyepiece position is much smaller.

For example, when our 10" Dob is pointed at the horizon, its eyepiece height is about 30". At zenith, the eyepiece height is about 48". But little observing is done at either horizon or zenith, and for the usual range of elevations the Dob's eyepiece height remains within the range of 37" to 47", only a 10" difference. That small difference is easily accommodated just by shifting observing position slightly, so a simple stool of appropriate height serves well for a small or mid-size Dob.

To determine the ideal height for a Dob stool, determine the minimum and maximum eyepiece heights of your scope at the extremes of elevation at which you normally observe. Then, seat yourself on any stool, lean slightly forward in the relaxed posture you would use when looking into the eyepiece, and measure the vertical distance from your eye to the ground, as shown in Figure 4-34. Also measure the upper and lower limits at which you can observe comfortably.

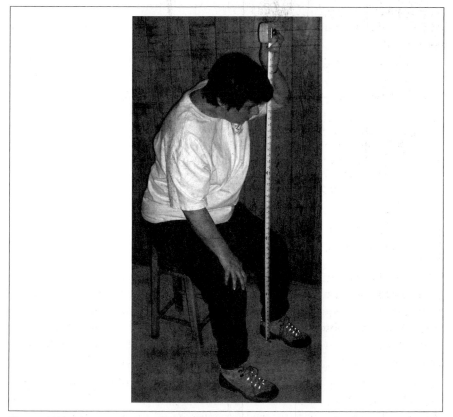

Figure 4-34. Measure your eye height to determine the proper seat height

For example, we started with an inexpensive oak barstool with a seat height of 24". Leaning slightly forward in observing position, Robert's eye was 51" above the ground, with a comfortable range of 47" to 57". That range is 10" higher than the range of eyepiece heights we measured for the scope, which means the seat height of the stool needs to be cut from 24" to 14".

> But Robert and Barbara share a scope and an observing stool. Although Barbara is 8" shorter than Robert, her eye height in a relaxed observing posture is only 2" less, giving an ideal seat height for Barbara of 16". So we compromised and trimmed the stool legs to give a seat height of 15".

SCTs, MCTs, short-tube refractors, and other tripod-mounted scopes are more problematic. On the good side, although the full arc from 0° to 90° translates to a moderately large vertical change in eyepiece position, the eyepiece height remains relatively constant within the usual range of observing elevations. Unfortunately, such tripod-mounted scopes have two problems of their own:

- Tripod legs often intrude, forcing you to sit farther from the scope and lean forward, which reduces the amount of vertical eyepiece shift you can accommodate comfortably.

- Instead of having the eyepiece always nearly horizontal or tilted slightly upward, as Dobs do, the eyepiece angle of a refractor or SCT changes with the elevation of the scope. When the scope is horizontal, you must look straight down into the 90° star diagonal, which is then located at the highest point of its arc. When the scope is vertical, you must look horizontally into the diagonal, which is then located at the lowest point of its arc.

We think the best type of observing chair for most tripod-mounted scopes is a drummer's throne, which is available from any music supply store. Look for a model that allows the seat height to be adjusted quickly over a relatively wide range, such as 16" to 28". Inexpensive models are available for as little as $25, and the best models cost $150 or more, but something in the $50 to $75 range generally suffices.

Long-tube refractors are the hardest type of scope to deal with because the eyepiece position changes dramatically with changes in elevation. For example, with its tripod legs fully extended, the eyepiece height of our 90mm long-tube refractor is 52" at horizon and only 20" at zenith, a difference of nearly three feet. Even within the usual range of elevations, the refractor's eyepiece height varies between 42" and 22", a difference of 20". That's much too large to accommodate from a fixed seat or even a drummer's throne without uncomfortable bending and stretching.

For long-tube refractors, the best seating choice is a fully adjustable astronomy chair, such as the model shown in Figure 4-35, which adjusts from a seat height of only a few inches to more than two feet. An adjustable chair such as this one makes observing much more convenient, particularly with telescopes whose eyepiece height change dramatically with changes in elevation. Commercial models are sold by astronomy retailers like Orion (*http://www.telescope.com*) and Company Seven (*http://www.company7.com*) for $200 or so.

Figure 4-35. An adjustable observing chair

If $200 is more than you want to spend on an observing chair, consider building your own. For $35 to $50 in materials and a few hours' work, you can assemble a fully adjustable observing chair that's the equal of any of the $200 commercial products. Web sites come and go, so we won't offer a specific URL, but a quick web search for "Denver observing chair" will turn up dozens of sites with detailed plans for the original Denver observing chair and many derivatives.

HACK #61 Stash Your Gear in a Photographer's Vest or Fanny Pack

You can never have too many pockets.

Astronomers carry a lot of stuff around with them. Red flashlight, eyeglasses, pens, eyepieces, Barlows, filters, lens caps, adapters—the list goes on and on. (Robert sometimes rattles.) Many of these items are much too expensive to risk dropping—everyone cringes when a $300 eyepiece hits the pavement—and all of them are hard to locate if you drop them in the dark.

Pockets are essential to hold all this stuff, and pockets with Velcro locking flaps are all the better. One excellent solution is to use a photographer's vest, which you can buy under different names from photography stores, astronomy retailers, and outfitters like L.L.Bean and Cabela's. Figure 4-36 shows Barbara wearing her fully loaded vest; this one is a model formerly sold by Orion Telescope and Binocular Center.

> Make sure to buy a vest large enough to fit over your bulky cold-weather clothing. It may be a bit loose in warm weather—although many vests are adjustable with Velcro tabs or similar arrangements—but better too large than too small.

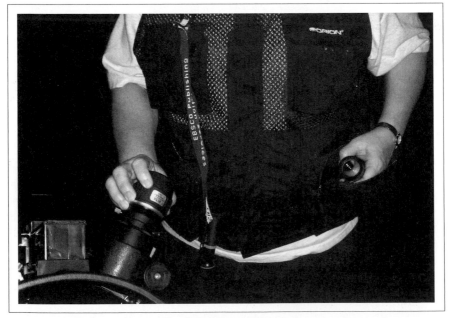

Figure 4-36. Barbara wearing a loaded photographer's vest

These vests are available in various materials, including canvas and mesh. Barbara prefers a mesh model, which is cool in summer and, she swears, warm in winter. (Robert doesn't understand how a mesh vest that is 95% holes could possibly help Barbara stay warm in cold weather, but he's been married too long to make the rookie mistake of pointing that out.)

If you use a vest, give some thought to how you'll organize it. Decide on a pocket for each of the items you regularly carry, and keep each item in its pocket. If you keep your red flashlight clipped to the left breast pocket, for example, you'll always know where to find it. If you just load items willy-nilly into the vest, you'll find yourself searching 20 pockets to find what you're looking for.

Robert finds a vest constraining, so he uses a fanny pack instead, which he wears to the front. Although a fanny pack doesn't offer nearly as many nooks and crannies as a vest, it is ideal for holding eyepieces, filters, and other small optical gear while you are actually working at the scope.

The model he uses was designed for photographers. It is well padded externally and provides padded interior separators that can be moved around and secured with Velcro. The top flap secures with Velcro sufficiently well that we are never concerned about dropping an eyepiece while working at the scope, and it can be zipped during transport or when it is otherwise not being used.

HACK #62 Use a Voice Recorder for Logging
Don't write it, speak it.

Traditionally, amateur astronomers have kept written logs of their observations, and most still do today. An increasing number, though, have started using a voice recorder to supplement or replace their written log sheets.

There are a lot of benefits to using a voice recorder:

- Most important, using a voice recorder encourages you to record more detailed observations.

- Because you can log your observations while you're right at the scope, looking through the eyepiece, your records tend to be more accurate and to include details you might overlook if you have to move to your chart table to write them down.

- A voice recorder can actually help you make better observations in an absolute sense because it encourages you to look more closely at the object as you search for details to speak into the recorder.

- Using a voice recorder saves precious observing time. What takes you 15 seconds to dictate into a voice recorder may take you 5 minutes to write down on a paper log sheet.

There are two types of voice recorder available:

Mini-cassette recorders

These devices are miniaturized versions of a standard tape recorder. They use small, dictation-size magnetic tape cassettes and are very inexpensive. You can find them in big-box stores selling for $25 or less. Unfortunately, that's about the best that can be said of them. The tapes they use are fragile, particularly in cold weather, which may cause the tape to break. Because they have a motor, they draw significant current from their batteries, which again is a problem in cold weather. The actual recording speed may vary with the temperature and the state of the battery, so your recordings may sound like Alvin the chipmunk or the Addams' Family's Lurch. Although they are miniaturized, they are still large enough to be inconvenient when you are working at your scope. Cassette-based voice recorders are a poor choice.

Digital voice recorders

A *digital voice recorder* (DVR) is all solid state. Instead of recording to a magnetic tape, it records to internal memory or a flash memory card. DVRs are small, light, and rugged. Because they record your voice in digital form, it's easy to transfer your voice logs to your computer, if the DVR provides a computer interface. (Not all do; look for one with a USB interface.) Transferring voice recordings to your computer allows you to keep them as a permanent record of your observing sessions. Because you can pause the file during playback, it's easy to transfer the data to your permanent observing ledger **[Hack #28]**. DVRs typically sell for $50 to $150, depending on brand name, features, recording time, and so on.

> If you own a portable MP3 player, you may already have a DVR you didn't know about. Many MP3 players, such as Barbara's Creative Labs MuVo N200, include a built-in microphone and voice recording capabilities. Many of them allow you to switch back and forth from voice recording mode to music playback mode, so you can use one device to serve both purposes.

H A C K
#63

Build or Buy an Equatorial Platform

The Dobsonian mount was a wonderful innovation. It's simple, inexpensive, smooth, extremely stable, and intuitive to use. The only downside is that a Dob doesn't track the motion of the stars. Or at least it doesn't unless you put it on an equatorial platform.

An *equatorial platform*, or *EQ platform* for short, is a squat, flat, motor-driven table that sits underneath the Dob base and gradually pivots the scope at the exact rate needed to counteract the apparent motion of the stars. When you use an EQ platform, you continue to point the Dob normally, moving it in altitude and azimuth to locate objects. The difference is, when you stop moving the Dob, it begins tracking and the object in the eyepiece remains centered rather than drifting out of view as Earth rotates. Figure 4-37 shows an EQ platform built by Steve Childers, underneath the rocker box of his 10" Orion XT10 Dob.

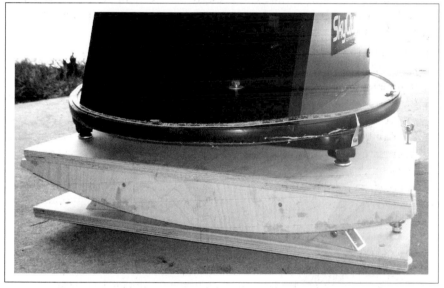

Figure 4-37. A typical equatorial platform

The equatorial platform pivots on the curved bearing shown in Figure 4-37, as well as a similar bearing at the far end of the platform, not visible in the figure. The bearing dimensions are calculated for the specific latitude at which the platform is to be used.

Using an EQ platform preserves all of the advantages of a Dob—large aperture; easy, intuitive motions, and so on—while adding all of the benefits of a driven equatorial mount. (Well, all but one; the Dob is still an alt-azimuth scope in terms of motions, which means you can't use the old EQ guys' trick of locating a bright object and then using the RA or declination slow motion controls to repoint the scope to a dim object on the same RA or declination line.) But the EQ platform is a real equatorial mount in every sense of the word. It just doesn't look like a standard GEM or fork equatorial mount.

Because an EQ platform tracks equatorially rather than alt-azimuthly (as, for example, a standard fork mount without an EQ wedge or a Dob Driver does), there is no field rotation during tracking. That means an EQ platform, if it is built precisely and polar aligned accurately, can be used for long-exposure astrophotography without requiring hardware or software field derotators.

Because the scope simply sits on top of the EQ platform rather than being attached to it, the range of motion is necessarily limited to prevent tipping. Most EQ platforms tilt from about -7.5° through vertical to +7.5°. This 15° range provides about one hour of tracking time. When the EQ platform reaches the end of its travel, you simply reset it for another hour's worth of tracking. Some EQ platform designs provide motorized rewinding, which takes a couple of minutes. Others allow you to reset the platform manually, simply by pivoting it back to the starting position, which takes only seconds. Figure 4-38 shows an EQ platform at one extreme of its motion, with the scope about 7.5° off vertical.

Like any equatorial mount, an EQ platform must be polar aligned if it is to track stellar motion properly. For visual use, a rough polar alignment works fine. You simply place the EQ platform with the north-south central axis pointing more or less north. Polar alignment within a couple degrees is sufficient for visual use. You can do an eyeball alignment on Polaris, or use a compass (adjusting for magnetic declination, of course) to align the platform accurately enough for visual use up to 300X or more.

For astrophotography, the polar alignment needs to be much more precise. You can use any of the standard methods that astrophotographers use for critical polar alignment, including drift alignment or a polar-alignment scope. Commercial platforms that are sufficiently precise for long-exposure astrophotography (which is to say platforms built by Tom Osypowski) offer optional polar-alignment tools.

Figure 4-38. An equatorial platform at one extreme of its motion, tilted about 7.5° off vertical

An EQ platform is simple in concept. The ground board, typically a sheet of stiff plywood, has adjustable feet that allow the platform to be leveled (an adjustable foot is visible at the lower right of Figure 4-39). The scope sits on top of the top board, which is usually another sheet of stiff plywood. Between the two boards sit the north and south bearings, a small motor, and some sort of drive mechanism that tilts the top board (and scope) to track the stars. The controller needed to adjust tracking speed may be a separate box that is cabled to the platform, or it may be built into the platform base. Figure 4-39 shows the ground board of Steve Childers' EQ platform.

The white box on the left contains the electronics that control the speed of the platform's motion. The stepper motor, which actually moves the platform, is visible in the oval cutout. It connects via a drive shaft to the drive wheel visible at right in the aluminum bracket assembly. That drive wheel bears against and drives the curved bearing surface connected to the top board, upon which the scope rests.

You can buy a commercial EQ platform or build your own EQ platform from scratch. Even commercial platforms are not off-the-shelf items, though, because a platform must be adjusted specifically to the latitude where you use it and the center of gravity of the scope you plan to mount on the platform.

Figure 4-39. The ground board, bearings, and drive mechanism of an equatorial platform

At least one commercial platform, the Johnsonian Type V, is adjustable for latitude within a wide range. Most platforms, commercial or home-built, are set for one specific latitude, although you can adjust them to ±5° or so by shimming up the north or south end. In practice, that means that most platforms are made for the owner's specific latitude, but are usable anywhere within 350 miles north or south of that location.

The most popular commercial platforms are made by Sam Johnson (*http://www.johnsonian.com*) and Tom Osypowski (*http://astronomy-mall.com/regular/products/eq_platforms/*). Both of these come highly recommended by their owners, but neither is inexpensive. The dual-axis, aluminum Johnsonian Type V platform sells for $900 and supports Dobs up to 16" or 140 pounds. The Johnsonian Type VI platform is an entry-level model that supports Dobs up to 10" or 60 pounds. The Type VI was announced in spring 2005 at an introductory price of only $399. Osypowski platforms range in price from $925 for a basic single-axis wooden model suitable for 6" to 11" Dobs to $4,250 for a dual-axis aluminum model suitable for 28" to 32" Dobs. Although these prices sound high, compared to the price of a traditional equatorial mount large enough to support scopes that size, a platform starts to sound like a bargain.

Many amateurs choose to build their own EQ platforms to save money, as well as to customize the platform to their own preferences. If you're comfortable with the idea of building a platform from scratch, you can probably do so for $150 to $200 in materials and two or three weekends' work. You'll need to calculate dimensions, center of gravity, bearing curves, and so on for yourself, but there are spreadsheets and other tools available to help with all that. Here are several online resources to help you get started:

- eqplatforms group (*http://groups.yahoo.com/group/eqplatforms*)
- Chuck Shaw's EQ platform page (*http://www.atmsite.org/contrib/Shaw/platform/*)
- Jan van Gastel's EQ platform page (*http://home.wanadoo.nl/jhm.vangastel/Astronomy/Poncet/e_index.htm*)
- Don Odegard's EQ platform page (*http://home.att.net/~segelstein/don/platform-1.html*)
- David Shouldice's EQ platform page (*http://members.tripod.com/denverastro/dsdfile/dspfile.htm*)
- Tom Hole's EQ platform page (*http://www.tomhole.com/EQ%20Platform.htm*)

An EQ platform may not be the best choice for large Dobs because it raises the Dob (and the eyepiece height) by 6" or so. For small Dobs, this may actually be welcome. Large Dobs already require a stepladder to reach the eyepiece when the scope is pointed near zenith, so another 6" of height might be the last thing you want. But for small and mid-size Dobs, those 12" or less, many astronomers think an EQ platform is the best single upgrade you can make.

Make Your Computer Work for You
HACK #64

Choose the best planetarium software and start planning observing sessions.

Planetarium software is an almost indispensable adjunct to traditional printed charts. Unlike static charts, planetarium software can provide a simulation of the night sky as viewed from any specified location, in real time or for some past or future date. Astronomers use planetarium software at home to plan future sessions, and in the field to provide updated charts of how the sky appears as of that moment and to provide detailed data about the objects they are observing.

The major advantage of planetarium software relative to printed charts is flexibility. With printed charts, what you see is what you get. With planetarium software, you can specify the field of view, the level of detail, the types of objects to be plotted, the limiting magnitude, and so on. You can zoom, rotate, and flip charts on-screen to correspond to the view in your finder or telescope, and then print them for use during the observing session.

You can pay hundreds of dollars for commercial planetarium software. Fortunately, you probably won't need to. Some of the best and most powerful planetarium programs cost nothing.

We won't say that free planetarium programs are superior to the commercial products in every respect, or even necessarily their equals. If that were true, no one would ever buy the commercial programs. All of these programs—free and commercial—have their own strengths and weaknesses. But we do think that nearly any amateur astronomer will be completely happy with one or more of the free programs. We suggest that you try the free alternatives first. If you find that none of them are completely suitable, then you can begin exploring the commercial alternatives.

Windows Planetarium Programs

The premier free planetarium program for Windows is the extraordinary Cartes du Ciel (Sky Charts), written by Swiss astronomer Patrick Chevalley. Cartes du Ciel is attractive, powerful, and immensely flexible. The complete program is a 15 MB download (*http://www.stargazing.net/astropc/*), which includes the Bright Star Catalog of 9,096 stars to magnitude 6.5; the Sky2000 catalog of 300,000 stars to magnitude 9.0; the full NGC catalog; the SAC (Saguaro Astronomy Club) 7.2 catalog of 10,000 nebulae; and ephemerides for the planets, comets, and asteroids, which can be updated online.

But the complete program is only the beginning. Cartes du Ciel makes it easy to download and install supplemental catalogs, of which dozens are available and many are indexed and searchable. For example, we've installed the full Tycho2 catalog (2.5 million stars to magnitude 12), the Hipparcos catalog (118,000 stars to magnitude 8.5 with very precise position data), the Catalog of Principal Galaxies, the Washington Double Star Catalog, and many more.

There are also many specialized catalogs available, including ones that correspond to the various Astronomical League observing programs. For example, Figure 4-40 shows Cartes du Ciel displaying a section of the southern summer sky with the Messier Catalog activated.

It's easy to switch among different catalogs when you want to chart different types of objects. For example, Figure 4-41 shows the same area of sky, but with the Messier Catalog disabled and the Herschel 400 catalog enabled.

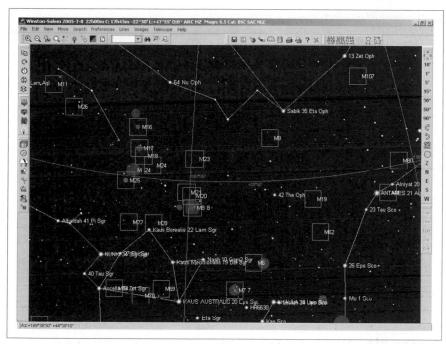

Figure 4-40. Cartes du Ciel, showing Messier objects in the southern summer sky

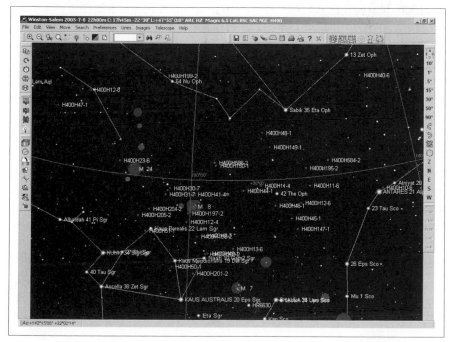

Figure 4-41. Cartes du Ciel, showing Herschel 400 objects in the same area of sky

Cartes du Ciel makes it easy to zoom in and out, set magnitude limits, insert finder circles, and do lots of other routine charting tasks. For example, Figure 4-42 shows a zoomed-in view of a small section of Sagittarius. We've increased the limiting stellar magnitude from 6.5 to 8.5 to correspond to the dimmest stars visible in our 50mm optical finder, and we've placed a Telrad finder circle to help us locate the Herschel 400 Object H.201-II, NGC 6569. (Note that the Messier open cluster M7, that group of stars visible at the lower right of the image, is no longer labeled.)

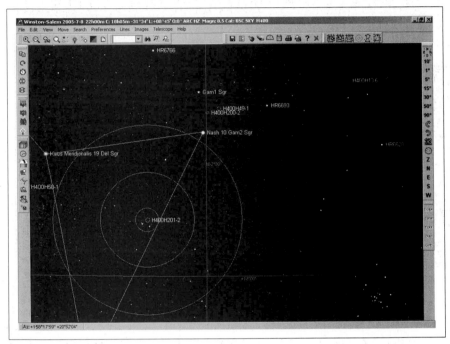

Figure 4-42. Cartes du Ciel, showing a detailed chart to locate Herschel 400 Object H.201-II

Cartes du Ciel also provides the advanced features you'd expect in a heavyweight planetarium program—e.g., controlling computerized telescopes, CCD integration, updating Solar system ephemerides online (including the moons of Mars, Jupiter, Saturn, and Uranus), using custom horizon maps, and so on. In short, Cartes du Ciel probably does nearly everything you need to do.

If the program has one weakness, it's in the area of observation planning. For example, you might want to plan an observing session by generating a list of objects of a particular type or subgroup that are visible on a particular evening. Cartes du Ciel provides only limited tools for this purpose. Fortunately, there's an integrated add-on utility available that serves that purpose well. RTGUI, shown in Figure 4-43, looks like a refugee from MS-DOS days, but it does the job.

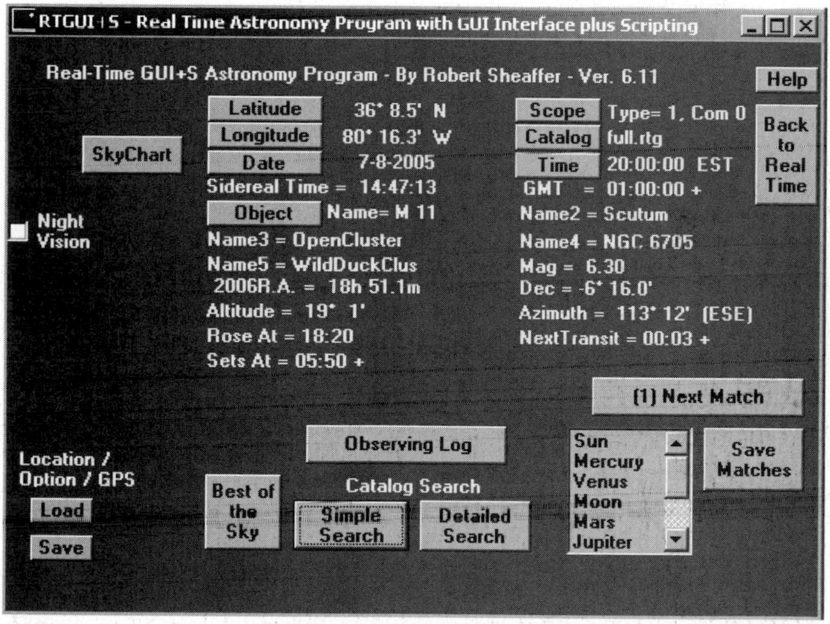

Figure 4-43. RTGUI displaying its Best of the Sky tour objects

RTGUI was actually designed as a lightweight real-time GUI link to automated scopes, small and fast enough to run on the obsolescent notebook computers that astronomers typically use in the field. But its feature list has grown to the point that it's very useful even for those who use manual scopes. We don't have room to discuss all of its features in detail, but there's no need to. For an excellent review of RTGUI, see Rod Mollise's article in the Fall 2004 issue of SkyWatch (http://skywatch.brainiac.com/swfall04.pdf).

> If you need an industrial-strength observation planning tool, choose the $40 Deepsky Astronomy Software (http://www.deepskysoftware.net) or the $100 SkyTools 2 (http://www.skyhound.com/skytools.html). Although both support other features, their real purpose is to generate customized observing lists, and both do an excellent job at this primary task. Most serious observers use one or the other, and many use both.

Cartes du Ciel also lacks support for Lunar observing. Although it shows a nicely detailed image of Luna, including an accurate terminator, that's as far as it goes. Fortunately, Patrick Chevalley and Christian Legrand have written a second program, Virtual Moon Atlas (VMA), that fills that hole. Full-time Lunar observers will still want a copy of Rukl's classic book, *Atlas of the Moon*, but even dedicated part-time Lunar observers will probably be happy with VMA, shown in Figure 4-44.

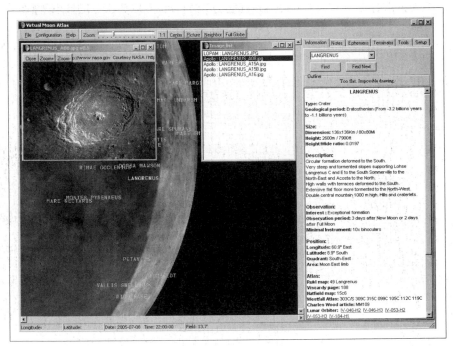

Figure 4-44. Virtual Moon Atlas, showing the crater Langrenus

> If VMA doesn't quite do the job for you, have a look at the $40 commercial product Lunar Phase Pro (*http://www.nightskyobserver.com/LunarPhaseCD/*). In general, LPP has Lunar mapping features similar to those of VMA, but superior observation planning features. There is another alternative, Lunar Map Pro, but we do not recommend it for various reasons, not least its intrusive copy-protection scheme.

Linux and Mac OS X Planetarium Programs

There are two standout planetarium programs for Linux, both of which can also be run on a Mac under OS X. The first program, the KStars Desktop

Planetarium, is a part of the Linux KDE desktop environment. In its early versions, KStars was a lightweight planetarium program—pretty enough, but useless for serious work. Beginning with the KStars 1.0 release with KDE 3.3, KStars has become, if not a heavyweight, at least a middleweight. We consider KStars 1.0, shown in Figure 4-45, to be suitable for casual use. In fact, it's what we generally use in the field on our elderly Compaq notebook.

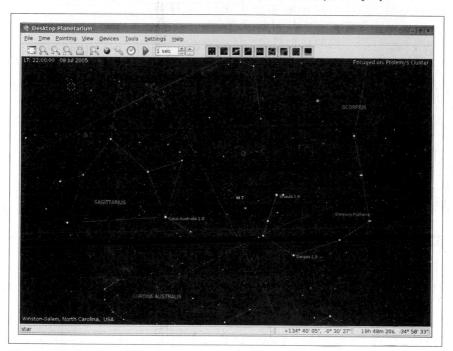

Figure 4-45. KStars Desktop Planetarium 1.0 displays the summer sky around Sagittarius

Although it's suitable for casual use in the field, KStars lacks the flexibility and configurability of heavyweight programs like Cartes du Ciel. (There is a Linux version of Cartes du Ciel in the works, but it's still in alpha as we write this.) In partial compensation, though, KStars adds some useful observation planning utilities for which Cartes du Ciel provides no direct equivalents. For example, the KStars Altitude vs. Time tool, shown in Figure 4-46, provides altitude/time curves for any objects you specify. In the image, we've entered several of the Messier Objects in and around Ophichus, Sagittarius, Scorpius, and Scutum, and graphed their altitude versus time. We learned that these objects culminate just before dawn local time, which allows us to schedule our observing sequence accordingly.

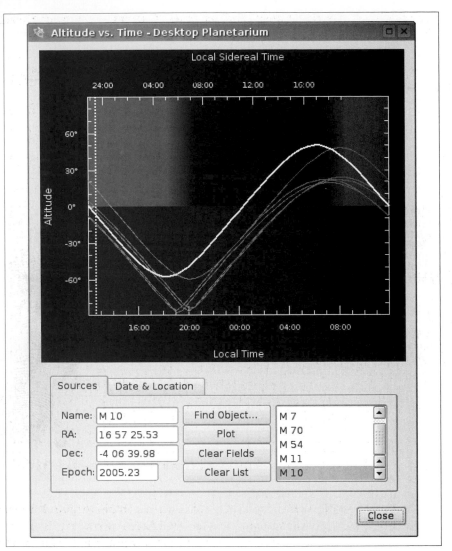

Figure 4-46. The KStars Altitude vs. Time tool allows you to schedule optimum observing times

The KStars What's up Tonight tool, shown in Figure 4-47, lets you go at it from the opposite direction. Rather than choosing specific objects and learning when they're best placed for observing, this tool lets you specify the time of your observing session and generate a list of objects that are well placed at that time. In the figure, we've told the tool to show us objects that are well placed for an evening observing session. We chose the Nebulae category, and the tool generated a list of available objects in that category. Selecting an object displays its rise, transit, and set times. Clicking the Object Details button provides more information about the object.

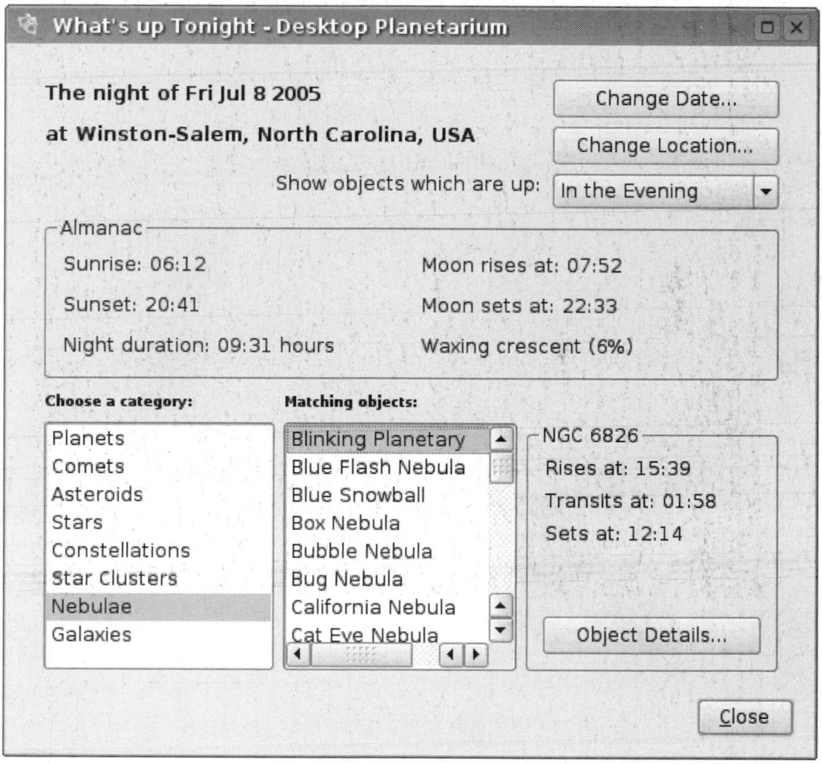

Figure 4-47. The KStars What's up Tonight tool lists well-placed objects for a specified observing session

If KStars isn't quite enough for your needs, there is an alternative. XEphem (*http://www.clearskyinstitute.com*), written by Elwood Downey, is a true heavyweight, by far the most powerful free astronomy program for Linux and Unix-like systems (including Mac OS X). XEphem is not as mature, polished, or feature-laden as Cartes du Ciel, but it has similar power and flexibility.

XEphem takes some getting used to, though. Just about every other planetarium program opens with a display of the night sky. Not XEphem. When you fire it up, the first thing you see is the control center dialog shown in Figure 4-48. You use this dialog to set time, date, location, and other parameters, and to load data files, generate ephemerides, set preferences, and access various views.

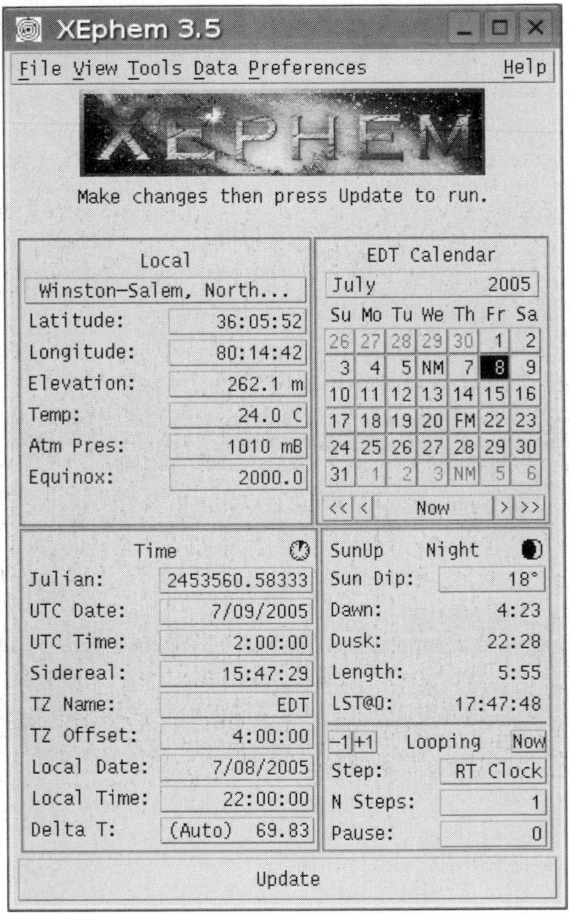

Figure 4-48. XEphem control center dialog

Choosing Sky View displays the planetarium screen shown in Figure 4-49. Although the X-Windows interface differs slightly from Windows, KDE, and Gnome, XEphem uses the standard conventions of a star map surrounded by icons on the left, top, and right, with status lines at the bottom. Anyone who's used other planetarium software will feel right at home. At first, that is.

As you begin to explore XEphem, you'll discover its oddities. XEphem definitely marches to its own drummer. For example, most other planetarium programs use the same method for zooming the display—draw a box on screen and click inside the box. XEphem instead uses a zoom slider, located at the left of the screen between the star map and the icons. Similarly, the right-hand slider controls right ascension or altitude (depending on which coordinate system you've enabled), and the bottom slider controls declination or azimuth. It takes some getting used to, but it's worth the effort. Once you understand its peculiarities, you'll find that XEphem can do just about anything you ask of it.

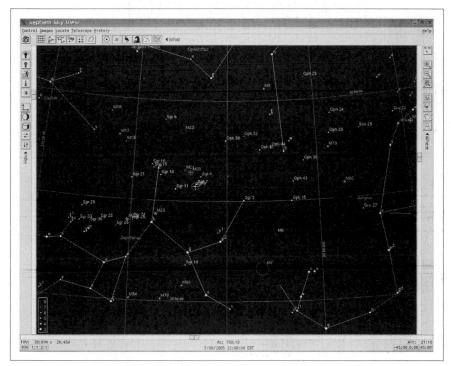

Figure 4-49. The XEphem Sky View window

 HACK #65 ## Astronomy Software in the Palm of Your Hand

Using the shareware Planetarium application for the Palm, you can keep your astronomy software right where you need it.

One of the fundamental problems for all amateur astronomers, especially beginners, is finding objects in the night sky. The sky is huge, and the field of view through a telescope eyepiece is extraordinarily small, so the likelihood of finding any specific deep-sky object without some specific observing aid is very low. One solution to this problem is a computerized GOTO scope, which has become an increasingly popular option for people who want to spend less time finding objects and more time looking at them. But many people also enjoy the challenge of finding objects manually. Indeed, my teenage son enjoys the "thrill of the hunt" even more than viewing the objects themselves. For finding objects on your own, you must have accurate maps of the night sky. Books and atlases like *Sky Atlas 2000.0* and *Uranometria* are wonderful resources, but they don't provide a real-time picture of the night sky: you have to contort the page at various angles to match the appearance of the stars. And while large atlases are nice to use on a table, they are not very convenient to hold at the eyepiece. Alternatively, you can print out detailed current star maps from astronomy software. But that can be cumbersome and limited to specific areas that you choose in advance. What if you change your mind and want to look at a different area? Or clouds cover up the area that you planned? Another alternative is to use astronomy software loaded on laptop computers. The computer will provide views to match what you will see in the sky. But it is also cumbersome: bringing the laptop to the eyepiece with you in the dark is neither convenient, nor healthy for the laptop.

One solution to these problems is to use astronomy software written specifically for the Palm and other handheld personal digital assistants (PDAs). If you're like me, you may already own a PDA to store addresses and run your daily calendar. If so, the main cost of this system is already paid for, since much of this software is modestly priced. While astronomy software on a handheld PDA might not have all the features of a high-end astronomy software package, the portability of a PDA is appealing. With a PDA, you can essentially hold a current map of the night sky in your hand at the eyepiece, with its screen zoomed in at whatever field of view you need, and you can access different screens by tapping a stylus instead of using a keyboard or mouse. One popular astronomy program for the Palm is Planetarium, shown in Figure 4-50. I have used this software for several years, and it has become one of the most indispensable parts of my observing package.

Planetarium is sold and downloaded online (*http://www.aho.ch/pilotplanets/*) as shareware. You download a demo version for free; if you like it, you pay the developer $24 for a code that activates the full version. The developer of this software is Andreas Hofer who has been helpful in creating upgrades and answering questions. In addition, you can join the online Yahoo group for astronomy PDA use, the PalmAstro group, *http://groups.yahoo.com/group/Palmastro/*.

Features of Planetarium

When you first use Planetarium, you enter your location and time zone, either from a large built-in database of cities around the world, or from latitude and longitude coordinates obtained elsewhere (the database at the Heavens Above web site is particularly useful: *http://www.heavens-above.com*). Your PDA needs to be set already with the correct time and date; these

Figure 4-50. Software at the eyepiece: using a Palm with Planetarium software to find objects

data provide Planetarium with the information it needs to give you accurate positions of celestial objects at any time from your location. If you change your location of observing, you simply tap on a new location from a pull-down menu; I have three custom locations set to the sites where I commonly observe.

Views. Planetarium views objects in two ways: a compass screen (Figure 4-51) and a sky view screen (Figure 4-52). The compass screen shows objects in two views: azimuth (as they would appear on a compass) and altitude. Tapping an icon switches over to the sky view screen, which shows objects as they appear in the sky relative to nearby constellations. Both views display current altitude and azimuth coordinates for selected objects. Figure 4-51 shows a Planetarium compass view. The altitude direction shows Mars at 20.3° from the horizon (H); the azimuth view shows Mars in the southeast, with an azimuth of 122.5°. Figure 4-52 shows a Planetarium sky view with a field of view of 60°, with Mars centered in the screen in the constellation of Aquarius.

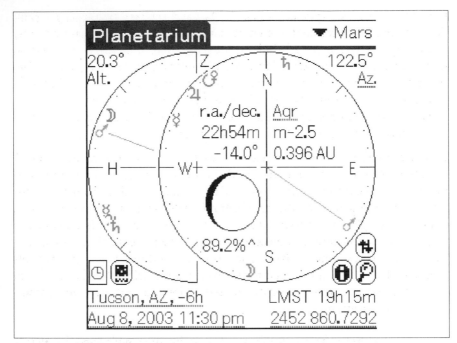

Figure 4-51. Planetarium compass view

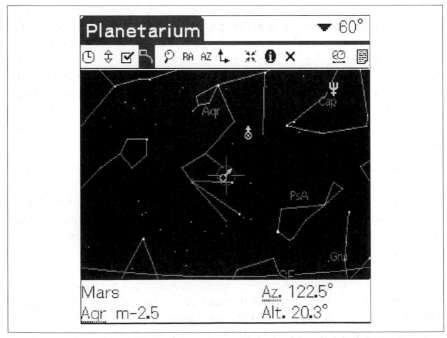

Figure 4-52. Planetarium sky view

Objects are grouped (planets, stars, DSOs), and selected by tapping a pull-down menu. Once selected, the object appears on the screen surrounded by circles representing a Telrad finder. If you like, you can tap an information icon that provides you with background information (e.g., distance, size, rise/set times). Tapping on another pull-down menu varies the field of view of the sky image, from 180° to 1°, with numerous increments in between. This way, you can see the object as it appears to the naked eye overhead, and then zoom to a smaller field of view as it appears in a finder scope.

Tapping another pull-down menu adjusts the time increment, from 1 minute to 90 days. You use the PDA's up/down button to show the position of the object at any time selected. Tapping a handy clock icon automatically resets the screen to the current time. The time increment button is important for planning observing sessions for months or days in advance, or in previewing positions of objects during an observing session. For example, I often want to know when a specific object will rise over a tree or roof-top, and this feature allows me to determine that easily. It's also useful to check out conjunctions, occultations, and eclipses in advance, all customized for your location.

Databases. Several different databases can be freely downloaded for Planetarium, depending on the level of detail you need. For example, the basic download contains either 9,096 stars, or if you are low on memory, 1,600 stars. I find the 9,096 version to be much more useful; it is similar to the number of stars you might see in a typical finderscope. Stars are listed by names or by the standard conventions (Yale, Bayer, and Flamsteed **[Hack #15]**). The standard database comes with the full lists of Messier and Caldwell objects, while an advanced download contains the Herschel 400 list of NGC objects, and an even more advanced list contains over 1,000 deep-sky objects. There is also a list of asteroids and comets. Several other custom download lists are available from the Planetarium web site.

Databases can be edited by the user. For example, if a new comet or asteroid is discovered, you can download the ephemeris (e.g., from *http://cfa-www.harvard.edu/iau/Ephemerides/index.html*) and add it to the list. This is a critical use of the software. After all, the locations of comets and asteroids are not listed on standard atlases, and if you want to do any comet hunting, it is extremely valuable to have that capability in your hand at the eyepiece.

Session planning. In addition to the object you selected, the compass view of Planetarium also shows the current phase of the moon, giving you a quick look to see whether the moon will be a factor in your observation plans. Tapping on the moon leads you to the twilight screen in a couple of taps, which gives you a quick run down on the vital statistics of the day, including rise and set times of the sun and moon, and exact times for twilight (civil, nautical, and astronomical). Moving the up/down button on the PDA provides you with this twilight information for other days, depending on the time increment you selected.

Log entries. Planetarium also contains an automatic log feature. If you want to log an object you've just observed, tapping the log icon will automatically add it as an entry into your personal logbook, including the object name and current time. Then you can add any other pertinent information if you like. I keep a simple one-letter code system where I can use the stylus to quickly enter the parameters of the observation. For lazy observers like me who just can't seem to get organized with a logbook, this feature is invaluable.

Locating Objects Manually with Planetarium

A primary goal of using PDA-based software is to find objects with minimum hassle. A PDA has a major advantage: you can take it with you to the eyepiece, where you have a real-time map of the sky to match the view you see in the finder. Whether you use a unit power finder like a Telrad or Rigel QuikFinder, or a finderscope, Planetarium is ideal for this purpose.

To accomplish this, you simply tap on a pull-down menu to select an object. You'll see the object with its relevant constellations in the Sky View screen, surrounded by Telrad circles (0.5°, 2°, and 4°). With the PDA in your hand, it's easy to match the position of the finder on your scope with the image on the screen. Starting with a wide view in the PDA (like 45°), you use the Telrad to point to the general area of the sky, and then quickly tap on the zoom view to narrow your field of view. Sometimes I end up with a custom view of 7° to precisely match the field of view of my finderscope.

Figures 4-53 and 4-54 show the process of finding M57 with Planetarium. First locate the general position of M57 in Lyra with a wide view of 45° (Figure 4-53), then zoom to 10° for a close-up view (Figure 4-54). Dotted circles represent a Telrad unit finder.

This is a quick and easy method to find objects. For example, my son and I have used it successfully in Messier Marathons to find over 100 objects in one night.

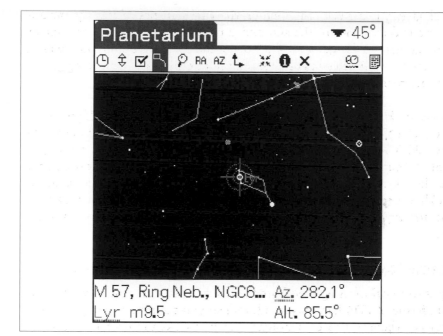

Figure 4-53. Locating the general position of M57 in Lyra

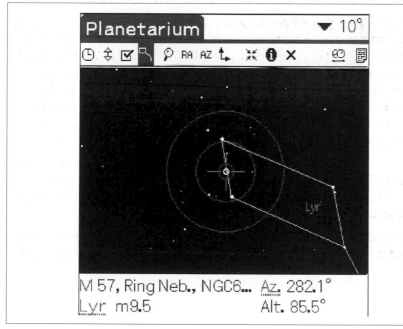

Figure 4-54. Zooming in to a detailed map for M57

Using Planetarium with Manual Setting Circles

For finding objects, many amateur astronomers find the setting circles on equatorial mounts to be virtually worthless because their small size precludes any degree of accuracy. But Dobsonians have large bases ideal for mounting setting circles. When used properly, these circles can quickly find virtually any visible object. Astronomy software provides the updated altitude and azimuth coordinates needed to find objects, and the most convenient way to do this is to use a PDA in conjunction with the circles.

Altitude and azimuth circles for Dobsonians are relatively easy do-it-yourself projects for Dob owners. Details for constructing and using these circles can be found on the Web (for example, see the Yahoo Skyquest forum and do a search under "azimuth ring": *http://groups.yahoo.com/group/skyquesttelescopes/*). The altitude circle is simply a protractor attached to the altitude bearing, while the azimuth circle is attached to the base of the scope with brackets so that it can be moved during alignment. Figures 4-55 and 4-56 show homemade manual setting circles on a 10-inch Dob. Figure 4-55 shows the altitude circle, and Figure 4-56 shows the azimuth circle. Using PDA-based software like Planetarium is a convenient way to provide updated alt-az coordinates for objects.

Figure 4-55. The altitude circle shows an altitude of about 44°

Figure 4-56. The azimuth circle shows an azimuth of about 180°

To align the circle, you find a known object (a bright star or planet), center it in the eyepiece, and look up its coordinates on the PDA. Then you simply move the azimuth circle around the base so that its pointer matches the azimuth in the PDA, and the circles are now aligned for the night. Throughout the night, you select an object in Planetarium to automatically provide its alt-az coordinates, then move the scope to these coordinates to locate the object quickly. One caveat: in order to work accurately, the base of the scope must be level. This is monitored with bubble levels glued at right angles on the base, then use either leveling feet or simple wooden wedges to level the scope. The leveling process only takes a couple of minutes each night.

Other Uses

The portability and features of Planetarium on the Palm give it a variety of other uses:

- It's a handy portable reference throughout the night. Need to know when Mars is rising tonight? Just pull it out of your pocket and you're set.

- As long as you can see the sun, you can use the compass view in Planetarium as a rough compass during the day. This is particularly useful when you're setting up a new observing site in daylight.

- Planetarium also provides you with an accurate value of magnetic declination for your observing location, so that you can calibrate magnetic compasses. This is handy for accurate polar aligning of equatorial mounts when Polaris is behind the trees, or during solar observing.

- The program can also interface with several commercial telescopes and GPS instruments via the PDA's serial port. I've never used this feature, so I can't comment on its utility. But it is a common item of discussion on the Yahoo PalmAstro group.

Bottom Line

Nothing is perfect, and there are a few problems using Planetarium and a PDA at the eyepiece. Most of these issues are associated with the PDA itself rather than the software. For example, some observers have trouble seeing the symbols and icons on the small screen of the PDA. Ultimately, this is the price you pay to have a screen in your hand at the eyepiece. Most annoying to your fellow observers is the light emanating from the PDA screen, which seems to be precisely situated to ruin your colleagues' night vision. Once again, this is not the fault of the software because Planetarium has a night vision mode in which the screen is lit with red light. Unfortunately, the Palm has an obnoxious white light that escapes from the edge of the screen. I solved this problem by attaching red film to the screen [Hack #44]; the Palm has a small gap between the frame and the screen where you can insert a sturdy piece of red film to block the white light. On some models, you can also turn off the backlit screen of the PDA and see the screen views with your red flashlight.

Other astronomy software will provide the basic features you find in Plane-tarium and include more advanced features as well. Certainly, Planetarium cannot replace the full features of a high-end astronomy program. But the advantage of PDA-based software is to use its portability to simplify the amateur astronomer's life. Armed with just a PDA, you can eliminate your star atlas, print-outs of star charts, logbook, laptop computer, and folding table. Your overall setup is faster and easier, and the reliability of the locat-ing system means that you can spend your time observing, not fumbling around in the dark.

—Steve Childers

Index

A

Aberrator, 240
absolute magnitude, 88
accessories, 257
achromatic refractors, 49
AFoV (apparent field of view), 280
airline-portable scope kits, 24
alcoholic drinks, 8, 19
altitude, 116
altitude-azimuth (alt-az) coordinate
    system, 116
altitude-azimuth (alt-az) telescope
    mounts, 43, 241
Anacortes Telescope and Wild Bird, 67
Antares RACI finders, 320
aperture, 29, 71
aperture masks, 220
aperture of telescopes, 40
apochromatic refractors, 49
apparent field of view (AFoV), 280
apparent magnitude, 88
Argelander, F. W. A., 106
asteroids, 150
astigmatism, eliminating from
    telescopes, 214–217
    causes, 214
Astromart, 38, 63, 66
astronomical twilight, 16
Astronomics, 67
astronomy
    learning process, 2
    observation (see observing)

astronomy chairs, 351
astronomy clubs, 3–5
    locating, 5
astrophotography, 197–202
    naked-eye viewing, versus, 154
averted vision, 79
azimuth, 116

B

backpacking kits, 22
Barlow lenses, 263–269
    2- and 3-element models, 268
    advantages, 265
    eye relief and, 266
    filter performance and, 267
    magnification, determining, 269–273
    magnification factors, 264
    model sizes, 264
    short versus long Barlows, 267
Barlow, Peter, 269
Barlowed lasers, collimation
    using, 233–236
Bayer catalog, 101–103
Bayer designation, 89
Bayer, Johann, 101
binoculars, 23, 26–40
    aperture, 29
    built-in digital cameras, 38
    buying used, 38
    collimation, 34
    diopter adjustment, 31
    exit pupils, 25, 30

H

Hands-On Optics, 67
headlights, 17
heat packs, 12
Henry Draper Catalog (HD), 108
High Point Scientific, 67
Hipparcos Catalog (HIP), 109
Historia Coelestis Britannica, 104
horizontal coordinate system, 116
Hubble Space Telescope (HST) Guide
        Star Catalog (GSC), 108
hydration, 11
Hydrogen-beta filters, 346
hypothermia, preventing, 8–14
    supplemental heat sources, 12–14
    wind and wind chill, 12

I

IC (Index Catalog) numbers, 175
instruments and visibility of objects, 94
integrated magnitudes, 96
interference coatings, 35
interference filters, 339–342
International Astronomical Union
        (IAU), 110
interpupilary distance, 32
Intes and Intes Micro, 63
intrinsic variable stars, 91
ITE, 63

J

Johnson, Sam, 358
jon-e Handwarmers, 13
Julian date, 107
junk scopes, 65

K

KStars Desktop Planetarium, 364
    night-vision mode, 258

L

L cones, 80
laptop computers (see notebook
        computers)
laser collimators, 222, 230
learning process, 2

Legrand, Christian, 364
lens coatings, 35–36
    quality, 36
lenses, cleaning, 304–308
LensPens, 307
light pollution, 160
    filters and, 163
    impact on entrance pupil, 26
    reducing, 166
Light Pollution Reduction (LPR)
        filters, 339–342
limiting magnitude at zenith (LMZ), 87
line filters, 345
Linux-compatible planetarium
        software, 364–369
LMZ (limiting magnitude at zenith), 87
logging, voice recorders, 353
logging systems, 173–176
    log sheets, design, 171
    spreadsheets, 173
long-period variable stars, 91
long-tube refractors, 24, 47
Lunar observing, 161

M

M cones, 80
Mac-compatible planetarium
        software, 364
magnification, 28
magnification and brightness, 95
magnitude, 86, 87–94
    ranges, 89
    of solar system objects, 92
    surface brightness and, 95, 96
    variable magnitudes, 90–94
    visibility, 94
Maksutov telescopes, 60
Maksutov-Cassegrain telescopes, 62
Maksutov-Newtonian telescopes, 63
Martian moons, viewing, 299
Mathematical Syntaxis, 86
McMaster-Carr, 252
mechanical Teflon, 252
mesopic mode of vision, 81
Messier, Charles, 176
Messier Marathons, 176–197
    date and time, 178
        Lunar phase, impact of, 178

V

variable stars, 90
virgin Teflon, 252
Virtual Moon Atlas (VMA), 364
visual magnitude, 87
voice recorders, 353

W

wanderers, 151
white light, 16
white-light checks, 21
wide-field eyepieces, 273–278

wind and wind chill, 12
Windows-compatible planetarium
    software, 360
Wratten color filter designations, 334

X

XEphem planetarium software, 367

Z

zenith, 116
ZHR (Zenithal Hourly Rate), 147
Zodiacal Catalog (ZD), 109

Colophon

Our look is the result of reader comments, our own experimentation, and feedback from distribution channels. Distinctive covers complement our distinctive approach to technical topics, breathing personality and life into potentially dry subjects.

The tool on the cover of *Astronomy Hacks* is a refractor telescope. Although magnifying glasses and burning glasses had been known since classical times, and eyeglasses were in use by 1300, it was not until the final decade of the 16th century that instrument makers first created optical instruments for scientific exploration.

The first simple microscopes were built by brothers Zacharias and Hans Janssen, about 1595. Although the telescope seems a logical follow-on to the microscope, no evidence exists that any telescope was built prior to 1608. The invention of the telescope is sometimes credited to Zacharias Janssen or James Metius, but evidence suggests that spectacle maker Hans Lippershey was the first to construct a telescope.

One day, while holding a spectacle lens in either hand, Lippershey happened to view a nearby church steeple through both lenses and was astonished to see that it appeared larger than before. He mounted the lenses in a tube to adjust and preserve their spacing, and thereby invented the refractor telescope.

Lippershey applied to the Dutch government for a patent, which was denied because he was unable to prove that he was the sole inventor. The government officials, however, recognized the value of Lippershey's invention. They bought his original telescope for 90 florins and paid Lippershey well to produce additional telescopes for them.

Opticians and instrument makers throughout Holland were soon producing telescopes, and within a year telescopes were being made throughout Europe. In 1609, Galileo Galilei, after reading a description of the telescope, constructed his own instrument and turned it to the heavens. Galileo first used his telescope to discover the moons of Jupiter, sunspots, the phases of Venus, and the craters and valleys on the Lunar surface. With his telescope, Galileo proved the Copernican heliocentric theory by establishing that the apparent motion of Jupiter's four moons could be explained only if those moons orbited Jupiter, and that the phases of Venus established that Venus must be orbiting the Sun.

Newton reinvented the second major type of telescope, the using mirrors rather than lenses to collect and focus light. Since th century, the craft of telescope making has been refined continually. We now have telescopes that enable us to see objects billions of light years away to the edge of our universe. There are also telescopes that capture energy such as radio wave emissions, gamma rays, and x-rays. But the refractor telescope, refined but essentially unchanged since the days of Lippershey and Galileo, remains a popular and useful scientific instrument.

Marlowe Shaeffer was the production editor and proofreader for *Astronomy Hacks*. Darren Kelly and Claire Cloutier provided quality control. John Bickelhaupt wrote the index.

Mike Kohnke designed the cover of this book, based on a series design by Edie Freedman. The cover image is an original photograph provided by Al Nagler. The background image is from Getty Images. Karen Montgomery produced the cover layout with Adobe InDesign CS using Adobe's Helvetica Neue and ITC Garamond fonts.

David Futato designed the interior layout. This book was converted by Keith Fahlgren to FrameMaker 5.5.6 with a format conversion tool created by Erik Ray, Jason McIntosh, Neil Walls, and Mike Sierra that uses Perl and XML technologies. The text font is Linotype Birka; the heading font is Adobe Helvetica Neue Condensed; and the code font is LucasFont's TheSans Mono Condensed. The illustrations that appear in the book were produced by Robert Romano, Jessamyn Read, and Lesley Borash using Macromedia FreeHand MX and Adobe Photoshop CS. This colophon was written by Lydia Onofrei and Robert Bruce Thompson.

Better than e-books

Search

over 2000 top
tech books

Download

whole chapters

Cut and Paste

code examples

Find

answers fast

Read books from cover
to cover. Or, simply click
to the page you need.

**Search Safari! The premier electronic reference
library for programmers and IT professionals**

Related Titles Available from O'Reilly

Hacks

Amazon Hacks

BSD Hacks

Digital Photography Hacks

eBay Hacks

Excel Hacks

Flash Hacks

Gaming Hacks

Google Hacks

Hardware Hacking Projects for Geeks

Home Theater Hacks

iPod & iTunes Hacks

Knoppix Hacks

Linux Desktop Hacks

Linux Server Hacks

Mac OS X Hacks

Mac OS X Panther Hacks

Network Security Hacks

PayPal Hacks

PDF Hacks

PC Hacks

Smart Home Hacks

Spidering Hacks

TiVo Hacks

Windows Server Hacks

Windows XP Hacks

Wireless Hacks

Word Hacks

O'REILLY®

Our books are available at most retail and online bookstores.
To order direct: 1-800-998-9938 • *order@oreilly.com* • *www.oreilly.com*
Online editions of most O'Reilly titles are available by subscription at *safari.oreilly.com*

Keep in touch with O'Reilly

1. Download examples from our books

To find example files for a book, go to:

www.oreilly.com/catalog

select the book, and follow the "Examples" link.

2. Register your O'Reilly books

Register your book at *register.oreilly.com*

Why register your books? Once you've registered your O'Reilly books you can:

- Win O'Reilly books, T-shirts or discount coupons in our monthly drawing.
- Get special offers available only to registered O'Reilly customers.
- Get catalogs announcing new books (US and UK only).
- Get email notification of new editions of the O'Reilly books you own.

3. Join our email lists

Sign up to get topic-specific email announcements of new books and conferences, special offers, and O'Reilly Network technology newsletters at:

elists.oreilly.com

It's easy to customize your free elists subscription so you'll get exactly the O'Reilly news you want.

4. Get the latest news, tips, and tools

http://www.oreilly.com

- "Top 100 Sites on the Web"—PC Magazine
- CIO Magazine's Web Business 50 Awards

Our web site contains a library of comprehensive product information (including book excerpts and tables of contents), downloadable software, background articles, interviews with technology leaders, links to relevant sites, book cover art, and more.

5. Work for O'Reilly

Check out our web site for current employment opportunities:

jobs.oreilly.com

6. Contact us

O'Reilly & Associates
1005 Gravenstein Hwy North
Sebastopol, CA 95472 USA

TEL: 707-827-7000 or 800-998-9938
(6am to 5pm PST)

FAX: 707-829-0104

order@oreilly.com
For answers to problems regarding your order or our products.
To place a book order online, visit:

www.oreilly.com/order_new

catalog@oreilly.com
To request a copy of our latest catalog.

booktech@oreilly.com
For book content technical questions or corrections.

corporate@oreilly.com
For educational, library, government, and corporate sales.

proposals@oreilly.com
To submit new book proposals to our editors and product managers.

international@oreilly.com
For information about our international distributors or translation queries. For a list of our distributors outside of North America check out:

international.oreilly.com/distributors.html

adoption@oreilly.com
For information about academic use of O'Reilly books, visit:

academic.oreilly.com

O'REILLY®

Our books are available at most retail and online bookstores.
To order direct: 1-800-998-9938 • *order@oreilly.com* • *www.oreilly.com*
Online editions of most O'Reilly titles are available by subscription at *safari.oreilly.com*